"This collection not only brings together issues in scientific understanding, representation and explanation but allows the authors the rare opportunity to directly engage with each other's works and thereby advance the debate."

Scientific Understanding and Representation

This volume assembles cutting-edge scholarship on scientific understanding, scientific representation, and their delicate interplay. Featuring several articles in an engaging 'critical conversation' format, the volume integrates discussions about understanding and representation with perennial issues in the philosophy of science, including the nature of scientific knowledge, idealizations, scientific realism, scientific inference, and scientific progress.

In the philosophy of science, questions of scientific understanding and scientific representation have only recently been put in dialogue with each other. The chapters advance these discussions from a variety of fresh perspectives. They range from case studies in physics, chemistry, and neuroscience to the representational challenges of machine learning models; from special forms of representation such as maps and topological models to the relation between understanding and explanation; and from the role of idealized representations to the role of representation and understanding in scientific progress.

Scientific Understanding and Representation will appeal to scholars and advanced students working in philosophy of science, philosophy of physics, philosophy of mathematics, and epistemology.

Insa Lawler is Assistant Professor of Philosophy at the University of North Carolina at Greensboro, USA.

Kareem Khalifa is Professor of Philosophy at Middlebury College, USA. He is the author of *Understanding, Explanation, and Scientific Knowledge* (2017).

Elay Shech is Associate Professor of Philosophy at Auburn University.

Routledge Studies in the Philosophy of Mathematics and Physics

Edited by Elaine Landry, University of California, Davis, USA and Dean Rickles, University of Sydney, Australia

For more information about this series, please visit: https://www.routledge.com/Routledge-Studies-in-the-Philosophy-of-Mathematics-and-Physics/book-series/PMP

Scientific Understanding and Representation
Modeling in the Physical Sciences

Edited by
Insa Lawler, Kareem Khalifa,
and Elay Shech

Routledge
Taylor & Francis Group

NEW YORK AND LONDON

First published 2023
by Routledge
605 Third Avenue, New York, NY 10158

and by Routledge
4 Park Square, Milton Park, Abingdon, Oxon, OX14 4RN

Routledge is an imprint of the Taylor & Francis Group, an informa business

Library of Congress Cataloging-in-Publication Data
Names: Lawler, Insa, editor. | Khalifa, Kareem, editor. |
Shech, Elay, editor.
Title: Scientific understanding and representation : modeling in
the physical sciences / edited by Insa Lawler, Kareem Khalifa,
and Elay Shech.
Description: First edition. | New York, NY : Routledge, 2023. |
Series: Routledge studies in the philosophy of mathematics and
physics | Includes bibliographical references and index.
Identifiers: LCCN 2022023904 (print) | LCCN 2022023905 (ebook)
Subjects: LCSH: Science—Philosophy. | Physics—Philosophy. |
Physics—Simulation methods.
Classification: LCC Q175 .S424135 2023 (print) |
LCC Q175 (ebook) | DDC 501—dc23 /eng20221006
LC record available at https://lccn.loc.gov/2022023904
LC ebook record available at https://lccn.loc.gov/2022023905

ISBN: 978-1-032-05495-7 (hbk)
ISBN: 978-1-032-06581-6 (pbk)
ISBN: 978-1-003-20290-5 (ebk)

DOI: 10.4324/9781003202905

Typeset in Sabon
by codeMantra

In memory of Margie Morrison (1954–2021)

Contents

Figures

Contributors

Sorin Bangu is Professor of Philosophy at the University of Bergen (Norway). He publishes in philosophy of science (physics and mathematics) and on Wittgenstein's later philosophy.

Julia R. S. Bursten is Associate Professor of Philosophy at the University of Kentucky. Her research has centered around building the philosophy of nanoscience. It is situated across the philosophical literature on kinds and classification, inter-theory relations, modeling, and scientific explanation.

Mazviita Chirimuuta is Senior Lecturer of Philosophy at the University of Edinburgh. She received her Ph.D. in visual neuroscience from the University of Cambridge, and she has published a number of articles on computational explanation in neuroscience. Her second book, on abstraction and idealization in neuroscience, will be published with MIT Press.

Finnur Dellsén is Professor of Philosophy at the University of Iceland, and Professor II at the Inland Norway University of Applied Sciences. He is the winner of the Nils Klim Prize and the Lauener Prize for Up-and-Coming Philosophers. Dellsén's current research focuses on the social epistemology of science, explanatory reasoning, and scientific and philosophical progress.

Henk W. de Regt is Professor of Philosophy of Natural Sciences at the Institute for Science in Society, Radboud University, The Netherlands. His research focuses on scientific understanding and explanation, and on public understanding of science. His monograph *Understanding Scientific Understanding* (Oxford University Press, 2017) won the 2019 Lakatos Award.

Roman Frigg is Professor of Philosophy in the Department of Philosophy, Logic and Scientific Method at the London School of Economics and Political Science, UK. He is the winner of the Friedrich Wilhelm Bessel Research Award of the Alexander von Humboldt Foundation. His research interests lie in general philosophy of science and philosophy of physics.

Kareem Khalifa is Professor of Philosophy at Middlebury College. He works on various topics in general philosophy of science, philosophy of social science, and epistemology. In 2017, his book, *Understanding, Explanation, and Scientific Knowledge,* was published with Cambridge University Press. He has published articles in journals such as *Philosophical Studies, Philosophers' Imprint, The British Journal for the Philosophy of Science,* and *Philosophy of Science.*

Daniel Kostić is Radboud Excellence Initiative Fellow in the Institute for Science in Society, at Radboud University. He was previously a Marie Skłodowska-Curie Fellow at the CNRS/ University of Paris 1 Panthéon-Sorbonne. His research interests range from issues in the philosophy of mind, metaphysics, philosophy of neuroscience and ecology, to scientometrics, with a special focus on topological explanations.

Jaakko Kuorikoski is Associate Professor of Practical Philosophy at the University of Helsinki. His current research interests include scientific explanation and understanding, the use of different kinds of evidence in the social sciences, and model-based social epistemology of science.

Insa Lawler is Assistant Professor of Philosophy at the University of North Carolina at Greensboro. She works on various topics concerning the epistemology and philosophy of science. Her work has been published in journals such as *Philosophy of Science, Synthese,* and *Noûs.*

C. D. McCoy is Assistant Professor at Yonsei University. His research focuses on the philosophy of science and the philosophy of physics. He addresses the conceptual and physical foundations of cosmology, the philosophy of thermodynamics and statistical mechanics, relativity theory, and quantum theory, the nature of motion, time, and chance, as well as prediction, explanation, and understanding.

Jared Millson is Assistant Professor of Philosophy at Rhodes College whose specializations include epistemology, logic, philosophy of language, and philosophy of science. He studies the nature of inquiry, including the norms governing psychological attitudes and speech acts associated with inquiry, formal modeling of these norms, and how they shape scientific representation and explanation.

James Nguyen is Assistant Professor in the Department of Philosophy at Stockholm University and Research Fellow at both the Institute of Philosophy, University of London, and the Centre for Philosophy of Natural and Social Science, London School of Economics and Political Science. His primary research interests are in the philosophy of science, and its intersections with epistemology and aesthetics.

Amanda J. Nichols is Professor of Chemistry at Oklahoma Christian University. Her research interests include inorganic materials and undergraduate research and philosophy of science with an emphasis on defenses of scientific realism.

Myron A. Penner is Professor of Philosophy at Trinity Western University and a Visiting Professor in the Department of Psychology at the University of British Columbia (until 2023). His research interests include epistemology, philosophy of science, and cognitive science of religion.

Christopher Pincock is Professor of Philosophy at The Ohio State University. His research focuses on the philosophy of science, the philosophy of mathematics, and the history of analytic philosophy. His current primary research project aims at developing a new defense of scientific realism that addresses worries about idealized modeling and historical change.

Angela Potochnik is Professor of Philosophy and Director of the Center for Public Engagement with Science at the University of Cincinnati. She is the author of *Idealization and the Aims of Science* (2017) and the coauthor of *Recipes for Science* (2018). She earned her Ph.D. from Stanford University in 2007.

Collin Rice is Assistant Professor of Philosophy at Colorado State University. His current research focuses on the epistemological and metaphysical issues surrounding scientific explanation and understanding, the use of highly idealized and abstract models in biology, and the positive roles idealizations play in scientific theorizing.

Mark Risjord is Professor of Philosophy at Emory University whose specializations include philosophy of science, philosophy of language, and epistemology, with a special interest in issues arising from the social sciences, medicine, and nursing. His recent work has concerned scientific modeling and developing an inferentialist-expressivist account of scientific representation.

Juha Saatsi is Associate Professor of Philosophy at the University of Leeds' School of Philosophy, Religion, and History of Science. He mainly researches philosophy of science, covering various topics in epistemology and metaphysics of science.

Elay Shech is Associate Professor at Auburn University's Philosophy Department, specializing in philosophy of science, physics, mathematics, and statistics, with additional developed research interests and projects in philosophy of biology and machine learning.

Emily Sullivan is Assistant Professor and Irène Curie Fellow at Eindhoven University of Technology and the Eindhoven Artificial Intelligence

Systems Institute. Her research explores how technology mediates our knowledge. She leads an NWO Veni project (2021–2024) on the explainability of machine learning systems.

Michael Tamir serves as Chief Scientist and Head of Machine Learning and AI for the Susquehanna International Group, and faculty for the UC Berkeley data science program. He has developed deep learning in natural language understanding applications, founded the faker-fact.org project, and worked as Head of Data Science for the Uber Advanced Technologies Group.

Acknowledgments

We wish to thank all our contributors, our anonymous reviewers, and the Routledge team, as well as Jack Trench for his editorial work on the index, and Auburn University's Department of Philosophy for funding the editorial work.

1 Introduction

Kareem Khalifa, Insa Lawler, and Elay Shech

Successful science, it would seem, is an effective means of both understanding and representing the empirical world. Although important antecedents can be found all the way back to the ancients, it is somewhat surprising that these themes have gained prominence in the philosophy of science only recently, with most monograph-length books on either scientific understanding or scientific representation being published in the past decade.[1] More surprising still is the sparse interplay between the literature on these two topics. Clearly, how a phenomenon is represented has far-reaching ramifications for how it is understood. Consider, for instance, idealizations, such as frictionless planes, infinite populations, ideal gases, and rational actors. Idealizations misrepresent their target systems, yet they frequently provide a deeper understanding than more accurate representations. However, to develop this idea, more detailed accounts of representation and of understanding must engage each other. Otherwise, it remains mysterious as to how a misrepresentation can provide a genuine understanding.

It was concerns such as these that animated a lively brainstorming session between Kareem Khalifa and Daniel Kostić in a Parisian cafe in March 2018: we would bring together leading scholars working on understanding and representation in the philosophy of science in the hopes of encouraging greater crosstalk between these two research areas. The fruits of those discussions led to the first annual Scientific Understanding and Representation (SURe) workshop held at the University of Bordeaux in February 2019, with subsequent meetings in Atlanta at Emory University (2020), in Nijmegen at Radboud University (2021), and in New York at Fordham University (2022).

The goal of this edited volume is to give readers a sense of the productive discussions that have emerged from the SURe workshops, with the hopes of advancing discussions concerning scientific understanding, scientific representation, and their delicate interplay. We also hope that readers get a sense of the intellectual energy and synergy that have characterized our workshops since their founding. To that end, this work includes both standalone chapters and "critical conversations" between pairs of contributors. In the critical conversation format, the contributors

DOI: 10.4324/9781003202905-1

each write an original target article on a related theme, and they then provide commentary on each other's work. What frequently emerges are unexpected rapprochements, deeper clarifications of earlier debates, and discoveries of new intellectual niches that were previously overlooked.

We have found it useful to group the contributions into four parts: (I) Understanding, Knowledge, and Explanation, (II) Understanding and Scientific Realism, (III) Understanding, Representation, and Inference, and (IV) Understanding and Scientific Progress. We discuss each in turn.

1   Understanding, Knowledge, and Explanation

Several interrelated debates concern the relationship between understanding, explanation, and knowledge. A longstanding position, which Khalifa (2017b) calls the "received view" of understanding, holds that understanding is knowledge of a correct explanation. Such a view is expressed, with varying degrees of explicitness, by philosophers working on scientific explanation (e.g., Hempel 1965, Kitcher 1989, Woodward 2003). However, the received view has faced several probing objections. Chief among these is that understanding requires skills or cognitive abilities that knowledge does not (de Regt 2017), and that understanding, unlike knowledge, can be achieved through what Elgin (2017) dubs "felicitous falsehoods." Both of these challenges to the received view concern scientific representation. Models are the paradigmatic kind scientific representation. The skills involved in the construction and interpretation of models are frequently associated with understanding. Similarly, scientific models frequently involve idealizations, approximations, and other departures from the truth that exemplify the kinds of felicitous falsehoods that yield understanding.

Part I, "Understanding, Knowledge, and Explanation," advances several of these discussions. It begins with an exchange between Henk de Regt and Kareem Khalifa that explores some of their longstanding disagreements in greater depth and with fresh perspectives. In earlier work, Khalifa (2012, 2017a, 2017b, 2020) criticized de Regt's contextual theory of understanding and advertised the advantages of his own, knowledge-based account. In Chapter 2, de Regt turns the tables, criticizing Khalifa's knowledge-based account of understanding and advertising the advantages of his contextual theory. De Regt's core arguments are twofold. First, understanding involves skills that are not required of knowledge. Second, knowledge is factive, whereas understanding is not. In Chapter 3, in the hope of reorienting his exchanges with de Regt, Khalifa considers the benefits of unifying their two accounts of understanding. He argues that de Regt's account substantially improves his own account of how considering different explanations contributes to understanding. Conversely, Khalifa argues that his account of understanding substantially improves upon de Regt's account of how explanatory evaluation contributes to understanding. Khalifa attempts to dissolve his apparent disagreement with de Regt about the so-called factivity of understanding through a voluntaristic approach to the scientific realism debates, in which theorists of understanding have considerable flexibility in how accurate an explanation must be in order to provide understanding.

In their replies, both de Regt (in Chapter 4) and Khalifa (in Chapter 5) explore the implications of this proposed "friendship" or synthesis of their respective views. Though broadly sympathetic to this synthesis, de Regt raises some questions about Khalifa's proposal that explanatory comparisons proceed by way of "empirical fitness," the requirement that good explanations perform at least as well as any rival in saving the phenomena. De Regt contends that this can be too restrictive, in no small part because scientists can reasonably disagree about the standards of empirical adequacy. By contrast, Khalifa shows how de Regt's objections to his view—concerning skills and factivity—can be answered using the hybrid of their two views developed in Chapter 3. Specifically, Khalifa argues that because the hybrid position includes all of the skills that de Regt underscores, it can meet de Regt's challenge in terms that the latter should accept. Further, Khalifa shows how voluntarism allows both realists and antirealists to accommodate the intuition that Newtonian mechanics provides understanding despite its inaccuracies.

While de Regt and Khalifa aim at dissolving debates concerning the factivity of understanding through their newfound commitment to voluntarism, Sorin Bangu and Mazviita Chirimuuta take different angles on this issue in Chapters 6 and 7, respectively. Bangu considers the question of whether past, false scientific theories such as the Newtonian gravitational theory provide some understanding of physical phenomena in our present times. Specifically, he asks whether de Regt's (2017) contextual theory of scientific understanding can support a non-factivist approach to understanding the gravitational deflection of light. Bangu argues that depending on how precisely we characterize gravitational light bending, de Regt's theory delivers two answers. Namely, characterized qualitatively, the Newtonian theory affords an understanding of the light bending phenomenon on de Regt's account. However, if this phenomenon is characterized quantitatively, then this account is shown to entail that no understanding is provided by the Newtonian theory. Hence, Bangu concludes that de Regt's account can only support a weak sense of non-factivism.

Chirimuuta's contribution, in Chapter 7, also broaches on the issue of understanding and truth, but she begins with the observation of a trade-off in models of complex occurrences. Specifically, increasing the predictive accuracy of the model often decreases intelligibility (the capacity of the model to provide scientific understanding). She argues that this trade-off lends support to non-factivist accounts of scientific understanding. However, non-factivism faces an important objection, concerning the hypothetical availability of de-idealized models that afford factive understanding. Chirimuuta replies to this objection by arguing that Potochnik's (2017) non-factivist account of understanding is overburdened by an ontological commitment to "real causal patterns." Chirimuuta shows that the account can be strengthened by replacing this commitment with the notion of an "ideal pattern," where the targets of model building are not patterns that are simply "out there" in nature, but they are, to some extent, dependent on the methods of data collection and processing chosen by the researcher. This proposal is then related to earlier accounts of scientific phenomena.

While de Regt, Khalifa, Bangu, and Chirimuuta all discuss the relationship between understanding and truth, Daniel Kostić focuses more squarely on explanation in Chapter 8. Rather than broaching larger debates about whether understanding requires explanation (Khalifa 2017b, Ch. 4) or not (Lipton 2009), Kostić discusses a particular kind of explanation: those that are driven by the resources of graph theory— so-called topological explanations. These more focused discussions provide the basis for seeing how different kinds of explanations produce understanding in different ways (Kostić 2019). To that end, Kostić provides a systematic and critical overview of topological explanations in the philosophy of science. Kostić's overview proceeds by presenting his account of topological explanation (presented in, e.g., Kostić 2020), and then comparing this account with some of its most prominent rivals. In the process, he highlights some problems with these alternatives and outlines his responses to some of the main criticisms raised by the so-called new mechanists.

Many of the threads in Part I come together in Juha Saatsi's contribution (Chapter 9). Saatsi identifies two dimensions of explanatory power. The first, "factive" dimension of explanatory power concerns how explanations relate to mind-independent reality. The second, "pragmatic" dimension of explanatory power concerns the relationship between explanations and us as epistemic agents who use explanations to gain understanding. Saatsi argues that keeping these two dimensions in mind helps us to better understand three key issues. First, the factive dimension of explanatory power immediately accords a role to truth and thereby provides yet another way of conceiving the relationship between understanding and truth, in addition to de Regt, Khalifa, Bangu, and Chirimuuta's accounts. Second, Saatsi argues that the distinction clarifies the explanatory role of mathematics in a manner that contrasts with Kostić's treatment of topological explanation. Kostić argues that topological properties are explanatory in themselves, whereas Saatsi argues that mathematics enhances the pragmatic dimension of explanatory power in a manner that is consistent with denying the explanatory power of mathematical properties along the factive dimension.

Third, Saatsi argues that scientific understanding may advance along either the factive or pragmatic dimensions of explanatory power, yet it is only when there is progress along the former that understanding amounts to an epistemic achievement. Pragmatic advances in explanatory power merely provide "cognitive scaffolding" for these epistemic achievements.

2 Understanding and Scientific Realism

In discussing the factive dimension of explanatory power and its implications for understanding, Saatsi also broaches upon the relationship between understanding and debates about realism and instrumentalism. In particular, scientific realism is, roughly, the position that what our best theories say about the world is approximately true (see Chakravartty (2017) for a discussion of the various dimensions of realist commitment). The realist holds that entities and structures posited by our best theories, including unobservable ones, exist and behave more or less as our theories say they do. Non-factivists about understanding, such as Potochnik (2017), hold that science aims at understanding instead of truth. So, on the face of it, non-factivism about understanding stands in tension with scientific realism. Part II, titled "Understanding and Scientific Realism," focuses on such issues. It begins with an exchange between Angela Potochnik and Christopher Pincock, who each offer a new defense of scientific realism. In Chapter 10, Pincock defends scientific realism while drawing on the requirements for successful scientific understanding and knowledge based on it. Whenever a scientist conducts a successful experiment that relies in part on theoretical beliefs about unobservables, grasps that this success is obtained independently of their actions and their scientific community, and understands the phenomenon in question well, then it is highly likely that they know that the unobservables exist. Pincock argues that no form of realism that involves context-dependent understanding or knowledge is feasible (such as Giere's (2006) perspectival realism or Potochnik's (2017) account of understanding), because they cannot explain how scientists extrapolate context-independent knowledge from their experimental findings. In Chapter 11, Potochnik defends a variant of selective realism. She argues that we can be realists, although scientists frequently work with models and idealizations that do not purport to be true. Her selective realism is based on the hypothesis that scientists do not aim at developing true theories. Instead, they aim at discovering causal patterns, that is, partial, simplified accounts of complex reality. These causal patterns exist in a mind-independent way. So, knowledge of causal patterns is knowledge about mind-independent reality. However, which patterns we detect depends partially on the research interests and the aims of the scientists who seek the explanations.

In his response in Chapter 12, Pincock insists that any form of selective scientific realism that construes the knowledge, explanation, or understanding involved as being context-dependent is not defensible. Knowledge of unobservables requires independence from the agent's situation, including their research interests. For instance, it requires us to know when an observable feature genuinely depends on something unobserved rather than being an artifact of the experiment. In her response in Chapter 13, Potochnik first argues that scientific realism should be about the truthmakers of theoretical claims (rather than the existence of unobservables or their character). Then, she addresses Pincock's concerns. She points out that the context-dependency of scientific knowledge only means that knowledge of some causal pattern could be unenlightening in a different research context, because it does not matter for the research goals in that context. The objects of scientific knowledge are selected based on context. The existence of the causal patterns is not context-dependent.

Writing also on the scientific realism debate in Chapter 14, Amanda Nichols and Myron Penner look at the evidence supporting the accuracy of current molecular models based on the data collected via infrared spectroscopy and high-magnitude microscopes. They argue that the best explanation for the correspondence between the observed data and that predicted by molecular models is that these models are correct with respect to the number and kind of atoms, bond connections, and orientation in three-dimensional space. This suggests that realism about molecular structures is justified. However, Nichols and Penner consider the possibility of withholding ontological commitment by reframing their abductive argument, such that the best explanation of correspondence between observed and predicted data is simply that molecular models are empirically adequate. Nichols and Penner discuss how the satisfactoriness of such reinterpretation depends on one's account of explanation and on how one characterizes the observable/unobservable distinction in light of technologically assisted observations such as infrared spectroscopy and high-magnitude microscopes.

Commitments to scientific realism and reductionism may suggest that in representing and modeling the world, we privilege those entities and structures that are the most fundamental. In Chapter 15, Julia Bursten looks at and expands on the work of Waters (2017, 2018) and Batterman (2021) as examples challenging reductive scientific metaphysics. Specifically, these are accounts that challenge the idea that the best guide to ontology is found in the so-called fundamental physical theory. She notes how these accounts highlight the important role that scale plays in defining ontological categories in scientific theories, thereby rejecting the reductionist perspective that the smallest scale is the most fundamental, general, or real. Consequently, Bursten urges philosophers to seek alternatives to the dominant part–whole approach to scientific metaphysics.

She identifies Massimi's (2022) perspectival realism and her own "conceptual strategies" (Bursten 2018) as such accounts, and in doing so brings into fruitful dialogue various anti-fundamentalist approaches to scientific metaphysics.

3 Understanding, Representation, and Inference

Scientific representations such as models afford understanding and facilitate surrogative reasoning and inference. Questions arise regarding how models represent, to what extent they are analogous to maps, and how they provide understanding and warrant inference. Part III, titled "Understanding, Representation, and Inference," includes essays that focus on such issues. In their contributions, Collin Rice and Jaakko Kuorikoski shed light on the social dimensions involved in scientific representations and the understanding that they produce. In Chapter 16, Kuorikoski defends a pragmatist concept of understanding; (explanatory) understanding is the ability to draw correct counterfactual what-if inferences about the object of understanding (rather than the possession of relevant correct mental representations). This inferentialist account of understanding is driven by the social epistemological function of attributions of understanding (drawing on Brandom 1994, Craig 1990). This ability-focused account of understanding can explain controversies within the sciences (which are concerned with what can be achieved with a given theory), it helps to defend the factivity of understanding (because it only matters whether the correct inferences are drawn), and it supports a moderate explanatory pluralism (because it allows for different kinds of counterfactual dependency). In Chapter 17, Rice argues that scientific representations and the understanding that they produce are a fundamentally social and dynamic enterprise and communal achievement. For instance, what scientific theories or idealized models represent can only be determined by how communities of scientists use these representations over time. Likewise, scientific understanding is gained by scientific communities using conflicting models, interests, and values over long periods of time. Rice uses a teleosemantic theory of mental representation to flesh out the historical components involved.

In his commentary in Chapter 18, Kuorikoski agrees that scientific representation and understanding are inherently communal and dynamic achievements. However, he is not convinced that the historical roots are decisive and points out, among other things, that grasping mature scientific representations does not require knowledge of their genesis. Kuorikoski offers an inferentialist analysis of communal and dynamic representation (based on Brandom 1994) as an alternative philosophical framework. In his commentary in Chapter 19, Rice acknowledges the virtues of an ability-focused account of understanding, but he raises various objections against Kuorikoski's analysis of counter-factual what-if inferences and the factivity of understanding. In addition, he argues against a purely pragmatist analysis of understanding. The possession of relevant correct mental representations seems to be necessary for having the abilities in question. As an alternative, Rice suggests that a proper account of the various social and epistemic functions of understanding involves both representationalist and pragmatist components.

In their exchange, James Nguyen, Roman Frigg, Jared Millson, and Mark Risjord discuss the nature of scientific representation and modeling, and that of cartographic representation. Specifically, in various publications (e.g., Frigg and Nguyen 2020), Frigg and Nguyen have explicated and argued for their preferred account of scientific representation, the "DEKI account," which consists of four main elements: denotation, exemplification, keying-up, and imputation. In Chapter 20, Nguyen and Frigg utilize the DEKI account to offer an account of how maps represent and facilitate surrogative reasoning. In doing so, they suggest that issues such as the nature of representational accuracy, the purpose relativity and historically situatedness of representations, and the possibility of total science, are further illuminated. For instance, they hold that the models-as-maps analogy, which entails that all scientific representations are partial, purpose relative, and historically situated, "pours cold water" on the dream of a complete, final scientific theory. In Chapter 21, Millson and Risjord criticize the DEKI account for not distinguishing between justified and unjustified surrogative inferences based on scientific representations. Specifically, they identify what they call "the fortuitous misuse of a map," where unjustified surrogative inferences from maps can sometimes accidentally result in true conclusions. The DEKI account's ability to distinguish use from misuse depends on what a representation denotes. However, looking at the methodology of archaeological site mapping, Millson and Risjord hold none of the common accounts of denotation enable DEKI to avoid fortuitous misuse.

In response in Chapter 22, Frigg and Nguyen distinguish between two different aspects of inferences from representations that might need to be justified, which they call derivational correctness and factual correctness. They then argue that although it is fair to demand that an account of representation provides justification of inferences having to do with derivational correctness, they maintain that factual correctness is beyond the scope of an account of representation (including DEKI). Moreover, Frigg and Nguyen explain that DEKI provides the needed justification in the context of derivational correctness via the notion of keying-up. However, in their reply in Chapter 23, Millson and Risjord argue that keying-up does not save DEKI from the problem of fortuitous misuse. They identify three general characteristics of mapping conventions and conditions of production, namely, relevance, measurement, and derivation, and hold that such characteristics are not circumscribed by the DEKI account in a way that accounts for derivational correctness. In so doing, they gesture at their own preferred account of scientific representation, the inferential-expressivist account (Khalifa et al. 2022).

Emily Sullivan's exchange with Michael Tamir and Elay Shech concerns the role of representation, understanding, and values in the context of machine learning (ML) models. In Chapter 24, Sullivan identifies three stages at which non-epistemic values determine the extent to which ML and deep learning models have an opacity problem. Specifically, she looks at ML model acceptance and establishing the link between the model and phenomenon, explanation with ML models, and attributions of understanding due to ML models. In all such contexts, Sullivan identifies how "questions of ML opacity, explanation, and understanding are entangled with non-epistemic values." Part of her argument is based on a claim argued for in Sullivan (2022) to the effect that the extent to which model opacity and complexity is a hindrance to understanding and explanation depends on how much "link uncertainty" the model has. Link uncertainty concerns the lack of empirical support linking a model to a target. In their contribution in Chapter 25, Tamir and Shech contrast how the term "model" in ML contrasts with traditional usage of scientific models for understanding and thus suggest the "Target of Machine Learning (TML) Hypothesis," namely, that the target phenomenon of understanding with ML models is the relationship(s) of features represented by the data. They then identify three modes of understanding given the said target, and disambiguate what they describe as the difference between implementation irrelevance and functionally approximate irrelevance in order to explore how this distinction impacts potential understanding with ML models. In so doing, Tamir and Shech address what they claim is an ambiguity in Sullivan's concept of link uncertainty, arguing that distinguishing empirical link failures from representational ones clarifies the role played by scientific background knowledge in enabling understanding with ML.

In her response in Chapter 26, Sullivan identifies various points of disagreement with Tamir and Shech, including issues having to do with the limits of implementation irrelevance, empirical versus representational link uncertainty, and the ostensible target of understanding with ML models. For instance, she notes that the TML hypothesis is model-centric and thus overlooks important ways in which ML may provide understanding of the aspects of the world, for example, by inducing explanations of phenomena. In contrast, Shech and Tamir maintain in Chapter 27 that we ought to moderate our expectations of the role that ML models play. For instance, they distinguish between the epistemic risk associated with making inferences based on an ML model that may turn out to be false and the practical risk associated with taking (or withholding) actions based on the inferences made using these models. They then argue that the notion of epistemic risk is not exclusively or especially problematic for ML models.

4 Understanding and Scientific Progress

Finally, issues of scientific understanding and representation figure prominently in philosophical discussions of scientific progress. A venerable view holds that science progresses as it achieves greater verisimilitude (Niiniluoto 2014, Popper 1972), which is clearly a kind of representational success. More pragmatically inclined philosophers characterize scientific progress in terms of its problem-solving capacity (Kuhn 1970, Laudan 1977). More recently, philosophers have proposed increased knowledge (Bird 2007) and increased understanding (Dellsén 2016) as alternative criteria of scientific progress. Finnur Dellsén and Casey McCoy explore the interrelations between several of these views in Part IV, titled "Understanding and Scientific Progress." In Chapter 28, McCoy argues that understanding is the principal epistemic or cognitive aim of science. He develops an understanding-based account of scientific progress with an eye on problem-solving practices in science. Scientists progress when they improve their theoretical understanding of the phenomena of interest. McCoy takes problem-solving abilities to be a measure of the understanding achieved or the potential understanding that could be obtained with a new scientific paradigm. In Chapter 29, Dellsén argues that scientific achievements (such as theories) need not be epistemically justified to be progressive (contra Bird's (2007) epistemic account of progress). Otherwise, scientific progress would be rarely achieved when scientists acknowledge that most previous theories about the phenomenon in question turn out to be false, when theories merely subsume or strengthen previous theories, or when scientific peers disagree on the correct theory for a given phenomenon. In all these cases, our scientific achievements are not justified, although they can be progressive, such as groundbreaking new work.

In his commentary in Chapter 30, McCoy submits that none of these cases need to be problematic when we analyze the progress made as occurring precisely at the point when we are adequately justified. He maintains that determining the beginning of a progressive period is subject to intuition—especially in Dellsén's cases. McCoy also points out that regardless of whether justification is necessary for progress, it plays such an integral part in the genesis of progress that it needs to be reflected in an adequate account of progress. In his commentary in Chapter 31, Dellsén argues that problem-solving does not constitute scientific progress. Following Laudan (1977), what a problem is and how it can be solved is determined based on the research tradition in question. Solutions to problems need not be (approximately) true. Dellsén suggests that this fails to make sense of the possibility of misunderstanding a phenomenon or misunderstanding when an achievement is progressive. If the role of problem-solving is merely to promote progress, it does not seem to be integral to scientific progress. On the one hand, problem-solving does not always promote progress, since solutions to problems might distract researchers or lead them astray in some cases. On the other hand, scientific achievements other than problem-solving, such as collecting new data, may also promote progress.

As we believe this introduction already suggests, scientific understanding, scientific representation, and their points of intersection stand as a touchstone for several related themes, including knowledge, explanation, scientific realism, reasoning, and scientific progress. As such, we hope that the chapters provide fresh insights on these perennial issues in the philosophy of science.

Note

1 See Frigg and Nguyen (2020) and Grimm (2021) for reviews of the scientific representation and understanding literatures, respectively.

References

Batterman, Robert. 2021. *A Middle Way: A Non-Fundamental Approach to Many-Body Physics*. New York: Oxford University Press. https://doi.org/10.1093/oso/9780197568613.001.0001.

Bird, Alexander. 2007. "What Is Scientific Progress?" *Nous* 41: 64–89. https://doi.org/10.1111/j.1468-0068.2007.00638.x.

Brandom, Robert. 1994. *Making It Explicit: Reasoning, Representing, and Discursive Commitment*. Cambridge, MA: Harvard University Press.

Bursten, Julia R. 2018. "Conceptual Strategies and Inter-theory Relations: The Case of Nanoscale Cracks." *Studies in History and Philosophy of Science Part B* 62: 158–165. https://doi.org/10.1016/j.shpsb.2017.09.001.

Chakravartty, Anjan. 2017. "Scientific Realism." In E. N. Zalta (Ed.), *The Stanford Encyclopedia of Philosophy* (Summer 2017 Edition), https://plato.stanford.edu/archives/sum2017/entries/scientific-realism/.

Craig, Edward. 1990. *Knowledge and the State of Nature: An Essay in Conceptual Synthesis*. Oxford: Clarendon Press.

Dellsén, Finnur. 2016. "Scientific Progress: Knowledge versus Understanding." *Studies in History and Philosophy of Science* 56: 72–83. https://doi.org/10.1016/j.shpsa.2016.01.003

De Regt, H. 2017. *Understanding Scientific Understanding*. New York: Oxford University Press. https://doi.org/10.1093/oso/9780190652913.001.0001.

Elgin, Catherine Z. 2017. *True Enough*. Cambridge, MA: MIT Press. http://dx.doi.org/10.7551/mitpress/9780262036535.001.0001.

Frigg, R., and Nguyen, J. 2020. *Modelling Nature. An Opinionated Introduction to Scientific Representation*. Berlin and New York: Springer. https://doi.org/10.1007/978-3-030-45153-0.

Giere, Ronald. 2006. *Scientific Perspectivism*. Chicago: University of Chicago Press. https://doi.org/10.7208/chicago/9780226292144.001.0001.

Grimm, Stephen R. 2021. Understanding. In E. N. Zalta (Ed.), *Stanford Encyclopedia of Philosophy*. https://plato.stanford.edu/archives/sum2021/entries/understanding/. Accessed July 24, 2022.

Hempel, C. G. 1965. *Aspects of Scientific Explanation, and Other Essays in the Philosophy of Science*. New York: Free Press.

Khalifa, Kareem. 2012. "Inaugurating Understanding or Repackaging Explanation?" *Philosophy of Science* 79(1): 15–37. https://doi.org/10.1086/663235.

Khalifa, Kareem. 2017a. "Review of *Understanding Scientific Understanding*, by Henk W. de Regt." *Notre Dame Philosophical Review*. https://ndpr.nd.edu/news/understanding-scientific-understanding/. Accessed December 22, 2017.

Khalifa, Kareem. 2017b. *Understanding, Explanation, and Scientific Knowledge*. Cambridge: Cambridge University Press.

Khalifa, Kareem. 2020. "Understanding, Truth, and Epistemic Goals." *Philosophy of Science* 87(5): 944–956. https://doi.org/10.1086/710545.

Khalifa, Kareem, Millson, Jared, and Risjord, Mark. 2022. "Scientific Representation: An Inferentialist-Expressivist Manifesto." *Philosophical Topics*. 50(1): 263–291. https://doi.org/10.5840/philtopics202250112

Kitcher, P. 1989. "Explanatory Unification and the Causal Structure of the World." In P. Kitcher and W. C. Salmon (Eds.), *Scientific Explanation* (Vol. XIII, pp. 410–506). Minneapolis: University of Minnesota Press.

Kostić, Daniel. 2019. "Minimal Structure Explanations, Scientific Understanding and Explanatory Depth." *Perspectives on Science* 27(1): 48–67. https://doi.org/10.1162/posc_a_00299.

Kostić, Daniel. 2020. "General Theory of Topological Explanations and Explanatory Asymmetry." *Philosophical Transactions of the Royal Society B: Biological Sciences* 375(1796): 20190321. https://doi.org/10.1098/rstb.2019.0321.

Kuhn, Thomas S. 1970. *The Structure of Scientific Revolutions*, 2nd edition. Chicago: University of Chicago Press.

Laudan, Larry. 1977. *Progress and Its Problems: Towards a Theory of Scientific Growth*. London: Routledge and Kegan Paul.

Lipton, P. 2009. "Understanding Without Explanation." In H. W. De Regt, S. Leonelli, and K. Eigner (Eds.), *Scientific Understanding: Philosophical Perspectives* (pp. 43–63). Pittsburgh: University of Pittsburgh Press.

Massimi, Michela. 2022. *Perspectival Realism*. Oxford University Press.

Niiniluoto, Ilkka. 2014. "Scientific Progress as Increasing Verisimilitude." *Studies in History and Philosophy of Science Part A* 46: 73–77. https://doi.org/10.1016/j.shpsa.2014.02.002.

Popper, Karl Raimund. 1972. *Conjectures and Refutations: The Growth of Scientific Knowledge*, 4th ed. London: Routledge & Kegan Paul.

Potochnik, Angela. 2017. *Idealization and the Aims of Science*. Chicago: University of Chicago Press. https://doi.org/10.7208/chicago/9780226507194.001.0001.

Sullivan, Emily. 2022. "Understanding from Machine Learning Models." *The British Journal for the Philosophy of Science* 73(1): 109–133. https://doi.org/10.1093/bjps/axz035.

Waters, C. Kenneth. 2017. "No General Structure." In M. H. Slater and Z. Yudell (Eds.), *Metaphysics and the Philosophy of Science: New Essays* (pp. 81–108). Oxford University Press. https://doi.org/10.1093/acprof:oso/9780199363209.003.0005.

Waters, C. Kenneth. 2018. "Ask Not 'What Is an Individual.'" In O. Bueno, R.-L. Chen, and M. Fagan (Eds.), *Individuation, Process, and Scientific Practices* (pp. 91–113). Oxford University Press. https://doi.org/10.1093/oso/9780190636814.003.0005.

Woodward, J. 2003. *Making Things Happen: A Theory of Causal Explanation.* New York: Oxford University Press. https://doi.org/10.1093/0195155270.001.0001.

Part I

Understanding, Knowledge, and Explanation

2 Can Scientific Understanding Be Reduced to Knowledge?

Henk W. de Regt

1 Introduction

Philosophers who have addressed the issue of scientific understanding have traditionally related it to the activity of explanation: scientists try to explain the phenomena they observe, and when they have found a satisfactory explanation, the phenomenon is understood. Thus, understanding appears to be the product of scientific explanation. But what exactly is the nature of that product? Long ago, when Carl Hempel's covering law model of explanation still ruled, the received view was that understanding is not much more than a psychological by-product of explanation, and accordingly uninteresting from a philosophical point of view. Although this idea is less popular today, it is still debated whether understanding is a cognitive achievement in its own right. The "friends of understanding" – as Kareem Khalifa calls us – argue that it is, but others, including Khalifa himself, deny understanding this special status, arguing that it can be reduced to the more traditional notions of knowledge and explanation. Thus, in *Understanding, Explanation, and Scientific Knowledge*, Khalifa (2017) develops an account of understanding as knowledge of explanations. Placing himself in the tradition of the received view founded by Hempel, he remarks that "the motivation for a radically novel conception of understanding will have to challenge understanding's status as a species of *knowledge*" (2017, 154; original italics).

My contextual theory of scientific understanding presents such a challenge, in the form of an argument that understanding crucially involves (context-dependent) skills and is thereby more than just a species of knowledge. A corollary of the theory is that scientific understanding is not factive. Khalifa's account, by contrast, denies both a role for special skills and the non-factivity of understanding. According to his Explanation-Knowledge-Science (EKS) model of explanatory understanding, scientific understanding is nothing more than having "scientific knowledge of an explanation" (2017, 11). This implies that understanding is factive, and that it does not require skills over and above those required for (scientific) knowledge.

DOI: 10.4324/9781003202905-3

In this chapter, I will critique Khalifa's claim that the skills involved in possessing an explanation are not special and his related claim that "grasping" an explanation is equivalent to having scientific knowledge of it. Section 2 argues that understanding is more than just a species of knowledge, and summarizes my contextual theory of scientific understanding. Section 3 focuses on the role of skills and replies to Khalifa's objections. Section 4 defends the thesis that (scientific) understanding is non-factive, and counters Khalifa's rival thesis that understanding is "quasi-factive".

2 Scientific Understanding: A Species of Knowledge?

The idea that the understanding produced by scientific explanations is nothing but a species of knowledge has been around for a while. Peter Lipton (2004, 30), for example, writes: "Understanding is not some sort of super-knowledge, but simply more knowledge: knowledge of causes." Consider, as an example, that we know that since 1900 the average temperature on Earth has increased; we have knowledge of global warming. When climate scientists explain this phenomenon, by pointing at the so-called greenhouse effect and the increase of CO_2 in the atmosphere, they have added more knowledge, namely knowledge of the *cause* of global warming. The understanding-as-knowledge view has recently been defended not only by Khalifa but also by Grimm (2006), Mizrahi (2012), Strevens (2017), and Kelp (2017). Although this view of understanding is more plausible than the idea that understanding is merely psychological, it turns out to be problematic when examined in more detail. I will present three concrete objections to it.

The first problem is that knowledge presupposes truth: we can only *know* something if it is true. So, if "understanding a phenomenon" is the same as "knowing the cause of a phenomenon", we can only understand a phenomenon if we know the *true* cause of it. At first sight, this appears quite reasonable. Consider, for example, the medieval theory about planetary motion, which assumes that the planets move because they are pushed by angels. This theory does not give us understanding, one might think, simply because it is not true. However, although this seems a plausible argument, it turns out that not much understanding is left if we make truth a requirement for understanding. Another look at the history of science makes this clear almost immediately: many scientific theories from the past were quite successful at the *empirical* level; they could describe and predict a host of observable phenomena in their domain. But most of these theories were later rejected, and they were replaced with new theories that describe and predict the phenomena even better. These new theories are often in direct contradiction with the old ones, which means that they cannot all be true. So, despite their empirical success, the old theories turned out to be false after all.

Does that mean that the scientists who developed and defended these theories lacked an understanding of the phenomena they could describe and predict so well with their theories? I think that this conclusion does not do them justice.

An example is Newton's theory of gravitation. This theory was, and still is, extremely successful in describing and predicting gravitational phenomena, such as the motion of falling bodies on the Earth, ballistics, the tides, and planetary motion. But from the viewpoint of current physics, Newton's theory does not give us a true account of the cause of gravitational phenomena. Since 1919, when Eddington's observations of the bending of light by the sun confirmed Einstein's general theory of relativity, the latter theory is regarded as the true explanation of gravitation. Einstein's theory differs fundamentally from Newton's theory: according to Einstein, Newton's force of gravitation is a fiction. Gravitational motion is caused by the curvature of space-time in the presence of masses. Does the fact that his theory turned out to be false imply that Newton did not have any understanding of gravitational phenomena? And that students in secondary school today, who still have to learn Newton's theory in their physics classes, do not acquire any understanding, but rather misunderstanding? I disagree: the case of Newton shows that also theories that we now know are false can still give us understanding, at least to some degree. In Section 4 below, this thesis will be explained and defended in more detail.

The second problem for the view that understanding is a species of knowledge is that the history of science teaches us that criteria for understanding and intelligibility vary considerably in the course of time. In the seventeenth century, for example, the generally accepted view was that only a mechanics based on contact action is intelligible, whereas in the eighteenth century (as a result of the success of Newton's theory of gravitation) action-at-a-distance was regarded as the paradigm of intelligibility (see Van Lunteren 1991). Of course, one might claim that only one of these positions is the correct one, but that would not do justice to the history of science. A more plausible option is to acknowledge that understanding is *contextual*: criteria for understanding and intelligibility depend on the historical context (and also on the disciplinary context). The strong contextual variation of such criteria does not easily align with the idea that understanding is a kind of knowledge.

The third problem for the idea of understanding as knowledge is that understanding appears to involve more than just knowledge. If you *know* what the cause of a particular phenomenon is, this does not imply that you *understand* that phenomenon. For example, merely knowing that global warming is caused by the increase of CO_2 in the atmosphere does not yet amount to understanding it. A student may be able to answer the question "Why does global warming happen"? correctly by answering "Because of the increase of CO_2 in the atmosphere". But this does not imply that she understands why global warming occurs – she merely knows what its cause is. This shows that understanding must be more than just having a particular type of knowledge. And this more, I submit, crucially involves *skills*. The student understands why global warming happens if she not only knows that it is caused by the increase of CO_2, but also "grasps" the causal, explanatory relation between cause and effect. In this case, she needs a theory or model of the climate system that includes the greenhouse effect. Moreover, she needs to be able to use that theory in order to derive the (causal) relation between the amount of CO_2 in the atmosphere and the mean global temperature. In sum, not only does she has to have knowledge, but she also needs the *ability* to use her knowledge. The latter is the crucial skill-component of (scientific) understanding.

The idea that understanding is an ability, a skill, is a central element of the contextual theory of scientific understanding that I have developed in my book *Understanding Scientific Understanding* (De Regt 2017). My theory is based on an analysis of episodes from the history of science, which shows that the criteria for what constitutes scientific understanding, and how to achieve it, can and do change in the course of history. What is more, even at one and the same time, scientists sometimes disagree about the criteria for understanding. Such variation is possible because understanding is context-dependent. This is the case because understanding involves a thinking subject, an "understander". It is human beings who understand, and a human being is always part of a context – a historical and social context, for example. For scientists, the context is, in important ways, determined by their disciplinary education, by their background knowledge, and by the state-of-the-art in their field. So, whether or not a scientific theory can provide understanding is partly dependent on who the understanders are, and what their context is.

It might be objected that this makes understanding a subjective affair that cannot be relevant to scientific explanation – which should be an objective and a generally valid relation between a theory and the phenomenon to be explained. However, as Cartwright (1983) and Morgan and Morrison (1999) have shown, a closer look at scientific practice teaches us that the relation between abstract theories and concrete observable phenomena is usually not a direct one, but one that is mediated by models. Reality is complex, and it is seldom the case that the behavior of a real system can be deduced directly from an abstract, general theory. The usual situation is one in which scientists first construe a simplified, idealized model of the system, and then apply the theory to it. Constructing such models is an art; it is a matter of choosing the right idealizations and approximations. There are no strict rules or algorithms for building scientific models: it is a human activity that involves skills, which can only be acquired in practice. So scientific explanation has a human face: scientists are human beings who should be able to build suitable models to explain phenomena. Whether or not a scientist is able to construct a model that relates the theory with the phenomena depends on two factors: on the one hand, the skills of the scientist, and on the other hand, the qualities of the theory. And there has to be a match between the two: scientists should have the right skills to work with the theory, to use it for building models of the phenomena.

In my contextual theory of scientific understanding, this idea is translated into the requirement that the theory used in the explanation has to be *intelligible* to the scientists, where intelligibility is defined as the value that scientists attribute to the cluster of qualities of the theory that facilitate its use (De Regt 2017, 40). Note that intelligibility, defined in this way, is not an intrinsic property of a theory, but is relative to (a community of) scientists. Accordingly, it does not make sense to say that, for example, the theory of evolution or quantum theory is intelligible, or unintelligible, in itself. Whether or not these theories are intelligible depends on the skills and background knowledge of the scientists who use them, that is, on contextual factors. However, this context-relativity does not entail that understanding is subjective: there are objective ways to test whether a theory is intelligible (to a scientist in a particular context).[1]

If my theory of scientific understanding holds water, understanding is more than just a species of knowledge. In particular, it is more than just knowledge of an explanation. Scientific explanations do not simply appear out of thin air; they are developed by scientists, who have to employ their expert skills to construct models of phenomena to which theories can be applied. These skills are at the core of understanding: without them, explanations would not have been developed in the first place, and what is more, without them explanations will not provide understanding to the understanders.[2] In the next section, we will look more closely at the skills involved in scientific understanding.

3 The Role of Skills: Khalifa's Criticism and a Reply

Let us start by presenting some examples of skills that may play a role in scientific understanding. First, even in the simplest cases of covering-law explanation, it appears that deductive reasoning skills are required. To use Hempel's original example of a DN-explanation: one can explain the fact that when a mercury thermometer is rapidly immersed in hot water, first the mercury column drops and then it quickly rises, by invoking "the laws of thermic expansion of mercury and of glass, and a statement about the small thermic conductivity of glass" (Hempel 1965, 246). With these laws and statements about the relevant background conditions, a DN-explanation can be constructed. However, merely *knowing* these laws and the background conditions is not equal to possessing an explanation: in addition, one should be able to use this knowledge in the right way to *deduce* the explanandum (De Regt 2017, 25). Thus, a student may have memorized the laws in question but may still be unable to use this knowledge to account for the behavior of the mercury column, when presented with it in an exam. The student also needs a specific skill to compose the explanation and thereby acquire understanding: the ability to construct a deductive argument from the available knowledge.

A second example of relevant skills, already briefly mentioned in the previous section, concerns idealization. Scientific understanding is often furnished via idealized models. Again, idealized models do not appear out of thin air: they have to be constructed. Idealization is accordingly not only a product but also a process, for which a particular skill is needed: the scientist should be able to choose the right idealizing assumptions. As mentioned above, this is not an algorithmic decision but involves human judgment. A scientist's ability to choose successful idealizing assumptions may be termed their "idealization skill".

As a case in point, consider the molecular models that have been constructed to achieve understanding of the behavior of gases. Such understanding results from application of the kinetic theory of gases to idealized models of the target systems in question. The most basic and most famous model is the ideal gas model, which represents gases as collections of elastic spheres (or centers of force) such that theoretical predictions of the relationship between pressure, volume, and temperature can be derived. The ideal gas model does not follow deductively from the kinetic theory nor from the target system: constructing it involves idealization skills. The same holds for more sophisticated models that have been developed to further explain the properties and behavior of gases. An example is the so-called dumbbell model that Ludwig Boltzmann advanced in order to solve the "specific heat anomaly". The story of the development and reception of this highly idealized model of diatomic gases nicely illustrates the central role of idealization skills in explanatory practice (see De Regt 2017, 205–216).

A third example of how skills figure in the process of achieving scientific understanding can be found in the history of Feynman diagrams, introduced by Richard Feynman to facilitate working with the abstract mathematical theory of quantum electrodynamics developed by Julian Schwinger and Sin-Itoro Tomonaga. Feynman diagrams are tools that are used to make quantum electrodynamics (QED) intelligible, and the successful use of these tools requires specific skills, as argued by David Kaiser in his book-length account of the history of Feynman diagrams. The acquisition of these skills is a matter of practicing under expert guidance. Although some physicists learned how to calculate with Feynman diagrams from published sources, they failed to use them as tools for dealing with completely new problem situations. According to Kaiser (2005, 13): "some form of understanding could be packaged and transmitted via texts, but not the more improvisational uses developed by those groups of physicists who shared informal contacts". The success of Feynman's diagrams was partly based on their visual nature – visualization is a ubiquitous skill that is widely used to enhance understanding, both in science and in other domains. In addition, however, effective use of Feynman diagrams required more specific skills, as Kaiser demonstrates – skills that were less widespread and more difficult to acquire, even by educated physicists.[3]

Examples such as the ones just given have not convinced Khalifa, who resists the idea that understanding is distinct from knowledge in that it requires specific skills (or abilities, as Khalifa prefers to call them). To be sure, Khalifa does not deny the fact that some skills are involved in understanding, but he adds that these are of a rather trivial kind and that also knowledge already "requires a fair amount ability" (2017, 54). What he denies is the thesis that understanding requires additional "special abilities" (ibid.). Against the popular thesis that understanding involves "grasping", a special ability that is not needed for mere scientific knowledge, Khalifa (2017, 64–65) argues that regardless of whatever ability is required for grasping, it is not special: grasping is nothing but having scientific knowledge of an explanation. Moreover, he admits that grasping in the latter sense requires possessing the concept of "explanation", which, in turn, requires the ability to answer what-if questions, but he suggests that also this ability is not special (2017, 59).

I submit, by contrast, that what-if reasoning is a special ability that exceeds believing, or knowing, an explanation. Thus, there is a difference between a student who simply *knows* the explanation that the greenhouse effect explains global warming and a student who *understands* why global warming occurs on the basis of this explanation. The former may have no clue when asked, for example, what would happen if the concentration of a particular greenhouse gas in the atmosphere increases, whereas the latter should be able to answer such what-if questions. In other words, the latter student has a special ability that the former lacks.

In his paper "Inaugurating understanding or repackaging explanation?", Khalifa (2012) directly addresses my arguments for a skill-based conception of understanding. First, he criticizes my analysis of the role of skills in covering-law explanation (first example above) and my interpretation of Hempel's account of it. To support his criticism, he cites a statement from the concluding section of Hempel's landmark essay *Aspects of Scientific Explanation*: "The understanding [that scientific explanation] conveys lies rather in the *insight* that the explanandum *fits into, or can be subsumed under*, a system of uniformities represented by empirical laws or theoretical principles" (Hempel 1965, 488; quoted by Khalifa 2012, 26, italics added). According to Khalifa, this shows that Hempel does not think that knowledge of explanans and explanandum suffices for possessing an explanation (and accordingly understanding in the Hempelian sense). In addition, one also needs "insight that the explanandum fits into, or can be subsumed under" the covering laws:

> De Regt includes knowledge of the explanans and explanandum, but he omits the student's "possession" of the information concerning their inferential connection. Curiously, this knowledge perfectly plugs the gap in the student's understanding. Moreover, it is just more propositional knowledge and, thus, need not be "know-how" or a skill. As a result, there is a Hempelian treatment of the example that makes no essential appeal to the skill condition.
>
> (Khalifa 2012, 26)

Khalifa calls my interpretation of Hempel "not entirely charitable" (25). Perhaps, but a more charitable interpretation would imply that Hempel here acknowledges a role for skills in understanding. While Khalifa assumes that Hempel's "insight" is merely "information" (or propositional knowledge), the quote does not imply this at all. On the contrary, the insight concerns an inferential connection, which clearly involves an ability: the student should not only *know that* the explanans entails the explanandum, but she must also *be able* to deduce the latter from the former, or else she still does not understand.[4]

Khalifa goes on to suggest that even if one grants that skills are needed for constructing deductive-nomological (DN) or model-based explanations, this is not very interesting: "just as we do not need a theory of understanding to tell us that DN understanding involves deductive-reasoning skills, it is not exactly newsworthy that constructing model-based explanations involves competence in approximating and idealizing" (2012, 28). Now, I am not sure whether newsworthiness is a strong argument for the philosophical importance of a claim (the news that Kanye West and Kim Kardashian are getting a divorce did not strike me as philosophically very interesting). But a competence is a skill, and accordingly differs from propositional knowledge, so the fact that idealization skills are involved in constructing model-based explanations should at least be newsworthy to those who favor an understanding-as-knowledge view.

4 Is Understanding Factive?

Accounts of scientific understanding that identify it with (a specific type of) knowledge imply that understanding – like any type of knowledge – must be *factive*. Knowledge that p is factive, because one can only know p if p is true.[5] If, following Lipton, "understanding why p" equals knowing that q causes p, then "q causes p" must be true, or else one cannot know that q causes p. Similarly, in Khalifa's view, if understanding why p amounts to knowing the explanation of p, then the explanation of *p* must be true. This, in turn, suggests that the theories and models on which the explanation of *p* is based are either (approximately) true or represent reality in an accurate way.[6]

However, plausible as this idea may look at first sight, it turns out to be problematic when examined in more detail. The reason is that truth is a very strong requirement: many scientific theories and models that are used in scientific practice are strictly speaking false. One important class of examples consists of models that involve idealizations. Science abounds with idealized models, which are inaccurate representations of their target systems in reality. A classic example was mentioned in the previous section: the ideal gas model. This model pictures a gas as a collection of randomly moving point particles without any intermolecular forces, which behaves according to the ideal gas law $PV = nRT$. The assumption that there are no intermolecular forces is an idealization: in real gases, such forces are always present. So, can we invoke the ideal gas model to understand the fact that in many circumstances the behavior of real gases conforms (approximately) to the ideal gas law? It seems that a factivist should answer this question in the negative: ideal gases are fictions that cannot be used to explain and understand how real gases behave. Some factivists have responded to this challenge by arguing for a more sophisticated version of their position, *quasi-factivism*, which asserts that not all of the elements of a theory or model have to correspond to reality.[7] As long as the "central" elements are true, the "peripheral" elements may be false (see Mizrahi 2012, 239, esp. figure 1). The quasi-factivist view is defended by Kvanvig (2009, 341), who claims that understanding of (a part of) reality "is related, presumably, to various pieces of information, and on the quasi-factive view, the pieces of information that are central to the understanding must be true".[8] A quasi-factivist approach to scientific understanding suggests that the idealizing assumptions in a model are relatively unimportant ("peripheral") and thus not essential for understanding. Still, Mizrahi (2012, 244) claims that de-idealization leads to an increase of understanding, with reference to the example of the van der Waals state equation.[9]

I submit, by contrast, that idealizations in models are typically not peripheral but essential to the enhancement of scientific understanding, because they enhance intelligibility. Therefore, I endorse a *non-factivist* view of scientific understanding. Next to the ubiquitous use of idealization in science, a different argument against factivism invokes the history of science. It regularly happens that an older theory, although superseded by a newer theory, is still used to achieve understanding in contexts where the new theory is not easily applicable. Sometimes, the old theory differs radically from the new theory, which implies that if the newer theory is (approximately) true, the old theory is false. A famous example is Newton's theory of gravitation. As has been discussed above, this theory was, and still is, extremely successful in describing and predicting gravitational phenomena. From the viewpoint of current physics, however, Newton's theory does not give us a true account of the cause of these phenomena. Nowadays, Einstein's general theory of relativity is regarded as the true explanation of gravitation. Einstein's theory differs fundamentally from Newton's theory: it states that gravitational motion is caused by the local curvature of space-time, rather than by a universal force of gravitation.[10] Hence, on a (quasi-)factivist view, Newton's theory could not give us any understanding of the phenomena: its central propositions are not even approximately true. But this seems absurd: when high-school students learn Newtonian mechanics, they acquire some degree of understanding – rather than misunderstanding – of natural phenomena.

A non-factivist approach to scientific understanding does justice to these intuitions. Non-factivism regarding understanding is defended by, for example, Angela Potochnik and Catherine Elgin. Potochnik offers a detailed investigation of the role of idealizations in achieving the aims of science, which leads her to deny that understanding is factive: "less truth can, in the proper circumstances, lead to greater understanding than would more truth" (2017, 103). Elgin (2017, 38) explicitly states that "a nonfactive explication of 'understanding' yields a concept that better suits epistemology's purposes than a factive one". It should be emphasized that non-factivists do not claim that facts, or truth, are irrelevant to scientific understanding. Surely, in order to provide understanding, a theory or model should at least get the relevant *empirical* facts right, although the degree to which it does so may vary with the context. In this sense, "understanding somehow answers to facts", as Elgin (2007, 37) observes.

My contextual theory of scientific understanding also entails a non-factive view of understanding, since it is the intelligibility rather than the truth of the theory that determines its success in providing explanatory understanding. A theory that is true but unintelligible cannot be used to construct satisfactory explanations, whereas a theory that is false but intelligible can in the right circumstances be useful for constructing explanations.[11] My emphasis on the importance of intelligibility (rather than truth) reflects the idea that understanding is a skill (rather than a type of knowledge). Since this skill can also be effective while using false theories, it implies that understanding is non-factive. Thus, my account accommodates both the use of idealizations in scientific practice and the idea that outdated theories (like Newton's) could produce understanding in the past and, in some cases, can still produce understanding today.

Khalifa (2017), by contrast, endorses the thesis that understanding is a species of knowledge and hence holds that scientific understanding is (quasi-)factive.[12] He has elaborated this in his EKS model of explanatory understanding. The EKS model is a comparative account of understanding, stating conditions for when "S1 understands why p better than S2" (2017, 14). Thus, there are degrees of understanding, and Khalifa suggests that these range from minimal understanding, via everyday understanding and a typical scientist's understanding, to ideal understanding. Minimal understanding of p is achieved when "S believes that *q explains why p*, and *q explains why p* is approximately true" (ibid.). Since even minimal understanding requires approximate truth, understanding is (quasi-)factive, according to Khalifa. So how does he deal with non-factivists' favorite examples of (alleged) scientific understanding without truth?

Against the historical argument that scientists of the past understood phenomena even in cases where we now know that their theories are false, Khalifa suggests that quasi-factivists can deal with those cases in two ways: either argue that the putative explanation in question is in fact (approximately) correct or argue that it is incorrect and, as such, does not yield genuine understanding. He illustrates these strategies via an analysis of the phlogiston case, which I have used to argue for non-factivism.[13] Phlogiston theory was an eighteenth-century chemical theory used to explain phenomena such as combustion and calcination. The theory assumed the existence of "phlogiston", an entity (or principle) that allegedly escaped from matter in the process of combustion. Around 1800, it was superseded by Lavoisier's oxygen theory, which marks the birth of modern chemistry. Today, phlogiston is regarded as fiction and the phlogiston theory as blatantly false. Nevertheless, I submit, eighteenth-century scientists did achieve (some degree of) understanding with the phlogiston theory, because they successfully used it to account for some aspects of combustion and calcination (see De Regt 2019). In this respect, there is no essential difference with the case of Newton versus Einstein: also, Newton's theory yielded understanding in the eighteenth and nineteenth century, whereas it has been superseded by Einstein's theory in the twentieth century. De facto, however, Newton's theory can still be used today to achieve understanding, because in certain contexts it is more intelligible than Einstein's, whereas the phlogiston theory is not intelligible anymore today.

Khalifa (2017, 162–163) first considers a possible quasi-factivist response that invokes the idea that phlogiston-theoretical explanations are, in some sense, correct. One way to do so is by adopting an antirealist stance: if phlogiston theory correctly predicts observable features of combustion and calcination, antirealists would allow it to figure in explanations while denying that phlogiston exists. Factivism can, indeed, be salvaged in this way, but it weakens factivism by relinquishing the intuition that scientific understanding requires truth across the board, rather than only with regard to empirical phenomena. Another way in which quasi-factivists might endorse phlogiston-theoretical explanations is by claiming that "phlogiston theorists were successfully referring to entities", for example, to free electrons, chemical potential energy, hydrogen, and oxygen (2017, 163–164). However, although some elements of phlogiston theory do seem to anticipate currently accepted theoretical terms (e.g., "dephlogisticated air" anticipates "oxygen"), the theory remains radically different from modern chemical theory (most notably because of its "principlist" nature; see Chang 2012). It appears too much of a stretch to interpret both theories realistically and consider them both to be true.[14]

But Khalifa has a second response to the historical argument, which has more force and is the one he seems to favor (see 2017, 162). Basically, the central idea of this response is that phlogiston theorists were at best "on the right track", but they certainly did not have "full-blown scientific understanding" (2017, 165). Accordingly, Khalifa characterizes their cognitive achievement as "proto-understanding", which falls short of explanatory understanding because it does not even meet the standards for minimal understanding (esp. the approximate truth of the explanation). One has proto-understanding if "one grasps a proposition's explanatory role but cannot situate that proposition within a correct explanation" (2017, 166). For example, although Joseph Priestley – a main proponent of phlogiston theory – discovered the conditions under which dephlogisticated air could be isolated, and thus grasped the explanatory role of the proposition that dephlogisticated air can be isolated under these conditions, he did not grasp "the larger explanation in which that role is played" (2017, 165). Khalifa does not tell us what this larger explanation is, but it seems clear that he takes it to be the modern chemists' explanation in terms of oxygen (indeed, he does not even mention dephlogisticated air, but he rather credits phlogiston theorists with the discovery of phenomena involving oxygen).

Although this may seem a plausible response for the case of phlogiston, there is a snag. For how does it apply to the case of Newton versus Einstein on gravitation, mentioned above? Here again, we have two radically different theories, both of which, therefore, cannot be approximately true (in the realist sense). Accordingly, on Khalifa's account, Newton's theory can at best give us "proto-understanding" of gravitational phenomena, and it falls short of genuine understanding. But that verdict is surely at odds with our intuitions: we would not want to say that Newton failed to meet the benchmark for minimal understanding, when he developed and published his theories in the *Principia*, his 1687 magnum opus that determined the course of physics for two centuries. Also, we would not want to deny high-school students a minimal understanding of gravitational phenomena when they have mastered Newton's theory. Finally, Khalifa's response to the historical argument is vulnerable to the pessimistic meta-induction, which implies that we do not have good reasons to believe that our currently accepted scientific theories are (at least approximately) true. Accordingly, if Khalifa ascribes mere proto-understanding to Priestley, in retrospect it may turn out that the same applies to today's scientists and that they should be denied "full-blown scientific understanding" as well.

The solution to this problem is to adopt a non-factivist perspective and reject the idea that there is an absolute benchmark for scientific understanding, such as (approximate) truth. This move does not imply giving up the possibility to distinguish between misunderstanding and genuine understanding, or between proto-understanding and full-blown scientific understanding. Such distinctions can be retained if the absolute benchmark is replaced with a *contextual* threshold for minimal understanding. With a contextual account of understanding, it is possible to ascribe genuine scientific understanding to someone in a particular context, while denying it to someone else in a different context, even if both employ the same theory or model. Applied to the historical argument and the phlogiston example, such contextualization entails that in the eighteenth-century context Priestley may have had a (perhaps minimal but still scientific) understanding of combustion with phlogiston theory; however, in the context of contemporary chemistry, phlogiston-theoretical explanations do not meet the benchmark for scientific understanding.

With respect to the second argument for a non-factivist conception of understanding – the ubiquitous use of idealizations in science – Khalifa (2017, 166–181) suggests that it can be rendered harmless via two different strategies. He calls the first the "Splitting Strategy", because it splits idealized explanations into idealizations and explanations, where the former are merely accepted whereas the latter are believed. The second strategy is the so-called "Swelling Strategy", which broadens the concept of knowledge such that it not only requires belief but also allows for mere acceptance as a condition for knowledge. Let me briefly evaluate Khalifa's views on idealization.[15] The Splitting Strategy assumes that idealizations play a non-essential role in the explanation, and that understanding is achieved not by virtue of the idealizations but rather in spite of them. However, as argued above, idealizations are often essential for understanding: they allow for the application of an intelligible theory to the model system. As regards Khalifa's alternative to deal with such cases, the Swelling Strategy, it seems that the thesis that acceptance of a claim that "q explains why p" is sufficient for understanding why p, as long as that claim is effective for achieving certain (context-dependent) scientific goals, is in line with a non-factivist approach to understanding. In particular, it agrees with the contextual theory, which states that scientific understanding requires intelligibility (rather than truth) of a theory, where intelligibility is associated with the qualities of a theory that facilitate its use. I conclude that the Swelling Strategy stretches the concept of knowledge in such a way that it becomes, indeed, conceivable that it covers understanding as well. Although I sympathize with this approach, I consider it to be a surrender to a (non-factivist) skill-based conception of understanding rather than a successful rescue attempt of the understanding-as-knowledge view.

5 Conclusion

In this chapter, I have argued that scientific understanding is more than just a species of knowledge, most notably because it involves skills, and I have defended the idea that understanding is non-factive as a corollary. These theses, which are associated with my contextual theory of scientific understanding, have been challenged by Kareem Khalifa, whose EKS model portrays understanding as a type of knowledge and accordingly as (quasi-)factive. While I have attempted to rebut Khalifa's view and defended my own, there may be a third way. Perhaps the two views can be reconciled if it is acknowledged that knowledge and skills are intricately related, and that science needs both. This would require a broadened conception of knowledge, on which "knowing" includes both explicit, propositional knowledge and implicit skills (know-how, or tacit knowledge). Both Khalifa's argument that knowledge requires abilities and his Swelling Strategy indicate that he might be open to such a reconciliation. I am happy to join him, but without abandoning understanding as a cognitive achievement in its own right.

Acknowledgments

I am grateful to Insa Lawler and Kareem Khalifa for their helpful comments on earlier versions of this chapter.

Notes

1 One such test is based on the criterion that a scientific theory is intelligible for scientists (in a particular context) if they can recognize qualitatively characteristic consequences of it without performing exact calculations (De Regt 2017, 101–108).
2 Some might object that my account conflates the process and product of explanation, or that it contains a circularity. However, the crucial point is that the skills required for constructing explanations are the same as those required for understanding and applying it. Hence, the circularity is a virtuous one; see De Regt (2017, 44–47).
3 See De Regt (2017, 251–255) for a more detailed account.
4 To justify his claim, Khalifa might opt for an intellectualist view of skills (know-how), which assumes that it can be reduced to propositional knowledge. But in that case, there would be no need for discussing the issue in the first place. The fact that he distinguishes between skills and (propositional) knowledge suggests that he does not take an intellectualist approach to know-how.
5 Knowledge is factive on traditional philosophical accounts of it, such as the view of knowledge as justified true belief. There may, of course, be alternative accounts of knowledge that do not entail factivity. A discussion of these alternatives falls outside the scope of the present chapter.

6 To be sure, this does not necessarily follow, and factivists have proposed alternative accounts on which the truth of "*q* explains *p*" does not require the truth of *q* (e.g., Khalifa 2017) or on which factive explanations can be gained from idealized models or theories (e.g., Lawler 2021).

7 The proponents of quasi-factivism are Kvanvig (2009), Mizrahi (2012), and Khalifa (2017). Stronger factivist views are defended by Strevens (2017) and Lawler (2021). Lawler classifies quasi-factivism as a weak form of non-factivism.

8 Kvanvig focuses on objectual understanding, which differs from explanatory understanding. For our purposes, this distinction is not essential.

9 There seems to be a tension between Mizrahi's view that idealization assumptions are merely peripheral and his claim that their removal increases understanding. I thank Kareem Khalifa for alerting me to this.

10 The fact that Einstein's theory reduces to Newton's theory in the limit of low velocities does not invalidate this point: this reduction concerns empirical predictions only; from an ontological perspective, the theories remain fundamentally at odds. Therefore, Mizrahi's (2012, 247) quasi-factivist interpretation of the Newton-Einstein is misguided.

11 Of course, not any intelligible theory will be useful for this purpose. Astrology, for example, will probably not do in any context (see De Regt 2017, 93).

12 Khalifa (2017, 154–156) defines quasi-factivism as the denial of non-factivism. Although he does not say so explicitly, it seems that quasi-factivism differs from factivism because it requires merely *approximate* truth.

13 See De Regt and Gijsbers (2017) and esp. De Regt (2019) for a detailed account of the case and a non-factivist analysis of it.

14 To be sure, some realist philosophers regard phlogiston as a term that is at least partially referring; see Ladyman (2011) and Radder (2012, 104). The discussion of their views is beyond the scope of this paper.

15 A detailed criticism of these strategies is given in Cornelissen and de Regt (2022).

References

Cartwright, Nancy. 1983. *How the Laws of Physics Lie.* Oxford: Clarendon Press. https://doi.org/10.1093/0198247044.001.0001.

Chang, Hasok. 2012. *Is Water H_2O?* Dordrecht: Springer. https://doi.org/10.1007/978-94-007-3932-1.

Cornelissen, Milo D., and Henk W. de Regt. 2022. Understanding in Synthetic Chemistry: The Case of Periplanone B. Forthcoming in *Synthese*.

De Regt, Henk W. 2017. *Understanding Scientific Understanding.* New York: Oxford University Press. https://doi.org/10.1093/oso/9780190652913.001.0001.

De Regt, Henk W. 2019. From Explanation to Understanding: Normativity Lost? *Journal for the General Philosophy of Science* 50, 327–343. https://doi.org/10.1007/s10838-019-09477-3.

De Regt, Henk W., and Victor Gijsbers. 2017. How False Theories can Yield Genuine Understanding. In S. R. Grimm, C. Baumberger, & S. Ammon (Eds.), *Explaining Understanding: New Perspectives from Epistemology and Philosophy of Science.* New York: Routledge, 50–75.

Elgin, Catherine Z. 2017. *True Enough.* Cambridge, MA: MIT Press. https://doi.org/10.7551/mitpress/9780262036535.001.0001.

Grimm, Stephen R. 2006. Is Understanding a Species of Knowledge? *The British Journal for the Philosophy of Science* 57, 515–535. https://doi.org/10.1093/bjps/axl015.

Hempel, Carl G. 1965. *Aspects of Scientific Explanation and other Essays in the Philosophy of Science*. New York: The Free Press.

Kaiser, David. 2005. *Drawing Theories Apart. The Dispersion of Feynman Diagrams in Postwar Physics*. Chicago: The University of Chicago Press. https://doi.org/10.7208/chicago/9780226422657.001.0001.

Kelp, Christoph. 2017. Towards a Knowledge-based Account of Understanding. In S. R. Grimm, C. Baumberger, & S. Ammon (Eds.), *Explaining Understanding: New Perspectives from Epistemology and Philosophy of Science*. New York: Routledge, 251–271.

Khalifa, Kareem. 2012. Inaugurating Understanding or Repackaging Explanation? *Philosophy of Science* 79, 15–37. https://doi.org/10.1086/663235.

Khalifa, Kareem. 2017. *Understanding, Explanation, and Scientific Knowledge*. Cambridge: Cambridge University Press. https://doi.org/10.1017/9781108164276.

Kvanvig, Jonathan L. 2009. The Value of Understanding. In A. Haddock, A. Millar, & D. Pritchard (Eds.), *Epistemic Value*. New York: Oxford University Press, 95–112. https://doi.org/10.1093/acprof:oso/9780199231188.003.0005.

Ladyman, James. 2011. Structural Realism versus Standard Scientific Realism: The Case of Phlogiston and Dephlogistcated Air. *Synthese* 180, 87–101. https://doi.org/10.1007/s11229-009-9607-8.

Lawler, Insa. 2021. Scientific Understanding and Felicitous Legitimate Falsehoods. *Synthese* 198, 6859–6887. https://doi.org/10.1007/s11229-019-02495-0.

Lipton, Peter. 2004. *Inference to the Best Explanation*. 2nd ed. London: Routledge.

Mizrahi, Moti. 2012. Idealizations and Scientific Understanding. *Philosophical Studies* 160, 237–252. https://doi.org/10.1007/s11098-011-9716-3.

Morgan, Mary, and Margaret Morrison, Eds. 1999. *Models as Mediators: Perspectives on Natural and Social Science*. Cambridge: Cambridge University Press. https://doi.org/10.1017/CBO9780511660108.

Potochnik, Angela. 2017. *Idealization and the Aims of Science*. Chicago: The University of Chicago Press. https://doi.org/10.7208/chicago/9780226507194.001.0001.

Radder, Hans. 2012. *The Material Realization of Science* (revised edition). Dordrecht: Springer. https://doi.org/10.1007/978-94-007-4107-2.

Strevens, Michael. 2017. How Idealizations Provide Understanding. In S. R. Grimm, C. Baumberger, & S. Ammon (Eds.), *Explaining Understanding: New Perspectives from Epistemology and Philosophy of Science*. New York: Routledge, 37–49.

van Lunteren, Frans H. 1991. *Framing Hypotheses*. Ph.D. Dissertation, University of Utrecht.

3 Should Friends and Frenemies of Understanding Be Friends? Discussing de Regt

Kareem Khalifa

In typical philosophical exchanges, one perfunctorily commends one's interlocutor before engaging in the intellectual equivalent of gladiatorial melee. Philosophical discussions about scientific understanding are not special in this regard. Some, like me, hold that understanding is derivative of explanation and knowledge. Call us the *frenemies* of understanding. Under the thrall of this gladiatorial paradigm, we frenemies have sometimes taken too much glee in deflating the ambitions of the *friends* of understanding, who, like my interlocutor, Henk de Regt, hold that something about understanding eludes explanation and knowledge.

These exercises in antagonism are tired, so I aim to disrupt them. I propose an alliance between these seemingly contradictory positions. Building bridges rather than burning them is the surer path to philosophical progress. Hence, the frenemy and friend of understanding should be friends![1] If you still pine for philosophical blood sport, then consider this a battle against debates that have outlived their usefulness.

1 Friends and Frenemies

Friendship between friends and frenemies requires a mutual understanding about understanding. So, I will begin by discussing de Regt's views and my own views.

1.1 My Frenemy Account

In my book (Khalifa 2017), I presented my frenemy view—the Explanation-Knowledge-Science (EKS) Model of understanding—as a traditional philosophical analysis, replete with necessary and sufficient conditions. However, as part of my kinder, gentler approach, I will be breezier here. I subscribe to three core principles, the first of which is the *Explanatory Floor*:

> Understanding why *P* requires possession of a correct explanation of why *P*.

DOI: 10.4324/9781003202905-4

The Floor's underlying intuition is simple. It seems odd to understand why P while lacking a correct answer to the question, "Why P?" For instance, the person who lacks a correct answer to the question "Why do apples fall from trees?" does not understand why apples fall from trees. Since explanations are answers to why-questions, the Floor appears platitudinous.

But what does it mean to "possess" an explanation? Since the Floor only concerns what understanding *requires*, explanatory possession is any representation of an explanation with a mind-to-world direction of fit.[2] So, understanding requires something like accepting or believing an explanation, but it does not require desiring or coveting an explanation. This still leaves the delicate issue of when an explanation is "correct." As we shall see below, this is one place where friends and frenemies might join forces.

My remaining principles allow understanding to rise above the Floor. Suppose that you can correctly identify two causes of a fire, and I can only identify one of those causes. *Ceteris paribus*, you understand why the fire occurred better than I do. This intuition is enshrined in the EKS Model's *Nexus Principle*:

> Understanding why P improves in proportion to the amount of correct explanatory information about P (= P's explanatory nexus) in one's possession.

However, two agents who possess the same explanatory information may, nevertheless, differ in understanding because of their abilities to use that information in illuminating ways. The last of my principles, *The Scientific Knowledge Principle*, does justice to this intuition:

> Understanding why P improves as one's possession of explanatory information about P bears greater resemblance to scientific knowledge of why P.

Now, for this last principle to have teeth, I owe you an account of the relevant kind of scientific knowledge. Behold its teeth!

> S has scientific knowledge of why P if and only if there is some Q such that S's belief that Q *explains why* P is the safe result of S's scientific explanatory evaluation (SEEing).

The core notions here are safety and SEEing. Safety is an epistemological concept that requires an agent's belief to not easily have been false given the way in which it was formed (Pritchard 2009). SEEing then describes the way in which beliefs in an explanation should be formed to promote understanding. SEEing consists of three phases:

1 *Considering* plausible potential explanations of why *P*;
2 *Comparing* those explanations using the best available methods and evidence; and
3 Undertaking *commitments* to these explanations on the basis of these comparisons. Paradigmatically, commitment entails that one believes only those plausible potential explanations that are decisive "winners" at the phase of comparison.

Thus, scientific knowledge of an explanation is achieved when one's commitment to an explanation could not easily have been false given the ways in which that explanation was compared with other potential explanations of the phenomenon.

So, when it comes to sentences of the form "*S* understands why *P*," the EKS Model suggests that in addition to the Floor's requirements, understanding requires having a "sufficiently good grasp" of the "relevant" explanatory information. Context determines "sufficiency" and "relevance." The Nexus and Scientific Knowledge Principles determine the sense of "good grasp."

As should be clear, my recipe for understanding consists of the two basic frenemy ingredients—explanation and knowledge. Give or take some niceties, the EKS Model suggests that we know a bunch of stuff, and we just happen to call the *explanatory* stuff that we know "understanding." So, frenemies take understanding to be a philosophical dish best served bland.

1.2 de Regt's Friendly Account

By contrast, de Regt's palate is subtler than mine. When he discusses understanding, he savors little hints of grasping and delights in morsels of skillful intellection. Since these goodies seem to go beyond knowledge and explanation, he is a *friend* of understanding. More precisely, de Regt offers the following Criterion of Understanding Phenomena (CUP):

A phenomenon *P* is understood scientifically if and only if there is an explanation of *P* that is based on an intelligible theory *T* and conforms to the basic epistemic values of empirical adequacy and internal consistency.

(De Regt 2017, 92)

Here, de Regt (2017, 38) construes empirical adequacy and consistency as values because "there may be variation in how these values are ranked and applied in specific cases." He takes them to be "basic" because every scientific explanation must exhibit these two values to some degree.

As might be expected, frenemies like me are not hostile to empirical adequacy and consistency. However, de Regt's master concept is that of intelligibility, which appears to be less at home behind frenemy lines. De Regt offers two complementary characterizations of intelligibility. First, he offers a "macro-level" characterization, intended to apply to science as a whole:

> *Intelligibility*: the value that scientists attribute to the cluster of qualities of a theory (in one or more of its representations) that facilitate the use of the theory.
>
> (De Regt 2017, 40)

Here, the theoretical "qualities" denote a wide variety of explanatory desiderata, including simplicity, scope, familiarity, causation, unification, mechanism, and visualizability. Moreover, "using" a theory is exercising the requisite judgment and skills needed to construct and evaluate explanatory models of the phenomena.

Different scientists place greater value on different clusters of qualities. Scientists' skills determine which of these clusters furnish intelligibility in a given context. However, in this skeletal form, one might worry that the contextualist catchall is a short path to triviality. De Regt artfully blunts this worry with a "meso-level" account of intelligibility,[3] the Criterion for the Intelligibility of Theories:

> **CIT$_1$**: A scientific theory T (in one or more of its representations) is intelligible for scientists (in context C) if they can recognize the qualitatively characteristic consequences of T without performing exact calculations.
>
> (De Regt 2017, 102)

Although de Regt offers no other criteria of intelligibility, he presents CIT$_1$ as providing only sufficient conditions (and subscripts it) to suggest that theories can be intelligible in other ways (De Regt 2017, 102, n.15). Moreover, CIT$_1$ appears intuitive—when scientists can effortlessly engage in qualitative reasoning with a theory, they have a "feel" for the theory. We frequently associate this "feel" for a theory with intelligibility and understanding.

Hence, like me, de Regt accords a prominent role to explanation in understanding. However, while I appeal to the frenemy concept of scientific knowledge to flesh out the remainder of understanding, he appeals to the friendly concept of intelligibility. With these details in hand, grab your popcorn and enjoy the show. Remember, however, that this is not a tale of a valiant hero defeating a dreaded rival—but rather a story of a budding friendship.

2 Consideration

Let's consider a case of understanding—in the service of understanding a case of (explanatory) consideration.[4] The late 1960s marked the dawn of the Standard Model in particle physics. Quarky ideas appeared in theoretical papers and on dusty chalkboards, but experimental evidence was scarce. The tides began to turn when experimenters at Stanford and at MIT discovered unexpectedly high cross-sections in deep inelastic electron-proton scattering experiments, suggesting a more rigidly structured proton than previously conjectured. In other corners of Palo Alto, theorist James Bjorken had been developing a model of hadronic structure. He asked his colleagues to plot two curves from the data of their already-interesting results to test his model. To their delight, his model correctly predicted these "Bjorken scaling curves," as they are now called.

However, there was one problem. Few of the experimenters *understood* Bjorken's model. Their exasperated cries fell on the ears of one Richard Feynman, who in August 1968 expanded on Bjorken's original explanation—the latter being heavily steeped in then-esoteric current-algebraic notation. Feynman's core idea is that hadrons are composed of hard, point-like entities that he called "partons" (and which are now called "quarks"). Mathematically speaking, Bjorken and Feynman's models are near equivalents, yet only the latter greatly facilitated experimenters in designing tests to compare this explanation with other explanations of Bjorken scaling.

Importantly, Feynman's *consideration* of the explanation of Bjorken scaling in terms of electron-parton interactions should be distinguished from the experiments done to *compare* that explanation with its rivals. This much the EKS Model gets right. However, my description of explanatory consideration has been paper-thin. De Regt's view has resources for a far richer account of explanatory consideration than mine. Our first opportunity for bridge-building arises.

2.1 Friendly Considerations

I propose the following:

A *potential* explanation E of phenomenon P *deserves consideration (in SEEing)* if and only if:

a E is based on an intelligible theory T, and
b E conforms to the basic epistemic values of empirical adequacy and
     internal consistency.

This differs from de Regt's CUP in being a criterion of *considering potential explanations* rather than one of *understanding phenomena*. However, taken at face value, this formulation undersells one of de Regt's key insights: intelligibility, empirical adequacy, and internal consistency all involve *value judgments*. After describing these judgments, I discuss their importance to explanatory consideration.

Recall that de Regt regards consistency and empirical adequacy as "basic values" in explanatory inquiry, meaning that any acceptable explanation must exhibit *some* degree of both. However, scientists must make value judgments about the degree, relative weight, and precise interpretation they attach to both these basic values and the explanatory desiderata that figure in intelligibility. We can get a feel for these kind of judgments when we turn to de Regt's (2020, 923) most precise characterization of empirical adequacy, which he borrows from Bhakthavatsalam and Cartwright (2017, 446):

> . . . a theory (or model or set of scientific claims) is empirically adequate when the claims it makes about empirical phenomena—or at least the bulk of these claims, or the central ones—are correct, or approximately correct enough.

Even if we grant an explanation's "central claim" about an empirical phenomenon is its explanandum, scientists still must make judgments about what counts as "a bulk of claims" and "approximately correct enough." To prevent crackpot explanations from deserving consideration, I will assume that these judgments must be made by scientists who are competent and well-informed (or "skilled" in the de Regtian dialect) about the phenomenon to be understood. Consistency and the various explanatory virtues that inform intelligibility (simplicity, unification, visualizability, etc.) admit of similar treatments.

De Regt's emphasis on expert judgments does not merely *describe* the inner workings of consideration that I so crudely black-boxed. It also highlights two of explanatory consideration's *normative* aspects. First, thinking of consistency and empirical plausibility as "basic values" to be balanced against intelligibility captures a kind of flexibility that seems appropriate at the phase of consideration. In this early phase of explanatory inquiry, scientists may tolerate some inconsistencies or incorrect predictions at an explanation's periphery if these inaccuracies are incidental to the phenomenon to be understood and the explanation appears especially promising. For instance, in considering Feynman's explanation, physicists discounted the absence of any empirical evidence for fractionally charged particles. Moreover, intelligibility—particularly the macro-level conception's "cluster" of theoretical qualities—is a good indicator of explanatory promise (Nyrup 2015).

Tolerance toward potential explanations with different bundles of consistency, empirical plausibility, and intelligibility is especially laudable at the phase of consideration. For instance, physicists' understanding of proton structure was enhanced by considering both Bjorken and Feynman's explanations. Moreover, scientists can be unjustified in accepting an explanation because they have failed to consider a deserving explanation. Failing to control for a confounding variable and overlooking auxiliary hypotheses about malfunctioning instruments are paradigmatic examples. In short, "underconsideration" makes SEEing less safe (Khalifa 2017, Ch. 7). The Scientific Knowledge Principle thereby indicates that understanding improves, in part, by more comprehensive consideration. Consequently, using de Regt's friendly account to unpack consideration is not merely *consistent* with my frenemy account—it *promotes* my preferred brand of understanding.

Synthesizing our two views highlights a second interesting facet of scientific practice, which has received meager philosophical attention. I call this *explanatory staging*: the representing of a plausible potential explanation so as to make it amenable to comparison with other plausible potential explanations of the same phenomenon, typically through empirical testing.

The effortless qualitative reasoning that de Regt prizes (in CIT_1) promotes explanatory staging. For instance, although it was formally identical to Feynman's model, Bjorken's current-algebraic model was not "staged" nearly as well, as evidenced by experimenters' greater facility with Feynman's model. The Stanford-MIT experimenters were highly proficient at designing scattering experiments involving electrons and *protons*. By piggybacking off Feynman's qualitative reasoning, they could represent Bjorken scaling as involving another kind of scattering experiment: one involving electrons and *partons*. This, in turn, allowed them to design further experiments by which to test the Bjorken-Feynman Frankenstein or "parton model" (the core ideas of which would later be incorporated into the Standard Model). Bjorken's largely quantitative reasoning did not generate these analogies with their previous experiments.

By leaning on de Regt's work, I hope to have set the stage for future adventures in explanatory staging. Our story of friendship, however, must turn to its next act—what scientists must do to advance their understanding *after* they have considered deserving candidates for explanations.

3 Comparison

To that end, observe that Bjorken and Feynman were not the only theorists to propose potential explanations of Bjorken scaling. Moreover, although these two theorists' explanations were quite similar, other explanations posited radically different hadronic structure. Sakurai's vector meson dominance (VMD) model was among these competitors to the parton model.[5] If Sakurai's explanation of Bjorken scaling was right, then the parton model was wrong.

Like the parton model, Sakurai's model deserved consideration as a potential explanation of Bjorken scaling in the late 1960s. The VMD models provided the most empirically successful and widely accepted explanations of scattering behavior in the high-energy physics community for most of the 1960s, so they appear empirically adequate and consistent in de Regt's more liberal sense.[6] Moreover, according to VMD models, ρ, ω, and φ mesons are the hadronic components of the photon. This lends itself to qualitative reasoning—by using Feynman diagrams, for example. Most relevant to Bjorken scaling, Sakurai held that proton-electron interactions were mediated by the aforementioned "vector mesons" instead of Feynman's partons. Since Sakurai was the foremost VMD theorist at the time, he certainly was competent and well-informed. Moreover, even defenders of the parton model thought enough of the VMD model to test it. Hence, per the criteria forged by our first foray into friendship, VMD deserved consideration as an explanation of Bjorken scaling.

But understanding is not simply an exercise in considering deserving contenders as explanations. Scientists march toward understanding by ascertaining which of these contenders is the champ. Sakurai's explanation of Bjorken scaling entails that the ratio of the proton's tendencies to absorb virtual photons longitudinally to its tendencies to absorb those photons transversely is quite large (in between 1 and 10). By contrast, parton models entail that this ratio should be quite small (in between 0 and 1). In August 1969, Richard Taylor led a run of "crucial" experiments at the Stanford Linear Accelerator designed to adjudicate between the parton model and Sakurai's VMD. The results showed the lower ratio predicted by the parton model to be correct. This was widely seen as a reason to reject Sakurai's VMD model. Hence, by the time that Taylor's experimental results were reported, it was clear that the empirical evidence favored the parton model over Sakurai's model. Consequently, Sakurai's model fails to provide understanding of Bjorken scaling precisely because of its incongruence with Taylor's findings.

As noted above, de Regt discusses the *evaluation* of explanatory models. However, he does not emphasize that models can be evaluated in different ways at different phases of explanatory inquiry. For instance, one can judge that an explanation, such as Sakurai's, is good enough to *deserve consideration*, but then make a subsequent comparative evaluation that another explanation, such as Feynman's, *better explains* the phenomenon of interest.

It is unclear how de Regt's approach can capture this distinction; for instance, de Regt's (2020, 931) discussion regarding Maxwell and

Boltzmann's disagreement about the latter's "dumbbell" model as one in which "neither . . . was irrational or unscientific: they only valued the empirical adequacy of the dumbbell model differently." This suggests that empirical adequacy is in the eye of the beholder. However, applying an analogous approach to Sakurai's case produces some counterintuitive results. Initially, Sakurai took the "main virtue" of his proposal as that of providing a clear experimental test by which to adjudicate between the parton and VMD models (Sakurai 1969, 981). However, in September 1969, Sakurai seemed to discount Taylor's findings, claiming that "we need better data," so he appears to have banished these results from his bulk (Riordan 1987, 164). Arguably, Sakurai merely changed *ad hoc* the degree to which he valued empirical adequacy to avoid refutation. However, this understandably struck most physicists at the time as a dubious way to achieve scientific understanding (Friedman 1991, 720; Riordan 1987, 165–166).

A less forgiving frenemy might conclude that de Regt's CUP runneth under: it fails to provide sufficient conditions for understanding phenomena. In the spirit of fostering friendship, this will not be my gambit. Rather, just as de Regt's account opened the black box in my account of consideration, I suggest that my account of explanatory comparison does the same for these aspects of his view. De Regt's account is consistent with, but does not entail, that different kinds of explanatory evaluation should be subject to different standards. Parts of my account provide de Regt with a framework for making these finer-grained evaluations.

3.1 Friendly Comparisons

Although I have not explicitly formulated my account of explanatory comparison using the language of empirical adequacy, it is no stretch to do so. When I have spoken of comparisons based on the best available evidence and methods, my examples consistently involve *empirical* evidence and methods apropos of *empirical* testing of explanatory hypotheses. With that in mind, I propose the following:

> An explanation Q of P is *empirically fit* in context C if and only if for all other explanations Q^* that also deserve consideration as an explanation of P in C, the judgment that Q^* *better explains why P than Q* is unsafe given the best evidence and best methods available in C.

Intuitively, this says that an empirically fit explanation is at least "tied for first" when it comes to saving the phenomena needed for methodologically sound comparisons. Hence, multiple explanations of the same phenomenon can be empirically fit. However, unlike de Regt's account of empirical adequacy, empirical fitness is demanding enough to do justice to the idea that Sakurai's model fails to provide understanding, for there is at least one deserving competitor, the parton model, that can safely be judged to be better than Sakurai's explanation given that Taylor's experimental results were available.

Empirical fitness seems largely congenial to de Regt's approach to understanding. For instance, the proposed alliance between frenemy and friend shows how empirical thresholds change at different stops on the road to understanding. De Regt's account of judgment-based empirical adequacy is still vital in explanatory consideration, so it should not be abandoned. However, it must be supplemented with the notion of empirical fitness if it is to avoid unduly awarding understanding to scientists who are mistaken about the phenomena they ought to save—as Sakurai was. Moreover, because an explanation can be empirically fit while still failing to accurately depict the unobservable, de Regt is not being saddled with the unpleasant task of saving the noumena. All of this resonates with de Regt's strong empiricist scruples.

The proposed synthesis of our views also refines de Regt's contextualism about understanding in three useful ways. First, we can think of explanatory consideration and explanatory comparison as two kinds of contexts in which scientists have different objectives. When scientists are considering explanations, their main goal is to avoid overlooking any deserving candidate, which favors relatively permissive standards. When they are comparing explanations, their main goal is to determine which of these explanations is the best, which favors more restrictive standards. Since these pull in seemingly opposite directions, these two contexts will involve different epistemic standards.

Second, empirical fitness more precisely identifies understanding's context-sensitive and context-invariant elements. Understanding-providing explanations must be empirically fit, that is a context-invariant requirement. However, how that requirement is realized depends on (a) which explanations are deserving of consideration and (b) which methods and evidence are available. I would call these two features part of a scientist's "context," though they are determined by scientific communities; not by individual scientists. For instance, what has been circulated within the relevant scientific community at a given time determines which methods and evidence are available. The disciplinary context determines what count as the *best* methods and evidence. In the case of Bjorken scaling, these context-infused standards involve different plausible potential explanations making competing predictions about what will happen in highly controlled experiments. However, in other contexts, such as the social sciences, explanatory comparisons frequently lack these features (Khalifa 2019). Since de Regt does not explicitly identify (a) and (b) as context-sensitive determinants of understanding—much less make them communal rather than individual—this seems like a fruitful elaboration of his view.

Third, stocking the contextualist cupboard in this way licenses more nuanced evaluations about a model's capacity to provide understanding.

For instance, subsequent revisions to Sakurai's work have resulted in so-called "generalized" VMD models that have had empirical success as effective descriptions of quantum chromodynamical models at low energies. Indeed, some recent VMD models even seem to fare much better than their predecessors in explaining deep inelastic scattering, which is precisely where Sakurai's explanation of Bjorken scaling faltered.[7] So, it makes little sense to throw out the entire VMD enterprise. This is why comparisons are tied to a specific explanandum and a particular context. These benefits of generalized VMD approaches can all be acknowledged while still claiming that the explanation that *Sakurai offered in 1969* did not provide an understanding of *Bjorken scaling* because it was empirically unfit.

Empirical fitness also invokes safety. Safety is my frenemy calling card, for it yokes understanding to knowledge. Despite this, I am hopeful that de Regt will not rebuff my attempt at friendship. Shorn of epistemological jargon, this is simply a requirement that the evidence and methods used to adjudicate between different candidate explanations are reliable tools for this task. Suppose that Taylor's experiments were unsafe. Then they would have been so poorly designed that, had the relevant ratio been greater than 1, they would have still indicated that it is less than 1. Consequently, they would have been too unreliable to assist in explanatory comparison. Surely, this would diminish their capacity to advance physicists' understanding of Bjorken scaling.

So construed, I see no reason for de Regt to deny the requirement that our empirical evidence and methods should be safe in this way. Moreover, intelligibility—the "friendliest" feature of his account—remains completely untouched by empirical fitness, including its inheritance of safety from the epistemologists down the hall.

4 Commitment and Truth

Suppose that only one explanation exhibits empirical fitness after comparison. What happens next? As noted above, the EKS Model holds that there is a third phase, *commitment*, where scientists adopt the appropriate cognitive attitude toward the different explanations they have considered and compared. The key question is whether scientists have, at this point, a correct explanation. If so, then they have cleared the Explanatory Floor. If not, then they are (at best) on the right track to understanding, but they still have work to do.

Thus, much hinges on how we define explanatory correctness. Some will decree that empirical fitness is enough. De Regt's CUP suggests as much.[8] Others will note that empirically fit explanations can be horrifically inaccurate, and horrifically inaccurate explanations frequently generate *mis*understanding. De Regt has recently suggested that we differ on precisely this issue (De Regt 2020, 931; De Regt and Höhl 2020). My dear friend is only half correct in this appraisal. Allow me to explain.

4.1 Setting the Record Straight

It is unclear to me whether de Regt has fully appreciated the flexible role of truth in my account of understanding. *Mea culpa:* although I have called my view of the role of truth in understanding "quasi-factivist," I have always been more "quasi" than "factivist." However, only the latter is a battle cry in the understanding literature, so quite reasonably, I am frequently portrayed as opposed to self-described "non-factivists," such as de Regt. Let me atone for my bad marketing choices and show how we are closer than he seems to think. As noted above, I am committed to:

> *The Explanatory Floor:* Understanding why *P* requires some *Q* such that *Q* correctly explains why *P*.

By Tarski's Convention T, we get:

> *Quasi-factivism*: Understanding why *P* requires some *Q* such that "*Q* correctly explains why *P*" is true.

Because of quasi-factivism's appeal to truth, I considered this view to be in the factivist family. More importantly, because frenemies hold that understanding is derivative of scientific knowledge of statements of the form "*Q* correctly explains why *P*," it is possible for quasi-factivism to be frenemies' *only* truth-requirement on understanding. It is certainly *my* only truth-requirement.

But why am I not a full-throated factivist but merely a half-hearted *quasi*-factivist? Primarily because adding truth-*talk* to the Explanatory Floor does not mean that I have strengthened the Explanatory Floor's truth-*requirements*. Specifically, quasi-factivism does *not* entail the following:

> *Explanatory Realism:* For all *P* and *Q*, if "*Q* correctly explains why *P*" is true, then '*Q*' is approximately true.

As I see it, full-throated factivists must add this realist[9] codicil to quasi-factivism. Consequently, they must deny that any explanans that falls short of approximate truth provides understanding.

Nor, however, am I a full-throated *non*-factivist. Like their dreaded realist rivals, these firebrands also add something to the quasi-factivist core:

> *Explanatory Antirealism*: There are some *P* and *Q* such that "*Q* correctly explains why *P*" is true and "*Q*" is not approximately true.

Unlike factivists, non-factivists can accept that explanantia that fall short of approximate truth provide understanding.

I add neither realist nor antirealist addenda to the quasi-factivist core.[10] This is possible precisely because the Explanatory Floor plays no favorites with respect to realism and antirealism. This is clearly expressed in the EKS Model's official "theory" of explanation:

> For all *P* and *Q,* if "*Q* correctly explains why *P*" is true, then *Q* satisfies your ontological requirements (so long as they are reasonable).[11]

In the hopes of dispelling confusion, let us give my longstanding position a new label: explanatory *voluntarism.*[12] On this line, some ontological positions will be more in keeping with explanatory realism; others, with antirealism. A frenemy who is also a voluntarist only demands that you should adjust the Explanatory Floor to be consistent with whatever ontological requirements you have settled on. If you impose more restrictive ontological requirements on explanation, as realists are wont to do, then adopt a similarly restrictive notion of understanding. If you have more relaxed ontological requirements on explanation, then adopt a similarly liberal notion of understanding. In short, your choices about explanation's truth-requirements should dictate your attributions of understanding. Indeed, as Section 4.3 illustrates, explanatory correctness is often overdetermined by both realist and antirealist ontological requirements, so realists and antirealists will frequently agree about *when* understanding has been achieved even if they disagree as to *why.* In such cases, I suggest that theorists of understanding should tolerate different stances toward the realism issue.

I propose that we theorists of understanding cheerfully abandon the word "factive," even when it is prefixed by a "non-" or a "quasi-." For those of us who stand firmly on the Explanatory Floor, positions such as realism, antirealism, and voluntarism more precisely characterize our differences than the dreaded F-word. Crucially, De Regt (2017, 23) stands firmly on the Explanatory Floor. Consider his pithy equation:

> understanding a phenomenon = having an adequate explanation of the phenomenon.

Moreover, I think he has suffered from similar ambiguities as I have. On one and the same page, De Regt (2017, 131) expresses a commitment to antirealism, when he writes that "we can have genuine understanding of phenomena on the basis of theories and models that defy realistic interpretation" and echoes the kind of quietism characteristic of voluntarism when he also writes, "understanding and realistic representation are independent."

As a final gesture toward friendship, I suggest that we both become voluntarists. To that end, I will show how I can accommodate explanatory antirealism and de Regt can accommodate realism without upsetting the synthesis effected in Sections 2 and 3. As such, intelligibility and safe SEEing—our most distinctively friendly and frenemy elements, respectively—remain intact.

4.2 Frenemy Antirealism

There has been little doubt that frenemies of understanding can be explanatory realists. However, it is simply mistaken to assume that frenemies cannot be antirealists. The basis for this judgment appears to be that frenemies take understanding to be derivative of knowledge, and since knowledge requires truth, frenemies must be realists. As the preceding lays bare, and as argued elsewhere (Khalifa 2011, 2017, Ch. 6), this argument is flawed. In a nutshell: antirealism entails that in some cases, "*Q explains why P*" is *true*(!), even though "*Q*" is false. To deny this modest truth-requirement is to fall off the Explanatory Floor. So, it is consistent with antirealism that in these cases, a scientist still knows that *Q explains why P* (though she does not know that *Q*). Since this is also compatible with the Nexus and Scientific Knowledge Principles, antirealists can also be frenemies of understanding.

4.3 Friendly Realism

As noted above, de Regt leans toward antirealism, but has suggested that, like me, he would like to put issues concerning realism to the side when discussing understanding. My goal here is to recast de Regt's antirealist stance as one of many permissible sets of ontological requirements. Realist ontological requirements that leave his account of intelligibility unsullied are also permissible.

To do that, I want to assuage De Regt's chief reservation about realism: that its truth-requirements on explanation will incorrectly disqualify perfectly good instances of scientific understanding. To that end, I begin by noting that de Regt's target is sometimes unclear. Is it the mild-mannered explanatory realism that only requires explanantia to be approximately true? Or is it "bogeyman realism," the position that every detail of an understanding-providing explanation must perfectly mirror reality's every nook and cranny to a divine level of resolution? Consider his admission that "while . . . there is a role for truth, this is not unqualified, objective truth but truth relative to a particular context and a particular purpose" (De Regt 2017, 136). I see plenty here to upset bogeyman realists, but no genuine friction with explanatory realism. Worrall's (1989) classic essay ushered in waves of realist positions that are "selective" in the posits that warrant ontological commitment. Hence, these positions are not "unqualified" about truth's role in explanation. Similarly, some realists have countenanced significant context-sensitivity through a contrastive theory of why-questions

(Lipton 2004). Generally, realists can comfortably claim that context and purposes play a central role in determining which truths are *relevant* or *significant* (Kitcher 2001). If this is all that de Regt means by the ways in which truth is "relative to a context," then all he has done is raise the hackles of realists who are not bogeymen. Moreover, he would not even disagree with the modest realists who ought to have been his genuine foils!

However, if de Regt really wants to put up a fight—not with me, but with the realists he has parodied—he may double down in two ways. First, on a more ambitious antirealist reading of his remarks, context and purposes do not simply make certain truths *relevant*, but instead make certain statements *true*. However, this incipient pragmatist theory of truth comes at a high cost: his claim that theories that are "strictly speaking false" are still "useful . . . in certain contexts" (De Regt 2017, 131) becomes inconsistent. Moreover, nothing in de Regt's work suggests such an exotic theory of truth, so I think this interpretation sits uneasily with his position.

Alternatively, de Regt may keep the debate going by making this modest realist thesis his foil. After all, if we read his CUP literally, an explanation that was *massively* false about *every* unobservable it posited could still provide understanding, so long as it was intelligible, consistent, and empirically adequate. Hence, it would seem that anyone who required correct explanations to be *approximately* true about *some* of their posited unobservables—that is, even the realists who are *not* bogeymen—would be too restrictive for his tastes.

However, so far as I can tell, de Regt's putative examples of antirealist understanding all leave interpretive wiggle room for this modest sort of realism. For example, all of the explanations offered for Bjorken scaling in the late 1960s assumed that electrons interact with hadrons.[13] However, realists can say even more than this and still operate within the confines of only saying that some parts of an explanation are approximately true. Consider a structural realist's take on the parton model's success. Bjorken's model yielded a novel prediction of his eponymous scaling curves. Structural realists regard novel predictions as evidence that Bjorken's mathematics latched onto a physical structure. Finally, they will note that Feynman's model piggybacked off of Bjorken's model. Consequently, parts of Feynman's model are isomorphic to the physical structures that Bjorken's model captured.

Crucially, realists can say all of this while still tolerating significant falsehoods in *some* parts of the parton model. Once again, this is essentially the "selective realist" approach that has been *en vogue* for more than three decades. For instance, Feynman's qualitative reasoning about partons more or less depicts them classically, as particles colliding in billiard-ball fashion. Yet we know that they are denizens of the quantum realm. However, the aforementioned suggestion—that the parton model is partially isomorphic to a physical (quantum) structure—is entirely consistent with its classical bits being false.

Once again, my point here is not that realism is correct and de Regt's antirealism is incorrect. It is simply that realists should not lose sleep over de Regt's arguments. As a voluntarist, I also think that the converse is true. De Regt should be unfazed by realist interpretations of his favorite episodes in the history of science. For instance, he is well within his rights to claim that the parton model provides an understanding of Bjorken scaling simply because it is empirically fit. Realists can heap similar praise upon the parton model, while adding that, for example, it is isomorphic to a physical structure. So, both parties will claim that the parton model provided understanding of Bjorken scaling. This accords well with voluntarism: in these sorts of cases, the choice between realism and antirealism makes no difference to understanding, so take your pick! More generally, I find that relitigating the entire realism debate simply to reach similar verdicts about understanding is not time well spent.

5 Conclusion

With this, our story of friendship comes to a curious close. If asked how explanatory inquiry must proceed in order to furnish understanding, it appears that de Regt and I can—and probably should—point toward the same things. Both of us ought to acknowledge that the skillful construction of explanatory models predicated upon expert judgments about which bundles of consistency, empirical adequacy, and intelligibility warrant consideration should be followed up by careful empirical comparisons to determine which of these models rises to the top. The remaining issue is how much the fruits of these labors must latch onto unobserved reality. On this point, I have suggested that de Regt and my friendship could be cemented by a more thoroughgoing commitment to a voluntarism with which we have both only flirted. So, the ball is in de Regt's court. He can either decline some of these invitations or cement the friendship by embracing voluntarism.

In conclusion, I note that the friendship discussed here is more accurately (but less colorfully) described as a synthesis of philosophical views. There is—I hope!—no *personal* animus between Henk de Regt and me. Ever since our first correspondences more than a decade ago, Henk has been generous and kind. I am fortunate to call him my friend.

Notes

1 In this chapter, I only look at de Regt's and my views; future work should forge connections with other friends and frenemies of understanding.
2 Khalifa (2017) restricted this only to belief. For reasons presented in Khalifa and Millson (2020), I have come to think of this as too restrictive. Moreover,

Khalifa (2017) characterized the Floor as an account of "minimal understanding." I thank Federica Malfatti for spotting some infelicities with that formulation.

3 Meso-level standards of intelligibility characterize a particular scientific community's norms regarding understanding. De Regt also describes individual or micro-level standards of intelligibility, but these do not figure prominently in his account.

4 See Khalifa (2017, Ch. 2) for further discussion.

5 The other main competitor was Arbanel et al.'s Regge exchange model. Most of my discussion of VMD has analogues to this model as well.

6 Prominent works include Gell-Mann and Zachariasen (1961), Ross and Stodolsky (1966), Sakurai (1960, 1962), and Stodolsky and Sakurai (1963).

7 Ironically, Sakurai and Schildknecht (1972) first proposed generalized VMD to explain Bjorken scaling in light of Taylor's results.

8 More precisely, empirically fit explanations deserve consideration (per Section 3.1). So, they will satisfy all of CUP's requirements (per Section 2.1).

9 Hereafter, I will use "realism" and its cognates as shorthand for "explanatory realism."

10 This, of course, echoes Fine (1986).

11 See, for example, Khalifa (2017, 7).

12 Chakravartty (2017) and van Fraassen's (2002) kindred positions suggest fruitful explications of ontological requirements' "reasonableness," for example, your requirements must be probabilistically coherent and must not undermine your epistemic goals.

13 Similarly, scientific realists have long sought to show that de Regt's go-to example for explanatory antirealism, Newtonian mechanics, approximates relativistic and quantum mechanics under suitably circumscribed boundary conditions. So-called correspondence results are their weapon of choice. I do not claim that these arguments are refutations of de Regt, only that he has not given us any reason to think that his antirealist interpretation of Newton is better than these realist interpretations. Consistent with my voluntarism, I take this as evidence that realists can enlist Newtonian mechanics as a repository of understanding just as ably as antirealists. See Chapters 5 and 6 (this volume) for further discussion.

References

Bhakthavatsalam, Sindhuja, and Nancy Cartwright. 2017. "What's So Special About Empirical Adequacy?" *European Journal for Philosophy of Science* 7 (3):445–465. doi: 10.1007/s13194-017-0171-7.

Chakravartty, Anjan. 2017. *Scientific Ontology: Integrating Naturalized Metaphysics and Voluntarist Epistemology.* Oxford: Oxford University Press.

De Regt, Henk W. 2017. *Understanding Scientific Understanding.* New York: Oxford University Press.

De Regt, Henk W. 2020. "Understanding, Values, and the Aims of Science." *Philosophy of Science* 87 (5):921–932. doi: 10.1086/710520.

De Regt, Henk W., and Anna E. Höhl. 2020. "Review of *Understanding, Explanation, and Scientific Knowledge,* by Kareem Khalifa." http://www.thebsps.org/reviewofbooks/kareem-khalifa-understanding-explanation-and-scientific-knowledge-reviewed-by-de-regt-hohl/.

Fine, Arthur. 1986. *The Shaky Game: Einstein, Realism, and the Quantum Theory.* Chicago: University of Chicago Press.

Friedman, Jerome I. 1991. "Deep Inelastic Scattering: Comparisons with the Quark Model." *Reviews of Modern Physics* 63 (3):615–627. doi: 10.1103/RevModPhys.63.615.

Gell-Mann, Murray, and Fredrik Zachariasen. 1961. "Form Factors and Vector Mesons." *Physical Review* 124 (3):953–964. doi: 10.1103/PhysRev.124.953.

Khalifa, Kareem. 2011. "Understanding, Knowledge, and Scientific Antirealism." *Grazer Philosophische Studien* 83 (1):93–112.

Khalifa, Kareem. 2017. *Understanding, Explanation, and Scientific Knowledge.* Cambridge: Cambridge University Press.

Khalifa, Kareem. 2019. "Is *Verstehen* Scientific Understanding?" *Philosophy of the Social Sciences* 49 (4):282–306. doi: 10.1177/0048393119847104.

Khalifa, Kareem, and Jared Millson. 2020. "Perspectives, Questions, and Epistemic Value." In *Knowledge from a Human Point of View*, edited by Ana-Maria Crețu and Michela Massimi, 87–106. Cham: Springer International Publishing.

Kitcher, Philip. 2001. *Science, Truth, and Democracy.* Oxford: Oxford University Press.

Lipton, Peter. 2004. *Inference to the Best Explanation.* 2nd ed. New York: Routledge. Original edition, 1991.

Nyrup, Rune. 2015. "How Explanatory Reasoning Justifies Pursuit: A Peircean View of IBE." *Philosophy of Science* 82 (5):749–760. doi: 10.1086/683262.

Pritchard, Duncan. 2009. "Safety-based Epistemology: Whither Now?" *Journal of Philosophical Research* 34:33–45.

Riordan, Michael. 1987. *The Hunting of the Quark: A True Story of Modern Physics.* New York: Simon & Schuster.

Ross, Marc, and Leo Stodolsky. 1966. "Photon Dissociation Model for Vector-Meson Photoproduction." *Physical Review* 149 (4):1172–1181. doi: 10.1103/PhysRev.149.1172.

Sakurai, J. J. 1960. "Theory of Strong Interactions." *Annals of Physics* 11 (1):1–48. doi: https://doi.org/10.1016/0003-4916(60)90126-3.

Sakurai, J. J. 1962. "Possible Existence of a $T=0$ Vector Meson at 1020 MeV." *Physical Review Letters* 9 (11):472–475. doi: 10.1103/PhysRevLett.9.472.

Sakurai, J. J. 1969. "Vector-Meson Dominance and High-Energy Electron-Proton Inelastic Scattering." *Physical Review Letters* 22 (18):981–984. doi: 10.1103/PhysRevLett.22.981.

Sakurai, J. J., and D. Schildknecht. 1972. "Generalized Vector Dominance and Inelastic Electron-proton Scattering." *Physics Letters B* 40 (1):121–126. doi: https://doi.org/10.1016/0370-2693(72)90300-0.

Stodolsky, Leo, and J. J. Sakurai. 1963. "Vector Meson Exchange Model for Isobar Production." *Physical Review Letters* 11 (2):90–93. doi: 10.1103/PhysRevLett.11.90.

van Fraassen, Bas C. 2002. *The Empirical Stance.* New Haven: Yale University Press.

Worrall, John. 1989. "Structural Realism: The Best of Both Worlds?" *Dialectica* 43 (1–2):99–124.

4 Frenemies or Friends?
A Reply to Khalifa

Henk W. de Regt

Kareem Khalifa has always been a personal friend, but whether he is a philosophical friend as well, a 'friend of understanding', was less clear to me. Although seriously interested in understanding – he wrote a whole book about it! – he did not regard it as something special: his account of (scientific) understanding reduces it to mere knowledge of an explanation. To him, the friends of understanding were offering old wine in new bottles.

It is always good to have a critical voice within a community of like-minded people, so Kareem's contributions have been extremely helpful in keeping the friends of understanding sharp and in preventing them from falling into a dogmatic slumber. In his contribution to the present volume, Kareem introduces the 'frenemy of understanding', an oxymoron that nicely describes his somewhat ambiguous but thought-provoking position. What is more, friends and frenemies can team up: he suggests that our opposing views of the nature of scientific understanding can be reconciled in such a way that a richer and more nuanced view of understanding results. Having read his chapter, I now regret the combative nature of my own chapter in this volume. In this reply, I consider whether I share his views on how such a reconciliation can be achieved.

Kareem's first step toward philosophical friendship is looking for common ground between our views. Thus, he observes that we share a commitment to what he calls the Explanatory Floor: the thesis that 'understanding why P requires possession of a correct explanation of why P' (this volume, 33). Now it is true that my contextual theory of scientific understanding concerns the understanding produced by scientific explanations; this is part of CUP (De Regt 2017, 92). However, I do not think that understanding is *reducible* to explanation: it is a cognitive achievement in its own right, rather than a 'derivative of explanation and knowledge' (this volume, 44). Accordingly, I leave open the possibility that understanding can be had in other ways, without explanation (as Lipton 2009 has argued, a view that Kareem attacks in Chapter 5 of his 2017 book). So, I hesitate to accept the Explanatory Floor as a dogma, but for the purposes of the present discussion I will focus on understanding as a product of scientific explanations.

DOI: 10.4324/9781003202905-5

A core element of Kareem's analysis of how explanations yield understanding is his notion of 'scientific explanatory evaluation', or SEEing (this volume, 34–35; cf. 2017, 12–13). The idea is, roughly, that someone understands a phenomenon P if they believe an explanation of P and the safety[1] of this belief is based on their SEEing, where SEEing is a three-stage process: first, *consideration* of plausible potential explanations of P; next, *comparison* of those explanations on the basis of the best available methods and evidence; and finally, *commitment* to the explanations on the basis of this comparison, which typically consists in believing only the explanation that has come out favorably. In his chapter, Kareem observes that my account sheds much light on the stage of consideration, whereas his own account has more to say about the stage of comparison. Accordingly, he suggests that our accounts are not opposed but compatible. His proposal for reconciliation consists in a proposal for the division of labor: apply my analysis to consideration and his analysis to comparison. With respect to the final stage, commitment, the bone of contention is the role of truth in understanding, the so-called factivity issue. Where I have argued for non-factivism, Kareem has previously defended (quasi-)factivism. Now, he proposes to bury the hatchet by taking a more tolerant stance in the form of a voluntarist attitude toward the realism-antirealism question that would leave room for both of us.

Let us have a closer look at these claims and suggestions. I will do so by focusing on the historical case that Kareem discusses: the explanation of deep inelastic electron-proton scattering in the 1960s (an episode analyzed in more detail in Chapter 2 of his 2017 book). In a nutshell, the story goes like this. James Bjorken's model of proton structure explained these scattering phenomena but was intractable and therefore difficult to test experimentally (this volume, 39). It was Richard Feynman's addition of the parton model to Bjorken's explanation that led to a testable model, which was subsequently vindicated in 1969, in crucial experiments that defeated rival models such as the vector meson dominance (VMD) model and the Regge exchange model (this volume, 40).

Kareem's friendly analysis of this episode emphasizes that Feynman's contribution nicely fits my account: the parton model allowed for qualitative reasoning and thereby enhanced the intelligibility of Bjorken's explanation. Indeed, here the case shows striking parallels with the examples I discuss in Chapter 7 of my book: the development of quantum mechanics and quantum field theory, where abstract theories were rendered intelligible with visual tools. Kareem argues that this aspect of the case belongs to the stage of *consideration*, and he concludes that my notion of intelligibility as an understanding-promoting feature of scientific theories is central to consideration. The 'plausible potential explanations' that deserve consideration are those that meet my criterion CUP, involving a balance between the values of intelligibility, empirical adequacy, and internal consistency.[2] Intelligibility allows for what

Kareem calls 'explanatory staging' (this volume, 39): the representation of rival explanations such that they can be compared empirically. I agree that this is how Feynman improved on Bjorken's explanation: by adding his parton model, he rendered Bjorken's original model more intelligible, such that it allowed for experimental testing.[3] So the two models were, in fact, different representations of one and the same explanation, where only Feynman's model deserved consideration because it was sufficiently intelligible.[4] The similarity with the genesis of quantum field theory two decades earlier is remarkable, when Feynman enhanced the intelligibility of Schwinger and Tomonaga's abstract formalism with his diagrammatic method (see De Regt 2017, 251–256).

The next stage is that of *comparison*, and here Kareem focuses on the experimental testing of the Bjorken-Feynman model against Jun John Sakurai's alternative VMD model. Here, it turned out that although the VMD model was good enough to deserve consideration, it failed when compared with Feynman's explanation (this volume, 40). Kareem suggests that here my account fails because it implies that 'empirical adequacy is in the eye of the beholder' (this volume, 41). By contrast, on his own account of comparison it can be asserted unambiguously that one explanation is to be preferred over another (and thus yields better understanding). Though discussed only very briefly in his book (2017, 12–13), Kareem now adds an extensive analysis centered around the notion of 'empirical fitness', which involves the comparative assessment of the safety of an explanation (this volume, 41). This is a valuable addition that affords a more precise comparative evaluation of rival explanations. In my work, I have perhaps overemphasized contextual variation in criteria for understanding, including the epistemic values empirical adequacy and consistency, and neglected the fact that scientists often reach a consensus about which model or theory affords the highest degree of understanding of the phenomena that it purports to explain. Kareem's criterion of empirical fitness is a welcome counterbalance to the relativistic tendencies in my account. I am grateful to him for offering an escape route here.

Still, I think that the assessment of empirical fitness is not always clear-cut. This is because the safety of the judgment that Q better explains why P than Q^* depends on "the best evidence and best methods available" (this volume, 41), and scientists do not always agree on what the best evidence and the best methods are. Here, it is instructive to compare the Feynman-Sakurai case with the Maxwell-Boltzmann case I analyzed in De Regt (2020). In the former case, scientists agreed on the relevant methods and evidence, apart from Sakurai's ad hoc rejection of the 1969 data that contradicted his model. Although Sakurai's attitude was clearly unreasonable and the VMD model was inferior to the parton model in 1969, this did not imply that it had to be buried once and for all. As Kareem observes (this volume, 43), later revisions of it achieved more success, and accordingly it had remained worthwhile to keep working on it, even in the face of its apparent empirical unfitness in 1969. So, contrary to what Kareem (this volume, 38) suggests, it is not only in the early stage of consideration that inconsistencies and empirical inadequacy may be tolerated.

This also applies to the Maxwell-Boltzmann case, in which there was less agreement about which empirical data were relevant, in particular whether spectral data should be considered as evidence against the model. But that case differs in that there were no alternatives to the dumbbell model: until the advent of the quantum theory, it was the only available explanation for the specific heat anomaly. So, a comparative evaluation of empirical fitness was impossible. It is not clear to me whether, and if so how, empirical fitness can be assessed in such cases (which are surely abundant in science). One might think that the only available explanation is, by definition, the best one, and is accordingly empirically fit. Or it may fail to meet the threshold for consideration, which, indeed, seemed to be Maxwell's verdict about Boltzmann's model.

Finally, Kareem addresses our apparent disagreement about the factivism issue: Does understanding require explanations to be true? This relates to the stage of *commitment*, in which scientists determine their doxastic stance toward the compared explanations. First of all, Kareem suggests that we stop talking about factivism (vs. non- and quasi-factivism). I happily agree: it too often leads to confusion. Moreover, I gladly accept the invitation to join the explanatory voluntarist camp (this volume, 45). I am no anti-realist but favor an agnostic position, and in my view understanding and realism are independent (De Regt 2017, 129–136).

Scientists are more flexible than philosophers. Although they often act and talk like realists, they may at times switch to an agnostic or outright anti-realist attitude toward their models and theories. But this does not entail that they are not committed to the explanations that those theories and models provide, as long as these explanations are correct. There is no better illustration of this than the words of Feynman, when he looked back on the development of the parton model:

> We have built a very tall house of cards making so many weakly-based conjectures one upon the other and a great deal may be wrong. . . . Finally it should be noted that even if our house of cards survives and proves to be right we have not thereby proved the existence of partons. . . . From this point of view the partons would appear as an unnecessary scaffolding that was used in building our house of cards. On the other hand, the partons would have been a useful psychological guide as to what relations to expect – and if they continued to serve this way to produce other valid expectations

they would of course begin to become 'real', possibly as real as any other theoretical structure invented to describe nature.

(Feynman quoted in Bjorken 1989, 58)

Notes

1 Khalifa (2017, 12): "A person's belief is safe just in case her belief-forming process could not easily have led to a false belief". Cf. Khalifa (this volume, 34).
2 Here, I ignore the distinction between theories and explanations, regarding Bjorken's and Feynman's models as explanations rather than theories.
3 See Bjorken (1989, 57–58) for a first-hand account.
4 This appears to be in line with Khalifa's (2017, 35) claim that Bjorken's explanation "did not hit the contextually relevant benchmarks" for understanding.

References

Bjorken, J. D. 1989. Feynman and Partons. *Physics Today*, 42/2, 56–59. https://doi.org/10.1063/1.881193.

De Regt, Henk W. 2017. *Understanding Scientific Understanding*. New York: Oxford University Press. https://doi.org/10.1093/oso/9780190652913.001.0001.

De Regt, Henk W. 2020. Understanding, Values, and the Aims of Science. *Philosophy of Science*, 87, 921–932. https://doi.org/10.1086/710520.

Khalifa, Kareem. 2017. *Understanding, Explanation, and Scientific Knowledge*. Cambridge: Cambridge University Press. https://doi.org/10.1017/9781108164276.

Khalifa, Kareem. this volume. "Should Friends and Frenemies of Understanding be Friends? Discussing de Regt."

Lipton, Peter. 2009. Understanding Without Explanation. In H.W. de Regt, S. Leonelli & K. Eigner (eds.), *Scientific Understanding: Philosophical Perspectives*. Pittsburgh: University of Pittsburgh Press, 43–63. https://doi.org/10.2307/j.ctt9qh59s.6.

5 Onwards, My Friend! A Reply to De Regt

Kareem Khalifa

In his contribution to this volume, Henk raises two important challenges to my thesis that understanding is derivative of explanatory knowledge.[1] First, he argues that my view accords an insufficient role to the skills that scientific understanding requires. Second, he argues that it wrongly disqualifies false theories that still furnish understanding.[2] While Henk was busy crafting these insightful comments, I did something sneaky: I abandoned my adversarial ways by synthesizing our two views in my contribution to this volume. While this "de Regt-Khalifa (DRK) synthesis" suggests that we now have less to disagree about, Henk's challenges still raise interesting questions against this new backdrop.

1 Skills

Henk's first argument against understanding as a species of knowledge rests on the premise that, unlike knowledge of an explanation, understanding requires the skills needed to grasp how an *explanans* relates to its *explanandum*. Fortunately, according to DRK, only an explanation that deserves consideration can provide understanding and any explanation that deserves consideration satisfies Henk's criteria of understanding. Consequently, Henk and I can provide the same responses to every objection he has raised to the *sufficiency* of my earlier accounts of understanding using resources from his work—including any of his skills-related objections.

But does this borrowing of Henk's work require me to abandon my position that understanding is a species of knowledge? I offer two reasons for sticking to my guns. First, DRK recruits Henk's account of understanding to elaborate the process of explanatory consideration that was *already* part of my knowledge-based account of understanding. Thus, DRK can follow my earlier work in treating explanatory consideration as an element of scientific knowledge of an explanation.

Second, Henk's objection to my view seems to assume that knowledge does not require skills. However, this clashes with what Pritchard (2012) calls the "ability intuition"—that knowledge requires cognitive ability. For instance, perceptual knowledge requires perceptual

DOI: 10.4324/9781003202905-6

abilities; inductive knowledge requires inductive reasoning abilities; memorial knowledge requires the ability to store and recall information; etc. Until Henk indicates what is wrong with the ability intuition, I am not sure why he draws the line between knowledge and understanding where he does.

For example, Henk claims that the deductive reasoning required in deductive nomological (DN) explanations is a skill that is not required to know such an explanation, for example, when one comes to know through rote memorization. However, if the ability intuition is correct, then deductive knowledge requires deductive reasoning skills. So, a younger, brasher Khalifa (2012, 27) rhetorically asked: "the big upshot is that understanding a DN explanation requires deductive reasoning skills. But do we really need a theory of understanding to tell us *that*?" I can now answer that non-rhetorically: no, the ability intuition applied to deductive knowledge can *also* tell us *that*.

The ability intuition also suggests that explanatory knowledge requires explanatory abilities. In my book, I argued that the ability to infer counterfactuals from explanations ("what-if reasoning") is among these abilities (Khalifa 2017, 57–61). Henk disagrees:

> . . . there is a difference between a student who simply knows the explanation that the greenhouse effect explains global warming and a student who understands why global warming occurs on the basis of this explanation. The former may have no clue when asked, for example, what would happen if the concentration of a particular greenhouse gas in the atmosphere increases, while the latter should be able to answer such what-if questions.
>
> (this volume, 23)

We seem to be talking past each other.[3] The argument in my book could have been clearer, so allow me to state it more explicitly:

1 Knowledge requires belief.
2 One can believe that p only if one can perform basic inferences with p.[4]
3 The basic inference that distinguishes explanations from near neighbors (e.g., conjunctions, correlations, conditionals) is: *a explains b, so had a been different, b would have been different* (= basic what-if reasoning).
4 So, one can know an explanation only if one can perform basic what-if reasoning with that explanation (From 1, 2, and 3).

So, I agree with Henk that a student who cannot perform basic what-if reasoning *does not understand* why global warming occurs. However, I disagree with Henk that this student *knows* why global warming occurs.

Premise 2 might be the source of our disagreement. Notably, the ability intuition provides one motivation for this premise. However, we can motivate this premise in other ways, too. Suppose that someone *claims* to believe that if it is raining then the streets are wet. However, he never believes that the streets are wet when he believes that it is raining, and he never believes that it is not raining when he believes that the streets are not wet. At this point, we have good grounds for denying his claim that he believes the conditional that claim is confused, insincere, etc.

Perhaps Henk might grant Premise 2 but deny Premise 3? For me, the motivation for Premise 3 is analogous to the previous example: it is unclear how someone could believe that *greenhouse gasses explain global warming* without also thinking that *changes in greenhouse gas levels would lead to changes in global warming*. So, if this is the true source of our disagreement, then a nice continuation of our conversation would have Henk proposing an alternative basic inference that believers/ knowers of explanations must be able to perform.

So, two outstanding issues suggest themselves. First, Henk and I may disagree about the ability intuition. Alternatively, we disagree about which abilities (if any) are required for explanatory knowledge. Having said all this, if both Henk and I hop aboard the DRK train, I do not know whether we should be agonizing over these issues. We agree on which skills are needed for particular explanations—deductive reasoning is required for DN explanations and what-if reasoning is required more broadly—and this might well eclipse any disagreements that we have about the relationship between understanding and knowledge.[5]

2 Truth

In constructing DRK, I was struck by how *voluntarism* dissolved a longstanding disagreement between Henk and me regarding truth and understanding. In this context, voluntarism entails that any good theory of understanding should be compatible with reasonable versions of both realism and antirealism. More precisely, take any sentence of the form "*Q* explains why *P*." For standard realists, "*Q*" must be at least approximately true; for structural realists, "*Q*" must be iso- or homomorphic to a physical structure; and for constructive empiricists, "*Q*" must only be empirically adequate. Voluntarism says that any of these options are fine in constructing your theory of understanding. As I suggested in my earlier contribution, this will sometimes result in corresponding differences in one's judgment as to which explanations provide understanding, but in many cases (such as my discussion of Bjorken scaling), it will not matter at all.

Henk rightfully presses me on a different class of cases than I have discussed: past theories that, by our current lights, still provide us with understanding even though they are known to be false. The shining example is Newtonian mechanics. I submit that a voluntarist treatment of Newtonian mechanics should parallel the one I offered of Bjorken scaling. Henk's chapter provides us with a nice antirealist interpretation of how Newtonian mechanics affords us understanding, so I will consider my job to be one of providing the realist alternative. As before, I do not claim that only one of these interpretations is reasonable; pick your poison.[6]

Henk's main reservation against realist interpretations of Newtonian mechanics is that "it appears too much of a stretch to interpret both [classical and relativistic] theories realistically and consider them both to be true" (this volume, 28). I tend to agree—at least if "interpreting a theory" means interpreting an *entire* theory. However, as I noted in my earlier contribution, to avoid conjuring "bogeyman realism," we should acknowledge contemporary scientific realists' more *selective* approach, in which the *parts* of the older theory responsible for high-grade empirical successes (e.g., novel predictions) must map onto some *parts* of that theory's successors.[7] As long as accurate parts of the theory are doing explanatory work, realists can account for understanding in these cases just as capably as antirealists.

Contemporary realists have applied this strategy to Newtonian mechanics (e.g., Saatsi 2019; Worrall 1989).[8] The *Principia* is packed with "use-novel predictions," that is, predictions of known facts that were not used to calculate free parameters in classical physics: Kepler's laws, the tides, etc. Further, the mapping between classical and relativistic mechanics—the so-called correspondence results—is among the most rigorous intertheoretic relations in science. So, a suitably nuanced account of how parts of a theory accurately represent parts of the world—rather than the kind of wholesale "realistic interpretation of a theory" suggested by Henk's remark—is both in lockstep with contemporary realism and allows realists to agree with antirealists that Newtonian mechanics provides understanding of several phenomena—including Kepler's laws, the tides, etc. So, we theorists of understanding are free to choose between Henk's antirealist interpretation of Newtonian mechanics and the realist one sketched here.

In conclusion, I am optimistic that this exchange nudges Henk, me, and others to pursue new questions about understanding. The relationship between understanding and knowledge, skills, and truth is a useful first step. We should now turn to exciting new topics such as explanatory staging, empirical fitness, and voluntarism.

Notes

1 For the sake of simplicity, this essay treats understanding as a species of knowledge. However, Khalifa (2017) only claims that understanding improves as it bears greater resemblance to scientific explanatory knowledge.

2 Henk also raises two other objections. First, he argues that my view sits uneasily with the contributions of idealizations to understanding. For replies, I refer readers to Khalifa (2020), Khalifa and Millson (2020), and Sullivan and Khalifa (2019). Henk also claims that my view cannot account for the historical and disciplinary variation of understanding. However, explanations and knowledge exhibit historical and disciplinary variation, so I am unclear what motivates this brief but suggestive idea.

3 The fancier what-if reasoning in Henk's example is not necessary for the minimal understanding I was discussing in Khalifa (2017, Ch.3) and to which Henk was responding. To my knowledge, Henk has not argued that minimal understanding requires fancier what-if reasoning.

4 I am amenable to other belief-like states figuring in understanding, for example, acceptance, credence, and endorsement. Slightly modified, this argument should work for those mental states, too.

5 Henk also mentions idealizing, approximating, and diagrammatic reasoning as other skills that my view will be hard-pressed to accommodate. For idealizations and approximations, see note 2. I see no special problem with diagrammatic reasoning. For example, we can know that A causes B from interpreting or drawing an accurate causal graph.

6 Here, I pursue what I called the "realist version of the Explanatory Objection" (Khalifa 2017, 163–164): the strategy of arguing "that the putative explanation in question is in fact (approximately) correct," as Henk puts it (this volume, 27). I tend to agree with him that the other strategies I develop are better suited for more thoroughly discredited theories such as the phlogiston theory than they are for Newtonian mechanics (though as a voluntarist, I will not rule them out.)

7 This need not presuppose that the new theory is true, as Henk suggests (28). Rather, continuity with successor theories provides evidence that the parts of the old theory were indispensable for generating the empirical success (Psillos 1996).

8 Saatsi cites several others.

References

de Regt, Henk W. this volume, "Can Understanding be Reduced to Knowledge?"

Khalifa, Kareem. 2012. "Inaugurating Understanding or Repackaging Explanation?" *Philosophy of Science* 79 (1):15–37. doi: 10.1086/663235.

Khalifa, Kareem. 2017. *Understanding, Explanation, and Scientific Knowledge*. Cambridge: Cambridge University Press.

Khalifa, Kareem. 2020. "Understanding, Truth, and Epistemic Goals." *Philosophy of Science* 87 (5):944–956. doi: 10.1086/710545.

Khalifa, Kareem, and Jared Millson. 2020. "Perspectives, Questions, and Epistemic Value." In *Knowledge from a Human Point of View*, edited by Ana-Maria Crețu and Michela Massimi, 87–106. Cham: Springer International Publishing.

Pritchard, Duncan. 2012. "Anti-luck Virtue Epistemology." *Journal of Philosophy* 109 (3):247–279. doi: 10.5840/jphil201210939.

Psillos, Stathis. 1996. "Scientific Realism and the 'Pessimistic Induction.'" *Philosophy of Science* 63 (s1): S306–S314. doi: doi:10.1086/289965.

Saatsi, Juha. 2019. "What Is Theoretical Progress of Science?" *Synthese* 196 (2):611–631. doi: 10.1007/s11229-016-1118-9.

Sullivan, Emily, and Kareem Khalifa. 2019. "Idealizations and Understanding: Much Ado About Nothing?" *Australasian Journal of Philosophy* 97 (4):673–689. doi: 10.1080/00048402.2018.1564337.

Worrall, John. 1989. "Structural Realism: The Best of Both Worlds?" *Dialectica* 43 (1–2):99–124. doi:10.1111/j.1746–8361.1989.tb00933.x.

6 Factivism in Historical Perspective

Understanding the Gravitational Deflection of Light

Sorin Bangu

1

Little Edith has an explanation for the appearance of presents under the Christmas tree: Santa Claus brought them! Emma, Edith's older sister, found out that (sadly) Santa does not exist, so her explanation is that their parents put the gifts there the evening before. It is intuitively evident that only Emma, and not little Edith, can be said to understand why there are presents under the Christmas tree. Scenarios like this lend credibility to the view that one's understanding as to why a phenomenon takes place must be based exclusively on truths, or at least on approximate truths – this is, henceforth, 'factivism' about understanding.

Yet, on reflection, and especially when taking into account what happens in many scientific contexts, a number of authors have embraced 'non-factivism' – the opposite view, that falsehoods *can* play principal roles in producing understanding.[1] Recent defenders of this conception include, among others, Elgin (2007, 2017), De Regt (2015, 2017), Rancourt (2017), and Doyle et al. (2019).[2]

There are currently two main lines of argument in favor of non-factivism. Its supporters have invoked both (i) the extensive employment of idealizations in science, and (ii) our inclination to say that past and false scientific theories, such as Newtonian mechanics, still provide understanding today.[3] In this chapter, I set the argument from idealizations aside.[4] My main concern will be the second argument, from history, as it appears in Henk de Regt's remarkable 2017 book, *Understanding Scientific Understanding*. I shall regard the sentence quoted below as representative for his take on the argument. The key-idea is that since there are prominent examples of past and false scientific theories that still generate understanding today, it follows that non-factivism is right:

> [A]lthough Newtonian mechanics has been superseded by Einstein's theory of relativity, and is therefore strictly speaking false, *it can be and indeed still is used* to explain phenomena, and in this way genuine understanding can be obtained.
>
> (2017, 131; my emphases)[5]

DOI: 10.4324/9781003202905-7

Although I retain my sympathy for non-factivism expressed in other works,[6] here I shall argue that this way of making the case for non-factivism, based on historical considerations (and focusing on the Newtonian theory), is quite weak. Central to my criticism will be an analysis of a famous episode in the history of physics, also discussed by de Regt (2017, 113–114): the fascinating astronomical-gravitational phenomenon typically presented as the 'bending' of light. Its observation, during a total solar eclipse in 1919, has corroborated Einstein's General Theory of Relativity (GR henceforth).[7]

In broad outline, the idea of the chapter is as follows. I will show that when applied to ('tested' on) this historically important example, de Regt's contextual theory of scientific understanding yields, in fact, *two* different answers (one affirmative and one negative) to our key-question here: do we, today, understand the bending phenomenon by using the Newtonian gravitational theory? We realize that de Regt's theory yields two answers in this case – and not only one (the affirmative answer) as he says above – as soon as we adopt a more precise characterization of the phenomenon of interest. Such a characterization distinguishes between a coarse-grained construal of the phenomenon-explanandum (that there is deflection *at all*) and a more fine-grained description of it (that there is deflection at *a specific angle*).[8] My main point will be that according to de Regt's theory we understand the former but not the latter; then, the negative answer weakens de Regt's support for non-factivism.

Before we move on, I need to spell out several assumptions on which the subsequent discussion relies.

To begin, I presume that the very point of explaining is to convey understanding.[9] Moreover, I agree with the view that understanding is the major aim of science, the primary epistemic good delivered by it[10]: for any natural phenomenon *p,* an explanation of it should be accepted *only* when it conveys understanding as to why *p* holds. This condition is meant to be uncontroversial, although it does rule out some types of explanations, in particular those explanations of the merely formal-mathematical kind often encountered in quantum mechanics.[11]

Another point I take for granted is that the 'deflection phenomenon' under study here refers to a full characterization of the observed behavior of light. As mentioned above, I will conceive of this phenomenon as having two facets: it can be presented under a coarse-grained description *and* under a fine-grained (quantitative) description. I take it that it is important to understand both dimensions of this phenomenon.

Finally, since, as I implied, we will unavoidably touch on the topic of scientific explanation, another caveat needed is that I will not be able to cover such a large theme properly. Most importantly, I will leave out all the various philosophical theories of explanation (deductive-nomological, causal, unificationist, counterfactual, difference-making, etc.) and I will talk about explanations only in generic terms, as (relevant) answers to why-questions.[12]

2

Let me introduce the basic ideas of de Regt's (2017) account. His avowed goal is to capture the gist of scientific understanding – an admirably ambitious project, since this notion is not only frustratingly elusive, but also contested, as the history of philosophy of science demonstrates.[13] In brief, according to his account, to say that we have a scientific understanding of a phenomenon it is necessary that there be an *explanation* of it based on an *intelligible* theory. In addition, such an explanation has to be *empirically adequate* and *internally consistent* (2017, 93).

When it comes to intelligibility, this is a feature that a theory possesses to the extent that the theory can be *used* by the scientists, that is, it can be employed in designing explanations, and explanatory models, of the phenomena of interest. A *pragmatic* feature (as is evident), intelligibility is also a *contextual* feature (hence his account is 'contextualist'). In agreement with Toulmin (1961) and Kuhn (1962), de Regt notes that a theory considered intelligible by a scientific community (in certain contexts of application) may be deemed unintelligible by a different community.[14]

Here is the first key-component of de Regt's account, the 'Criterion for Understanding Phenomena':

> CUP: A phenomenon P is understood scientifically if and only if there is an explanation of P that is based on an intelligible theory T and conforms to the basic epistemic values of empirical adequacy and internal consistency.
>
> (2017, 92)

Before I take up the other central component (intelligibility), let me first signal, and sort out, a potentially confusing terminological issue. In the formulation of the CUP above, and throughout his 2017 book, de Regt takes empirical adequacy to be a feature of an *explanation* (a positive feature: an epistemic value). However, this usage of the term 'empirically adequate' is, it seems to me, non-standard. Van Fraassen, for instance, the philosopher from whom I assume de Regt (and virtually everyone else) borrows the term, always refers to the empirical adequacy as a feature of a *theory*.[15] To stress, van Fraassen, unlike de Regt, attributes this feature systematically not to explanations, but to theories (or to models) on whose basis we formulate explanations. So, although I am not entirely sure what it is for an *explanation* to be empirically adequate in de Regt's view, I will be charitably assuming here that by the empirical adequacy of an explanation it is meant the empirical adequacy of the *theory* on which the explanation is based.[16]

Thus, to claim that a certain explanation (of a certain phenomenon) is empirically adequate is to say that the explanation draws on an empirically adequate theory. This is a theory that 'saves the appearances', in the sense that it makes correct predictions about what we observe. As is well known, to believe that a theory is empirically adequate is van Fraassen's alternative to believing that a theory is true (full stop), which involves the more committing belief that what the theory says about the unobservable part of reality is also true. The theory we will be focusing on here is the Newtonian theory of gravitation, which, as de Regt acknowledges repeatedly, is not a true theory; today, we know that the central unobservable structure postulated by Newton – the force of gravity – does not exist.

The next concept to discuss is *intelligibility*, the other pilar of de Regt's account. It is described as

> the value that scientists attribute to the cluster of qualities of a theory (in one or more of its representations) that facilitate the use of the theory.
>
> (2017, 40)

In addition to being context-sensitive, intelligibility is a multifaceted, multiple-realizable (as one may say) feature of a theory: different scientific communities, co-extant or even across history, may take different aspects of a theory to contribute to facilitating its use (e.g., its visualizability, its appeal to causal mechanisms, its unified character, its mathematical simplicity). De Regt describes this feature in more detail as follows:

> Criterion for the Intelligibility of Theories (CIT$_1$): A scientific theory T (in one or more of its representations) is intelligible for scientists (in context C) if they can recognize the qualitatively characteristic consequences of T without performing exact calculations.
>
> (2017, 102)[17]

With both CUP and CIT$_1$ on the table, we are finally in the position to assess the verdict that his theory of understanding gives about the phenomenon of gravitational deflection of light. So, let us ask the key-question once again: do we, today, understand this phenomenon by using the (false) Newtonian theory? As announced, I shall argue that the answer is both 'yes' and 'no' – so not only 'yes' as de Regt assumes. More precisely, I claim that de Regt's theory entails both that we understand it and also that we do not. This is so because the first answer has to be given when we consider the phenomenon in its coarse-grained construal, whereas the second has to be given when we take into account the finer-grained construal of the phenomenon. In and of itself, this is of course not a problem; but the negative answer *is* a problem for a non-factivist like de Regt.

3

Let us begin by focusing on the coarse-grained construal of the phenomenon of interest here, that there is *some* deflection of starlight when passing near the Sun. One can see right away that the application of CUP and CIT_1 criteria to the Newtonian theory produces the desired result: we do understand this phenomenon (again, in its coarse-grained version). According to these two criteria, we understand the coarse-grained version of the deflection phenomenon: the theory enables us to understand why there is deflection *at all*. This is so because there is an explanation of it based on an intelligible (in this context[18]) theory, the Newtonian gravitational theory. Moreover, this theory conforms to the basic epistemic values of empirical adequacy and internal consistency. As has been shown by, for example, Henry Cavendish (1731–1810) and Johann Georg von Soldner (1776–1833), we can explain (and predict) the deflection of light near a massive object if (i) we model light, just as Newton seemingly did, as composed of a stream of massive corpuscles[19] and (ii) we take these particles, in so far as they have mass, to be subject to the (large) gravitational pull of the massive object, according to the law of universal gravitation.

So, what the classical theory of gravitation tells us in this context is empirically adequate (recall that we consider the deflection phenomenon from a coarse-grained perspective) and internally consistent. Thus, two (of the three) conditions that CUP mentions are satisfied. What about the third condition, the intelligibility of the theory (in this context)? One can make a compelling case for it, too. Indeed, as CIT_1 requires, the physicists can recognize consequences of the Newtonian gravitational theory without performing exact calculations. One way they can recognize them, I submit, is via an analogy between a cannon shooting projectiles horizontally and a flashlight emitting light particles horizontally as well. Based on the law of gravitation, the Newtonian can infer without calculation that the light corpuscles, just like the projectiles, will be deflected ('downward') toward the larger body – because they will also feel the force of attraction (as given by Newton's law) due to the much larger mass near them (the Earth and the star, respectively). Note that this is a phenomenon described exclusively in coarse-grained terms: we are interested in why there is *some* deflection. There is no fine-graining involved here, no quantitative-numerical aspects to take into account (i.e., no question is raised as to why the angle of deflection is what it is observed to be).

However, if we consider the phenomenon under its fine-grained (quantitative) aspect – namely, if we want to understand why light is deflected at *a certain angle* – then it turns out that the Newtonian theory is *not* empirically adequate anymore. The same physicists mentioned above, von Soldner and Cavendish, were the first to calculate the values of the deflection angle from derivations based on the classical gravitational theory.[20] Von Soldner's paper was published in 1804 – exactly 100 years after Newton's publication of his *Opticks* – and the value he reported was 0."84. This value is close to a calculation done by Cavendish 20 years earlier, who found a similar number (0."875) but did not publish the result.[21] Crucially, both these results are seriously inaccurate. They are about *half* of the true value of 1.7" observed by the 1919 expedition, value calculated (and predicted) by Einstein himself from GR in 1915.

That the Newtonian theory is not empirically adequate when it comes to the fine-grained construal of the phenomenon of interest means that one of the conditions of CUP (empirical adequacy) is not satisfied. Since this condition is not satisfied, we have to conclude that according to de Regt's theory the Newtonian physicist (today) does *not* understand the phenomenon when construed fine-grainedly (that is, she does not understand why the angle is 1.7 seconds of the arc). To clarify, I accept that the other two conditions, the intelligibility of the theory in this context and the internal consistency of the explanation, *are* satisfied. Moreover, let me dispel the impression that by pointing out that the Newtonian theory may not be empirically adequate when the bending phenomenon is construed fine-grainedly / *quantitatively*, I may be talking past de Regt, who requires that the scientists "recognize the *qualitatively* characteristic consequences of the [Newtonian] theory" (2017, 102; my emphasis). This is *not* so, because this is the intelligibility condition, which, again, I accept it is satisfied; what I claim is not satisfied is another condition, empirical adequacy.

Recall that all this holds for the Newtonian gravitational theory. Importantly, this does not mean that according to de Regt's theory we lack, today, an understanding of this phenomenon – construed as both coarse-grainedly and fine-grainedly. By applying his theory (specifically, the CUP and CIT_1 criteria) to Einstein's GR this time, we shall see that, despite some doubts I raise below, de Regt's theory does yield the intuitively right result – namely, that we do understand this phenomenon, and under both construals. To see why this is so we need to take up these three key-questions: Is it possible to formulate, on the basis of GR, an explanation of the phenomenon considered both coarse-grainedly and fine-grainedly? Is GR (and the explanation based on it) empirically adequate and internally consistent? Is GR an intelligible theory (in this context)?

It is not hard to accept that the first two questions should receive affirmative answers. To begin with the first one (about the explanation), the

Einsteinean physicist will tell us that the phenomenon (considered in its coarse-grained form) is, ultimately, a consequence of the 'equivalence principle'. Imagine, along the lines of Einstein's original *gedankenexperiment* (Einstein 1917), an elevator floating in outer space, far away from any celestial body. Suppose that we drill a little hole into one of its walls and flash a horizontal light beam through it. When the elevator is at rest, the beam will follow a straight path parallel to the elevator's floor. But when the elevator accelerates rapidly 'upward', the beam traveling across the elevator will appear to an observer inside as bending toward the floor, that is, 'downward' (whereas to an observer outside the elevator it will continue to appear straight). By the principle of equivalence, such an accelerated frame of reference can also be regarded as being stationary and affected by a gravitational field. Hence, Einstein concluded, light moving in a 'real' gravitational field will also follow a bent path; so, we understand why there is deflection. Note, however, that although this is correct, the complete explanation and prediction of the phenomenon (which includes its fine-grained version) has to be formulated within a fully worked out GR.[22] (As well-known, this explanation states that light moves along null geodesics of the spacetime, and once they are warped by a large amount of mass-energy due to the presence of the Sun, the path of light will appear curved as well.)[23]

As for the other two features required by the CUP, the empirical adequacy and the internal consistency of GR, it is sufficient to recall that (a) GR has been extensively probed by some of the sharpest minds in theoretical physics in the last 100 years or so, and found to be free of inconsistencies,[24] and also (b) that it has been comprehensively tested empirically since its proposal in 1915. Needless to say, it has met both these kinds of challenges admirably well. From the derivation of the anomalous precession of the perihelion of Mercury (more on this later), to the observed gravitational redshift, and to accounting for gravitational 'lensing', the theory has delivered impressively accurate empirical results. Hence, I take it that there is virtually no doubt that GR explains the bending of light (construed both coarse-grainedly and fine-grainedly), and that this theory (and the explanation based on it) possesses the two required key-features.

What about the intelligibility of GR? Although the answer to this question involves some complications, I am inclined to accept that in the end it is affirmative. Recall that CIT_1 requires us to assess whether GR is such that the physicists working with it (in a particular context) can recognize characteristic qualitative consequences of it without performing exact calculations. Is this so?

On the one hand, GR is mathematically very complex, and it may be prudent to say that a definitely affirmative answer is hard to give. It may depend not only on *which* particular consequences one is interested in recognizing without calculating, but also on *who*, that is, which particular physicist, tries to recognize these consequences in that way. (A genius like Einstein may be able to recognize them, whereas a merely competent practitioner may not.) On the other hand, it seems reasonable to say that after acquiring sufficient familiarity with the theory, even an ordinary physicist *may* be able to derive such consequences (in a given context) based on nothing else than just grasping the point of the left-side and right-side expressions appearing in the field equations – a point captured by John Wheeler's well-known catchphrase that 'spacetime tells matter how to move; matter tells spacetime how to curve'. Thus, a phenomenon like the one of interest here, the curving of light passing near a large amount of mass-energy, may be derivable in this fashion.[25]

Now, let me reiterate the intermediate conclusions we have reached so far. De Regt's theory allows us to say that we understand the deflection phenomenon (both as coarse-grained and as fine-grained) when it is explained by using Einstein's GR. But – and this is the relevant finding – such an unequivocal answer is *not* available when the explanation is based on the Newtonian gravitational theory. When we use this theory, we see that de Regt's account entails that we obtain at best a partial understanding of the phenomenon: we can be said to understand it when it is presented in a coarse-grained way (i.e., we understand why light is deflected *simpliciter*), but not when it is presented in a fine-grained way: we do not understand why the value of the deflection angle is 1.7″.

With this, we have reached the main, and worrisome, conclusion announced at the outset: that de Regt's theory is a weak ally in fighting the cause of non-factivism. Contrary to what de Regt assumes, CUP, the central pillar of his theory, simply does not let us claim without qualification what a non-factivist would like to claim, namely that (an explanation grounded in) a false theory (the Newtonian gravitational theory) still provides an understanding of phenomena. (And, to stress, this is so since the phenomenon of gravitational deflection of light serves as a counterexample.) The Newtonian theory provides only an understanding of the phenomenon conceived in a coarse-grained way; that is, it provides only partial understanding.

4

Let me now address two potential responses to the challenge I raised here. Both maintain that even if the worrisome conclusion drawn above holds, the present criticism still misses the target.

First, de Regt could concede that although the Newtonian physicist is today entitled to claim that she understands this phenomenon only when it is conceived in a coarse-grained way (and thus partially), this 'amount' of understanding is *enough* to score decisive points in favor of non-factivism. In suggesting this defensive strategy on de Regt's behalf, I draw on a certain gradualism that can be detected in his approach. He uses phrases such as 'some understanding' (2017, 92) and he says that he finds it 'awkward' to accept that 'eighteenth-century chemists like Stahl and Priestley, did not possess *any* scientific understanding at all' (2017, 132; my emphasis). Hence, the obvious implication is, I take it, that they had 'some degree of understanding' (2017, 132) – not as much as we have today but, by comparison, *less*.[26]

In reply, I have to first acknowledge that the idea to adopt a comparative-gradualist approach to understanding is quite natural. But there is a rather obvious problem with this line of defense, namely that it remains an open question as to how much is enough: is having an understanding of the coarse-grained version of the phenomenon enough to convince a factivist? And not only a factivist, but anyone who believes that the quantitative, fine-grained aspects of science are essential? I am afraid that on de Regt's account these will always remain open questions, since his theory does not seem to offer guidance as how to go about answering them. My own inclination (admittedly not an argument) is to insist that lacking understanding of the quantitative, fine-grained aspect of a phenomenon is a pretty serious flaw – and hence a factivist will not be swayed.

The second response de Regt may make is to bite the bullet, and accept that even if the bending of light poses a problem, this is just one recalcitrant counterexample and the objection does not generalize. I reply that it very likely does generalize; other such phenomena can be found.

Consider another historically important astronomical phenomenon, the precession of the perihelion of Mercury.[27] It can be shown that according to de Regt's theory of understanding, Newton's theory of gravitation provides (even today) an understanding of the coarse-grained construal of this phenomenon, viz., that there is precession *at all*. (As such, the Newtonian theory derives it as due to the various gravitational pulls of the planets around Mercury.) However, as is well known, the theory fails to account for the quantitative / fine-grained aspect of the phenomenon, since it gives the incorrect *amount* of precession. This is measured to be $5600''$ of arc per century, whereas Newton's equations give only $5557''$, with the remainder of $43''$ obtained by using GR.

In addition to this, there are other such cases, and much more general. Take, for example, the composition of velocities. Suppose that v_m is the velocity of a missile relative to the spaceship from which it is shot; and suppose that while shooting, the spaceship also moves with velocity v_s (relative to an observer at rest) in the same direction as the missile. Then, the resultant velocity of the missile relative to the observer at rest is, for the Newtonian, $v_N = v_m + v_s$. For the Einsteinean relativist, on the other hand, it is

$$v_E = (v_m + v_s)/[1 + (v_m v_s/c^2)],$$

where c is the speed of light.

As is evident, both the Newtonian theory and Special Relativity entail that the resultant velocity is *larger* than each component, just as observation confirms: $v_N > v_m$ (and $v_N > v_s$), and the same holds for v_E: $v_E > v_m$ and $v_E > v_s$. Crucially, however, note that this formulation – 'the resultant velocity is larger than each of the component velocities' – amounts to a coarse-grained construal of the observed phenomenon. Thus, when the phenomenon is conceived in this fashion, the Newtonian theory explains (and predicts) it correctly. But when it comes to telling us *how much larger* the resultant velocity is – that is, when we consider a fine-grained construal of the same phenomenon – the Newtonian theory gets this quantitative aspect wrong. We do not observe the ratio v_N/v_m, but the relativist ratio v_E/v_m. Moreover, the difference between these ratios[28] increases as the component velocities v_m and v_s approach the maximum allowed speed (the speed of light c).

5

To close. In a paper predating the 2017 book, de Regt writes:

> Newton's theory is, from today's perspective, false and does not describe reality: the attractive forces that Newton postulated do not exist according to Einstein. But does this mean that Newton had no understanding of gravitational phenomena? That seems an unacceptable claim.
>
> So we face the dilemma of either giving up the idea that understanding requires truth or allowing for the possibility that in many if not all practical cases we do not have scientific understanding. I will argue that the first horn is preferable: the link between understanding and truth can be severed. This becomes a live option if we abandon the traditional view that scientific understanding is a special type of knowledge. While this view implies that understanding must be factive, I avoid this implication by identifying understanding with a skill rather than with knowledge.
>
> (2015, 3782)

I agree with the point made at the end of the first paragraph above, when read this way: if we apply de Regt's theory *in Newton's time*, then it delivers a unique (affirmative) answer indeed. So yes, as we saw, we can say that Newton understood (in de Regt's sense of 'understood') the deflection of light. In *that* epoch, no fine-grained / quantitative aspects of the deflection phenomenon were known; hence, in their absence, and if – presumably – *some* bending had been observed experimentally, de Regt's theory gives a unique affirmative answer. However, in so far as de Regt sees himself as arguing for non-factivism, this is ultimately irrelevant. In order to evaluate his position, we have to make the assessment *today* – because it is today when we know that the Newtonian theory is false – and not in Newton's time, when the theory was taken to be true.[29] And, from today's perspective, when the quantitative aspects of the phenomenon are known (and problematic), the answer is, as argued, no longer unique.[30]

Finally, let me also note that the weakness of de Regt's historical case for non-activism cannot be eliminated by appealing to the role of skill (mentioned in the second paragraph of the quotation above). When we consider our understanding of a whole range of (gravitational) phenomena within the classical (gravitational) framework, passages like this second paragraph fail to take into account the whole picture (once again). Indeed, a Newtonian physicist does need skill to extract a prediction and an explanation of a phenomenon out of a false theory – when this phenomenon is construed coarse-grainedly. (And yes, the exercise may be pedagogically useful today, even illuminating in certain respects, as de Regt and Gijsbers (2017) note.) But, on the other hand, no amount of skill can turn $0.''84$ into $1.7''$, nor $5557''$ into $5600''$, and so on – such that the Newtonian gets the fine-grained (quantitative) aspects right. With this missing, I am afraid that one is entitled to maintain that the non-factivist case for the claim that we can obtain scientific understanding from a false theory is incomplete, and hence rather weak.

Acknowledgments

First of all, I am grateful to the editors for the invitation to contribute to this volume. For very helpful critical feedback on the various drafts, I especially thank Insa Lawler and Elay Shech, as well as Andrei Marasoiu, Magdalena Małecka, and Gabriel Tarziu. The responsibility for the final version is mine.

Notes

1 When it comes to these labels, I follow Khalifa's (2017, 156) definition: a non-factivist holds that 'understanding why p does not require belief in any approximately true explanations of p'. Thus, for a factivist, understanding does require such belief; hence, no understanding can be obtained from a false explanation, or false theory. Another label in circulation is 'quasi-factivism'; for clarifications on it, see Khalifa (2017, Ch. 6).

2 Some of these authors may not self-identify, explicitly, as non-factivists; for instance, the label does not appear in de Regt (2017). However, in so far as de Regt and Gijsbers aim to show 'How false theories can yield genuine understanding' (the title of their 2017 paper), the label is, I believe, justified. Moreover, in the presentation of his 2017 book at LSE (available here https://www.youtube.com/watch?v=SY1vySjRFdA; accessed 26 January 2022), de Regt explicitly calls his view 'non-factivist'. Also, Khalifa (2017, 154) includes him among non-factivists.

3 As we will see in section 5, there are some complications with this line of argument, arising because one has to be careful as to which historical point of reference one takes when judging the status of a theory.

4 Elgin (2007, 2017) is the most dedicated advocate of this idea – for a recent take on her views, see Lawler (2021) and Frigg and Nguyen (2021) – but de Regt also follows this line of thought (2017, Ch. 6). On the opposite side of the philosophical fence, we find Kvanvig (2003). There are other authors, among them Potochnik (2017) and Strevens (2017), who hold views on the relationship between idealizations and understanding different from Elgin (although they also agree that idealizations facilitate understanding).

5 Strictly speaking, there is no such thing as 'Einstein's theory of relativity'. Einstein produced two theories of relativity, the 'special' and the 'general' one. With the exception of a brief mention of Special Relativity at the end of the paper, my focus here will be on the General theory.

6 See Bangu (2017b).

7 F. W. Dyson and A. S. Eddington organized two expeditions to observe the eclipse on 29 May 1919. Eddington and E. Cottingham went to the West African island of Principe, whereas A. Crommelin and C. Davidson were in charge of the other expedition, to Sobral (Brazil). The goal was to measure the deflection angle of starlight passing near the Sun, which was found to be in accordance with the predictions of GR. For philosophical discussion, technical and historical details, see, for example, Earman and Glymour (1980).

8 De Regt makes no mention of the fine-grained (quantitative) aspects of bending in his (2017) work.

9 Relatedly, I assume that there is no other way to gain understanding. Thus, I side with Strevens' *simple view* (2013) here, *contra* for example, Lipton (2009).

10 In another work (Bangu 2015), I advanced the idea that the progress of science can be cashed out in terms of increasing our understanding of the world. See also Dellsén (2016).

11 There are many examples of this – viz., the explanation of quantum tunneling – but presenting even one of them would take us too far afield.

12 I shall not discuss the factivity of knowledge (one cannot know p unless p is true), and in particular of scientific knowledge. Since understanding has been claimed to be a kind of knowledge (i.e., knowledge 'of causes'; Grimm (2014)), it would follow that understanding itself must be factive. However, since I doubt that understanding is a kind of knowledge, I will put this issue aside.

13 Hempel pretty much dismissed it, and Salmon followed in his footsteps. Friedman (1974) tried to rehabilitate it, but his (unificationist) account of explanation and understanding fell prey to technical errors, as shown by Kitcher (1981). However, not everyone is convinced that these technical flaws make Friedman's view irredeemable. See Bangu (2017a), and Sober and Roche (2017).

14 Another important aspect of the account is that it disregards the 'phenomenology of understanding' (2017, 20–23). Thus, it is not necessary that scientists 'feel' they understand the phenomenon of interest, or experience the 'aha' moment, etc. For a nuanced view on phenomenology, see Trout (2007).

15 See, for example, his classic (1980). The phrase appears quite often, on pp. 4, 10, 12, 13, 18, etc. As far as I can tell, there is not a single place where he talks about an *explanation* as being empirically adequate. The Stanford Encyclopedia of Philosophy entry on 'constructive empiricism' (Monton and Mohler 2021) follows the same usage.

16 My reading is supported by passages like this one, where he uses the term for both theories and explanations: 'Scientific theories (and explanations for

that matter) should conform to the observable world; a desideratum that in its minimal form can be called empirical adequacy' (2017, 37–38).

17 The index signals that this is only one version of intelligibility, as it covers theories formulated mathematically. For other kinds of theories, other, similar criteria can be given.

18 Here I take 'context' to refer to the context of application of the theory, namely to the effects of gravitation on light; this would be what de Regt calls a 'theoretical context' (2017, 97, 99). Thus, applying the theory to explaining the tides would be to use it in a different context (for the discussion of tides, see his 2017, 131).

19 Newton did suggest – see, for example, Query 1, at the end of his *Opticks* (1952, 1704) – that light is composed of particles; and that, as such, even if of very small mass, light is affected by gravity just as ordinary matter is. Another (false) assumption that these Newtonians – which also include John Michell (1724–1793) – made was that light coming from a distant star slows down near the Sun.

20 For a presentation and analysis of these calculations, see Treder and Jackisch (1981) and Ginoux (2021).

21 As Will (1988) documents.

22 See Holton and Brush (2005, Ch. 30.10). In fact, Einstein presented *two* derivations of the deflection phenomenon. As Pais (1982, 303) notes, the first dates from ca. 1911, before he completed GR, and relied only on the principle of equivalence; the second was presented in 1915, when he used, in addition, his own approximate solution to the GR field equations. It is this latter derivation that yielded the correct value 1.7″ of arc. (The first derivation of the deflection got only the coarse-grained aspect of the phenomenon right, but not the correct value of the angle. He obtained 0.″83, value whose similarity to von Soldner's prompted Lenard's antisemitically motivated, and false, accusation that Einstein plagiarized von Soldner. For details, see Wazeck (2014).

23 I leave aside some subtleties (e.g., the difference between local and global light deflection; for this, see Ehlers and Rindler (1997)). De Regt mentions Eddington's derivations of this phenomenon (Eddington 1920, Ch. 6). They are done within the GR framework, but also mix in Newtonian elements. One helpful classical analogue is the bending of light when passing in between two media, as illustrated by the proverbial straw appearing kinked in water.

24 To make something of an argument from authority here, if people of the stature of Eddington, Wheeler, Penrose, Hawking, or Weinberg have not found any internal inconsistency within GR, then it is almost certain that there is not any. Although it is true that GR and quantum field theory are in conflict when both are applied to black holes, this is not an inconsistency of GR alone. Vickers (2013) gives a thorough discussion of inconsistency in physics, but GR is not among the possible examples.

25 Note that strictly speaking, light is radiation, not matter; but photons have energy, energy is equivalent to mass (as Special Relativity shows), and mass has inertia, so light behaves like matter, hence (curved) spacetime does tell it how to move.

26 However, as he specifies rightaway, this amount of understanding is not sufficient to claim that their theory helps us in understanding the phenomenon of combustion today (2017, 132).

27 Mercury orbits the Sun on an ellipse, but the point where it is the closest to the Sun (the perihelion) does not remain fixed. It slowly moves around the Sun, giving rise to an inclination, then rotation, of the plane of the orbit. This motion is called *precession*. It takes 250,000 years for the perihelion to go all the way around the Sun and come back to where it started.

28 Of course, these two ratios are equal (within experimental error) if the two speeds v_m and v_s are much lower than c.

29 Importantly, de Regt himself realizes that the assessment has to be done from the contemporary perspective, as we saw from the passage cited at the outset: '[A]lthough Newtonian mechanics has been superseded by Einstein's theory of relativity, and is therefore strictly speaking false, *it can be and indeed still is used* to explain phenomena, and in this way genuine understanding can be obtained' (2017, 131; my emphases).

30 This is the complication I alluded to in footnote 3. When it comes to factivism, it matters from which historical perspective we judge.

References

Bangu, S. 2015. "Scientific Progress, Understanding and Unification." In G. Sandu, I. Parvu, and I. Toader (eds.), *Romanian Studies in Philosophy of Science*. Series Boston Studies in the Philosophy and History of Science. Pp. 239–253. Dordrecht: Springer. http://dx.doi.org/10.1007/978-3-319-16655-1_15.

Bangu, S. 2017a. "Scientific Explanation and Understanding: Unificationism Reconsidered." *European Journal for Philosophy of Science* 7(1): 103–126. http://dx.doi.org/10.1007/s13194-016-0148-y.

Bangu, S. 2017b. "Is Understanding Factive? Unificationism and the History of Science." *Balkan Journal of Philosophy* 9(1): 35–44. http://dx.doi.org/10.5840/bjp2017913.

Dellsén, F. 2016. "Scientific Progress: Knowledge versus Understanding." *Studies in History and Philosophy of Science* 56: 72–83. http://dx.doi.org/10.1016/j.shpsa.2016.01.003.

de Regt, H. 2015. "Scientific Understanding: Truth or Dare?" *Synthese* 192: 3781–3797. http://dx.doi.org/10.1007/s11229-014-0538-7.

de Regt, H. 2017. *Understanding Scientific Understanding*. New York: Oxford University Press. http://dx.doi.org/10.1093/oso/9780190652913.001.0001.

de Regt, H., and Gijsbers, V. 2017. "How False Theories can Yield Genuine Understanding." In S. Grimm, C. Baumberger, and S. Ammon (eds.), *Explaining Understanding*. Pp. 50–75. New York and London: Routledge. http://dx.doi.org/10.4324/9781315686110.

Doyle, Y., Egan, S., Graham, N., and Khalifa, K. 2019. "Non-factive Understanding: A Statement and Defense." *Journal for General Philosophy of Science* 50: 345–365. http://dx.doi.org/10.1007/s10838-019-09469-3.

Earman, J., and Glymour, C. 1980. "Relativity and Eclipses: The British Eclipse Expeditions of 1919 and Their Predecessors." *Historical Studies in the Physical Sciences* 11(1): 49–85. http://dx.doi.org/10.2307/27757471.

Eddington, A. 1920. *Space, Time and Gravitation*. Cambridge: Cambridge University Press.

Ehlers, J., and Rindler, W. 1997. "Local and Global Light Bending in Einstein's and Other Gravitational Theories." *General Relativity and Gravitation* 29: 519–529. http://dx.doi.org/10.1023/A:1018843001842.

Einstein, A. 1917 [1955]. Über *die spezielle und die allgemeine Relativitätstheorie*. Braunschweig: Vieweg; translation by R.W. Lawson as *Relativity: The Special and the General Theory*. New York: Crown Publishers.

Elgin, C. 2007. "Understanding and the Facts." *Philosophical Studies* 132: 33–42. http://dx.doi.org/10.1007/s11098-006-9054-z.

Elgin, C. 2017. *True Enough*. Cambridge, MA: MIT Press. http://dx.doi.org/10.7551/mitpress/9780262036535.001.0001.

Friedman, M. 1974. "Explanation and Scientific Understanding." *Journal of Philosophy* 71: 5–19. http://dx.doi.org/10.2307/2024924.

Frigg, R., and Nguyen, J. 2021. "Mirrors without Warnings." *Synthese* 198: 2427–2447. http://dx.doi.org/10.1007/s11229-019-02222-9.

Ginoux, J.-M. 2021. "Albert Einstein and the Doubling of the Deflection of Light." *Foundations of Science*. http://dx.doi.org/10.1007/s10699-021-09783-4.

Grimm, S. 2014. "Understanding as Knowledge of Causes." In A. Fairweather (ed.), *Virtue Epistemology Naturalized. Bridges between Virtue Epistemology and Philosophy of Science*. Synthese Library. Pp. 329–345. Dordrecht: Springer http://dx.doi.org/10.1007/978-3-319-04672-3_19.

Holton, G., and Brush, S. 2005. *Physics, the Human Adventure*. New Brunswick: Rutgers University Press.

Khalifa, K. 2017. *Understanding, Explanation, and Scientific Knowledge*. Cambridge: Cambridge University Press. http://dx.doi.org/10.1017/9781108164276.

Kitcher, P. 1981. "Explanatory Unification." *Philosophy of Science* 48: 507–531. http://dx.doi.org/10.1086/289019.

Kuhn, T. 1962. *The Structure of Scientific Revolutions*. Chicago: University of Chicago Press. http://dx.doi.org/10.7208/chicago/9780226458144.001.0001.

Kvanvig, J. 2003. *The Value of Knowledge and the Pursuit of Understanding*. Cambridge: Cambridge University Press. http://dx.doi.org/10.1017/CBO9780511498909.

Lawler, I. 2021. "Scientific Understanding and Felicitous Legitimate Falsehoods." *Synthese* 198: 6859–6887. http://dx.doi.org/10.1007/s11229-019-02495-0.

Lipton, P. 2009. "Understanding without Explanation." In H. De Regt, S. Leonelli, and K. Eigner (eds.), *Scientific Understanding: Philosophical Perspectives*. Pp. 43–63. Pittsburgh: Pittsburgh University Press. http://dx.doi.org/10.2307/j.ctt9qh59s.6.

Monton, B., and Mohler, C. 2021. "Constructive Empiricism." *The Stanford Encyclopedia of Philosophy* (Summer 2021 Edition), Edward N. Zalta (ed.). https://plato.stanford.edu/archives/sum2021/entries/constructive-empiricism/.

Newton, I. 1952 (1st ed. 1704). *Opticks or A Treatise of the Reflections, Refractions, Inflections & Colors of Light*. New York: Dover Publications.

Pais, A. 1982. *Subtle Is the Lord: The Science and the Life of Albert Einstein*. New York: Oxford University Press.

Potochnik, A. 2017. *Idealizations and the Aim of Science*. Chicago: University of Chicago Press. http://dx.doi.org/10.7208/chicago/9780226507194.001.0001.

Rancourt, B. 2017. "Better Understanding through Falsehood." *Pacific Philosophical Quarterly* 98(3): 382–405. http://dx.doi.org/10.1111/papq.12134.

Sober, E., and Roche, W. 2017. "Explanation = Unification? A New Criticism of Friedman's Theory and a Reply to an Old One." *Philosophy of Science* 84: 1–24. http://dx.doi.org/10.1086/692140.

Strevens, M. 2013. "No Understanding Without Explanation." *Studies in History and Philosophy of Science* 44: 510–555. http://dx.doi.org/10.1016/j.shpsa.2012.12.005.

Strevens, M. 2017. "How Idealizations Provide Understanding." In S. Grimm, C. Baumberger, and S. Ammon (eds.), *Explaining Understanding*. Pp. 37–49. London: Routledge. http://dx.doi.org/10.4324/9781315686110.

Toulmin, S. 1961. *Foresight and Understanding*. London: Hutchinson.

Treder, H.-J., and Jackisch, G. 1981. "On Soldner's Value of Newtonian Deflection of Light." *Astronomische Nachrichten* 302: 275–277.

Trout, J. D. 2007. "The Psychology of Scientific Explanation." *Philosophy Compass* 2: 564–591. http://dx.doi.org/10.1111/j.1747-9991.2007.00081.x.

van Fraassen, B. 1980. *The Scientific Image*. Oxford: Clarendon Press.

Vickers, P. 2013. *Understanding Inconsistent Science*. New York: Oxford University Press. http://dx.doi.org/10.1093/acprof:oso/9780199692026.001.0001.

von Soldner, J. G. (1804). "Ueber die Ablenkung eines Lichtstrahls von seiner geradlinigen Bewegung. *Berliner Astronomisches Jahrbuch*." Pp. 161–172. English translation: "On the Deflection of a Light Ray from Its Rectilinear Motion." Available at https://en.wikisource.org/?curid=755966 (accessed 12 October 2021).

Wazeck, M. 2014. *Einstein's Opponents: The Public Controversy about the Theory of Relativity in the 1920s*. Cambridge: Cambridge University Press. http://dx.doi.org/10.1017/CBO9781139084185.

Will, C. M. 1988. "Henry Cavendish, Johann von Soldner, and the Deflection of Light." *American Journal of Physics* 56: 413–415. http://dx.doi.org/10.1119/1.15622.

7 Ideal Patterns and Non-Factive Understanding

Mazviita Chirimuuta

1 Background: A Trade-Off between Prediction and Understanding

The perfect model would be all things to all scientists: it would accurately predict the behaviour of a natural system, it would represent it accurately (so that familiarity with the model would furnish us with true beliefs about the target), and it would enhance our understanding of it. As has been observed, at least since Levins (1966), models are like human beings in that the superb all-rounder is a comic book fantasy. And yet, as with human beings, the complementary strengths of a community of models can make up for the deficiencies of each one by itself. As I have argued elsewhere (Chirimuuta 2020), in some modelling projects, especially of very complex neural systems, a trade-off obtains between the ability of the model to make accurate predictions of the system's behaviour, across an acceptably broad range of conditions, and its *intelligibility* (de Regt 2017: chapter 4) – its capacity to yield an understanding of the targeted system.

That there can be an incompatibility between the scientific goals of prediction and understanding, or explanation,[1] is a point made long ago by Nancy Cartwright (1983). In her examples, the fundamental laws of physics are said to be "untrue" (not predictively accurate of the observed phenomena) but explanatory; whereas cobbled together ensembles of phenomenological laws and models derived from fundamental laws are needed to predict the behaviour of most of the concrete physical objects around us, but they do not help to explain them. In cases from computational neuroscience, I examined the performance of hand-coded, essentially linear models of the responses of neurons in primary visual and motor cortex (V1 and M1) in comparison with new generations of models produced via machine learning (ML) techniques – recurrent or convolutional networks that use learning algorithms to fit the V1 or M1 data. The new models have shown impressive predictive accuracy across a broader range of cases than managed by the older classes of models. However, this performance comes at a cost. The complex non-linear function learned by the network to fit the physiological data

DOI: 10.4324/9781003202905-8

is not available explicitly to the neuroscientist. The relative opacity of the ML-derived models means that they are less intelligible, and hence afford the scientist less understanding of the brain, in contrast to the classic hand-coded models. Conversely, the older models have less predictive power but do yield more understanding of the cortex.

The commonality between these examples from neuroscience and Cartwright's from physics goes deeper than the observation of the trade-off. One might be tempted to say that the classic models in neuroscience are simply false, now rendered obsolete with the arrival of the more mathematically sophisticated ML alternatives. However, it should be emphasised that the older models of the visual system can accurately predict the responses of neurons when the brain has been *rendered simple by experimental means*: that is, when the animal is anaesthetised and the visual stimuli are artificial patterns like bars and blobs, which have none of the rich detail of naturally occurring images.[2] This is a parallel to Cartwright's observation that the fundamental laws of physics, such as Newton's universal law of gravitation, *can* accurately predict the behaviour of physical systems in contrived experimental conditions; for example, when circumstances are controlled so that the dominant force on the two bodies is only due to gravity, and others such as air resistance and electrical attraction/repulsion are made negligible. The shielding around a physics lab and artificial conditions in neurophysiology are techniques used to produce a simplicity in the experimental occurrences that would not normally occur. Thus, we should not be tempted to say that the old linear models of the cortex were just cases of "unsuccessful" science, whereas the famous idealisations of physics are part of "successful" science. The thing to recognise on both counts is that there is predictive success, but in the limited domain in which reality is engineered to conform to ideal conditions, such as the low-pressure, low-temperature regime in which the ideal gas law makes accurate predictions.[3]

In a recent paper that discusses the application of ML to physics, Hooker and Hooker (2018) make a case that scientists' capacity to understand physical interactions is restricted to the simple cases described adequately by fundamental laws like Newton's equations, but that predictive power (enhanced by ML or just techniques of numerical approximation) holds up for a wider range of more complex systems, with the downside that we do not have "interpretable knowledge" (i.e., understanding) of them (p. 180). This observation leads these authors to contest the value and relevance of understanding as a goal for science. Instead, I retain understanding as a central epistemic aim of scientific research: awareness that a goal cannot always be met is not, by itself, a reason to give up on it.

The existence of the trade-off between prediction and understanding provides a compelling reason to consider that at least in some cases scientific understanding is *non-factive*, because the cases show how departures from accuracy make the scientist's representations more and not less likely to confer understanding. Non-factivism is the idea that the achievement of scientific understanding is not essentially a matter of finding out the facts about the system and acquiring true beliefs about it (e.g., the discovery of a true explanation), but that it depends on the creation of scientific representations that aid human cognition – especially through the strategies of abstraction and idealisation – involving departures from representational accuracy. The non-factivist assertion is that idealised models offer epistemic benefits, in their conferring of understanding, that are not matched by more realistic counterparts.

In our neuroscience cases, it is assumed that the more recent ML models are more predictively accurate over a wider range of inputs because they more accurately represent the complex, non-linear functions computed in the visual or motor cortex.[4] A question that occurs is, *how can a model provide more understanding when it is less true of its target system than the more realistic model?* The answer is that the models that provide understanding are a compromise between the mind-boggling complexity of the natural system and the human capacity to grasp an array of relations. A model that just delivered the Truth of Nature would be no more comprehensible than Nature by itself; in order to produce understanding, models must simplify, and this means departing from the facts. Putting more "truth" into a model can make it better at predicting but need not make it better at providing understanding, because more truth will often make the model more cognitively opaque.

Recently, a number of philosophers have defended non-factivism about understanding, treated as a separable topic from explanation, in particular Elgin (2017), de Regt (2017), and Potochnik (2017). Consistent with the observed trade-off, and the implication that understanding and predictive accuracy are separable goals in science, Elgin has argued that understanding is conveyed by "felicitous falsehoods." In this chapter, I examine an important objection to non-factivism from Sullivan and Khalifa (2019), and I show how Potochnik's (2017) version of non-factivism would need to be modified in order to resist it. Sullivan and Khalifa argue that only the truthful features of idealised models confer an epistemic benefit, and that departures from truth act only as a psychological crutch. I will show how Potochnik's account succumbs to these objections because of an ontological commitment to "real causal patterns" (Section 2), and that the account can be strengthened by replacing this with the notion of an "ideal pattern," where the targets of model building are not patterns that are simply "out there" in nature, but they are to some extent dependent on the methods of data collection and processing chosen by the researcher (Section 3). I also show how ideal patterns relate to the notion of scientific phenomena. Phenomena are regularities indicated by datasets, which are the target of modelling and theory building. Against Bogen and Woodward (1988), and in agreement with McAllister (1997) and Massimi (2011), I reject a strongly realist construal of phenomena. Phenomena – taken as ideal patterns – are not pre-existing regularities discovered by scientists, but regularities that are to some extent brought about by the practices of experiment and data analysis. This rejection of the real patterns' account of phenomena is prefigured by Cartwright (1983: 19), who asks,

> But where are the patterns? Things happen in nature. Often they happen in regular ways when the circumstances are similar; the same kinds of causal processes recur; there are analogies between what happens in some situations and what happens in others. . . . But there is only what happens, and what we say about it. Nature tends to a wild profusion, which our thinking does not wholly confine.

2 Potochnik's Non-Factivism, and an Objection

2.1 Real Causal Patterns

Angela Potochnik presents a compelling picture of causal complexity as the reason for the prevalence of idealised models in science: observable nature presents us with countless occurrences, some of which are relevant to human interests, and some are more reproducible than others. For scientific representations to be useful for understanding and action, they must be judicious in their choice of what is represented, and in their mode of representation. Like Cartwright's, Potochnik's notion of "truth" or factivity of a scientific representation is one of accuracy in the depiction of empirical observations – what Potochnik calls the "phenomena." As will be discussed below,[5] these phenomena are "surface phenomena," what appears to scientific observation and instrument readings, not what is inferred from datasets. Potochnik's discussion is centred on model-based science, and not on laws. Idealised models are "departures from the truth" (Potochnik 2017: 95), just because they misrepresent the phenomena. Phenomena themselves are characterised as embodying very many "causal patterns" (Potochnik 2017: chapter 5). On the positive role of idealisations, we are told that idealisations aid the discovery of causal patterns:

> idealizations contribute to understanding by representing as-if to the end of depicting a causal pattern, thereby highlighting certain aspects of that phenomenon (to the exclusion of others) and revealing connections with other, possibly disparate phenomena that embody the same pattern . . . .
>
> (Potochnik 2017: 97)

In order to diffuse the worry that non-factive understanding might just be "merely apparent," a projection of a pattern into the phenomenon, which confers on the person projecting no more than a subjective feeling of understanding, Potochnik employs the notion of "real causal pattern," inspired by Dennett's (1991) account of "real patterns" (Potochnik 2017: 115–116). Real causal patterns are said to be "embodied" in phenomena, and in Potochnik's use of the notion – in a departure from Dennett – patterns have a reality that is *not* dependent on there being an agent capable of discerning them (Potochnik 2017: 28). The shift from Dennett's "semirealism" towards a more robust realism can be made clear with a discussion of the famous "bar code" pattern.

Dennett and Potochnik concur that Figure 7.1 *A-E* (which, for Potochnik, are each analogous to an observed phenomenon) all "embody" the bar code pattern of alternating black and white squares. While making the point that phenomena can embody many different patterns, Potochnik introduces the new pattern "bc-algorithm," which is made according to the recipe, "produce a dot matrix of pattern bar code, but change the color of a specified number of randomly selected dots" (Potochnik 2017: 28). The "bc-algorithm" pattern is "embodied" in each instance of Figure 7.1, including *F*. Image *F* does *not* embody "bar code" because, as Dennett states, it is not possible to transmit image *F* more efficiently than otherwise, by representing it as a combination of bar code with a certain amount of random noise. Indeed, for Dennett F is not a pattern at all because it contains no regularity, being a (pseudo) random array of dots – there is no representation of it that is more compressed than a pixel-by-pixel specification (a bitmap). In contrast, Potochnik embraces the idea that there may exist patterns that are hidden from the observer by noise, as "bc-algorithm" is in *F*.[6]

Figure 7.1 The pattern "bar code" is an alternating series of solid black and white squares obscured by varying amounts of noise. Dennett (1991: Figure 1), reproduced in Potochnik (2017: Figure 2.1).

According to Potochnik, observed phenomena comprise real causal patterns mixed with "noise" (which may be other causal patterns, not currently of interest to the researcher). The job of the idealised model is to separate the wheat from the chaff, making the real pattern of interest more salient, comprehensible, and useful. As Potochnik (2017: 43) puts it, "[i]dealizations provide a way to set aside complicating factors to discern a causal pattern of interest." This allows Potochnik to assert that "understanding is achieved by *revealing* causal patterns, patterns that are imperfectly embodied by phenomena and that relate to only some elements of phenomena" (p. 91, emphasis added). Without doubt, this is of a piece with Potochnik's commitment to scientific realism. Potochnik states that scientists are seeking "truths about causal patterns" (p. 119), and she emphasises that causal patterns are real in the robust sense of "out there in nature independently of human recognition," when contrasting her own account with Cartwright's "unruly" picture of Nature. Here, Potochnik says that the success of science is best explained by its being able to discover some of these "complex and variable causal relationships" (p. 41), which is an echo of the "No Miracles" argument for scientific realism. Thus, understanding is non-factive in the sense of being "not true to the surface phenomena," but at the same time we have here a notion of idealised models achieving agreement with real patterns not readily apparent within phenomena, and thereby "revealing" them to the scientist.

This is a view about the role of idealisation as a means to understanding, via the revelation of truths masked by complexity, that is sometimes expressed by scientists themselves, for example in a textbook appendix on modelling in neuroscience, Abbott et al. (2013) state:

> The purpose of modeling is to illuminate, and the ultimate test of a model is not simply that it makes predictions that can be tested experimentally, but whether it leads to better understanding. No matter how detailed, no model can capture all aspects of the phenomenon being studied. As theoretical neuroscientist Idan Segev has said . . . modeling is the lie that reveals the truth.

We must now consider whether this view leads to a defensible version of non-factivism about understanding.

2.2 An Objection to Non-Factivism

There are important arguments against the proposal that understanding is non-factive, because it is conferred by idealised models (Khalifa 2017: chapter 6). In particular, I will consider recent objections to the claim that idealisations have epistemic value because they aid understanding. Sullivan and Khalifa (2019) present the case that, on the one hand, "idealizations' false components only promote convenience instead of understanding," and on the other that, "only the true components of idealizations have epistemic value." Although these authors do not discuss Potochnik's account of understanding, in this section I will show that versions of their criticisms can be levelled against it.

Using the familiar example of the ideal gas model, Sullivan and Khalifa (2019: 676) argue that non-factivist interpretations can be rebutted in three stages: (1) by showing how the idealisation brings convenience ("makes certain calculations easier to perform or certain features of the phenomenon more salient"); (2) by showing how a de-idealised or approximately true version of the model could deliver the same epistemic goods as the idealised version, though in a less convenient manner; and (3) by showing that the epistemic benefits of the idealised model are only conferred by those component parts of the model that approximate to a de-idealised counterpart – what Sullivan and Khalifa call the "demythologising" move.

One initial obstacle in the way of the application, to Potochnik, of these three moves is that the second and third seem to depend on the availability of a de-idealised counterpart to the idealised model, such as the virial equation in relation to the ideal gas model. Potochnik frequently emphasises that idealisations are "rampant and unchecked" in scientific practice, which is to say that scientists do not pursue de-idealisation. This ensures that de-idealised counterparts are not available for the majority of models. However, some modifications of stages (2) and (3) suggest themselves, which would reinvigorate the factivist argument. The crucial point for the factivist is not the existence of de-idealised models but *the thought that the supposedly idealised models are in fact not idealised with respect to real causal patterns.*

In Potochnik's use of the terminology, a model counts as idealised only if it misrepresents an observed, surface phenomenon. This seems to suggest that the only way a model can be de-idealised is if it sticks more closely to the phenomenon, for example by making more accurate predictions over a wider range of circumstances. But the typical way to think about idealisation, and hence de-idealisation, is in terms of accuracy of representation of both observed phenomena and unobserved processes in nature, including processes and entities that are, in principle, unobservable. Only an anti-realist (empiricist) would reject this characterisation, and Potochnik is not an anti-realist. Thus, it follows naturally from her account to say that *hypothetically* there is a class of de-idealised, or rather non-idealised,[7] models that accurately represent the real causal patterns. In fact, most of her examples of idealised models – *with respect to the surface phenomena* – would count as *non*-idealised, *with respect to real causal patterns.* The idea is that in the natural system, other causal patterns interfere with the one that is the target of the model, making its manifestation ("embodiment") in the observed phenomenon "imperfect." But that does not render the model any less true of the target causal pattern, which is really operational in nature, even if obscure. With these pieces in place, the factivist can make a case that the truthful modelling of causal patterns is what confers the epistemic benefit of the model, though it is idealised with respect to the surface phenomena. This benefit turns out to be the product of truth after all.

In case this seems an uncharitable reading of Potochnik, note that she does herself tell us that idealised models have "features in common" with the target system, understood as a composite of numerous, interlaced, causal patterns:

> Models and the target systems they are used to represent bear some features in common, while other features of the systems are neglected or falsified in the models. This is accomplished via the use of abstractions and idealizations in characterizing the model, in order to render it simpler or easier to deal with in some other way.
> (Potochnik 2017: 43)

This is almost an admission that the epistemic benefits of the idealised models are conferred to the extent that something about them is an accurate representation of the target – by sharing common features with it – and that departures from accuracy are for convenience only. This is an important parallel to Sullivan and Khalifa's claim (3) that it is only the approximately true components of an idealised model that confer epistemic benefits. And so, the factivist can argue, the use of abstraction and idealisation is merely to make things "simpler and easier" by bracketing the extraneous causal patterns, whereas the epistemic value of the idealised model is conferred by its accurately representing something there in the natural system – one of its causal patterns.

3 Ideal Patterns

It is worth summarising the situation so far. Non-factivists note that idealised models – ones not providing accurate representations of observations – confer understanding nonetheless. Potochnik says that this is due to the idealised models' representation of a causal pattern that is only imperfectly embodied in the surface phenomena (p. 91). But as shown in the previous section, the factivist can insist that the representation of a real pattern is enough to make the understanding factive after all. My proposal for non-factivists is that they replace the notion of real patterns with that of *ideal patterns*. In the next sub-section, I will explain what these are, before showing how they can successfully defuse the factivists' objections.

3.1 Characterising Ideal Patterns

The notion of an *ideal pattern* is actually closer to Dennett's "semirealist" intent than Potochnik's "real causal pattern," but I name it "ideal" in order to mark its distinctness from the commonly used sense of "real" as "just out there in nature" and "independent of the scientist." Deeper into the 1991 paper, Dennett presents another version of the bar code pattern. The point is that patterns like the ones in Figure 7.1 can be generated without first making a "clean" version of the pattern and then corrupting it with noise. Instead, repetitions of a somewhat stochastic (but *not* completely random) process will yield an observable regularity, and "cleaner" versions of the patterns can be generated by processing the observed data in order to filter out the differences between the repetitions (Dennett 1991: 44). These ideal patterns do *not* arise from the superposition of a perfect regularity with noise, or some interfering regularities.

In order to employ this notion of pattern to our case, I will draw an analogy between Dennett's contrast enhancer and the data processing operations that are ubiquitous in science, and which generate the actual targets of modelling and theory building. (The target is almost

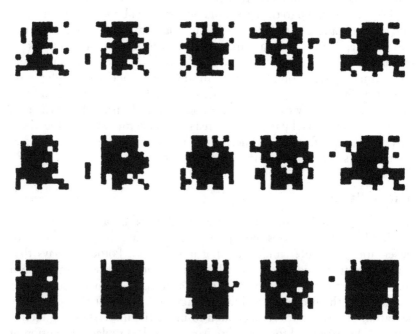

Figure 7.2 Illustration of an "Ideal Pattern." The top row is generated by a stochastic process. Pixels are drawn from normal distributions, evenly spaced along the row. Hard edges appear in the second and third row through application of a contrast enhancer – such results of averaging are what I call "ideal patterns." From Dennett (1991: Figure 4).

never the raw data.) A process as simple as averaging across trials creates "hard edges" that are not there in the raw data. For instance, neural recordings show trial-to-trial variability regarding the exact onset time and rate of activity in response to a stimulus. Traditional neural models target average data – the simplified and regularised patterns that are created through data processing. In this sense, the target of the model is an "ideal pattern," like the cleaned-up pattern in the third row of Figure 7.2, a pattern that is partially human dependent, and not a "real pattern" in the sense of a pattern that is just out there in nature.

To illustrate, consider the neurophysiology experiment performed by Georgopoulos et al. (1982), in which spike data were reported for five trials of one motor cortex neuron's responses to eight directions of movement. Onset time and level of excitation (or inactivation) in response to movements vary across trials but the pattern that is the actual target of modelling is a trial averaged one that is, strictly speaking, a creation of the scientist. The brain itself does not average across trials and each instance of variation in trials may be "meaningful" to the brain – reflecting different modes of attention or posture – not pure noise (Stringer et al. 2019). However, an ideal pattern is not synonymous with "fiction," something "made up" in a way that is disconnected from the observation

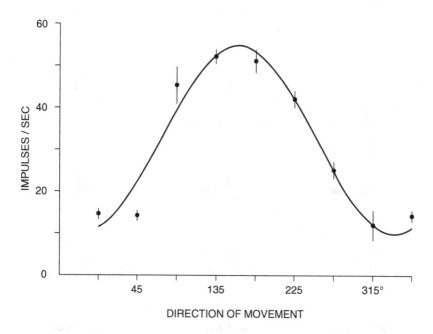

Figure 7.3 The Cosine model of a neuron's tuning to direction of movement. The cosine function is fit to trial average data. From Georgopoulos et al. (1982).

of a concrete system. Rather, an ideal pattern is a *regularised* version of a pattern that is in nature, but should *not* be thought of as an *underlying regularity*. The pattern is "ideal" in the sense of being more perfect than what is to be encountered in our messy, and – to the scientist – less than ideal, reality.[8]

Figure 7.3 illustrates how ideal patterns become the targets of models that then confer understanding. The many thousands of recorded spikes are condensed down into trial averaged means and standard deviations of the peak response for each direction of movement – nine data points that can be modelled with a cosine curve, though not without some error. The end result is the classic cosine model of motor cortex responses, which was later used in brain computer interface technologies to decode motor intentions from neural activity (Koyama et al. 2010). Figure 7.4 is provided in order to show that there can often be latitude, for the scientist, in deciding what the pattern in the data is supposed to

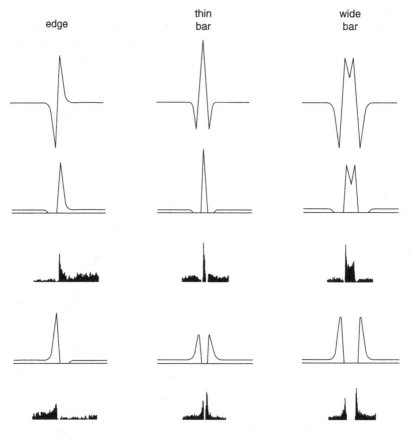

Figure 7.4 Comparison of model prediction and neurophysiological data, after Marr and Ullman (1981: 165); Marr (1982: 65).

be. It shows how in choosing the Laplacian of Gaussian function (line traces) to model responses of visual neurons, David Marr treated the responses as if they were completely symmetrical, whereas there is a noticeable pattern of asymmetry in the peaks of the neural data (solid black shapes).[9] The transformation of patterns that are asymmetrical in the data into ones that are symmetrical in the model is the kind of regularisation that typifies ideal patterns.

3.2 Responding to the Objection

On Potochnik's account, the idealised model has features in common with a causal pattern out there in nature, but only imperfectly embodied in the observed phenomena because of causal complexity. Instead, I propose that the idealised model represents an "ideal pattern," which is suggested by the data but does not have an existence independently of the scientists' activity. This blocks the move of the factivist who supposes that any epistemic benefits of the idealised model could just as well be conferred by non-idealised models that depict the underlying causal pattern. On my account, it is a mistake to posit these hypothetical, non-idealised models, since I do not presuppose the existence of these simpler causal patterns, independently of scientific activity.

The factivist might insist, however, that the idealised model is an approximately true representation of the natural system, by virtue of it representing an ideal pattern that originates from a pattern in the raw dataset. In response to this, I emphasise, first, that the ideal pattern contains features that cannot be found in the raw dataset, but are introduced with data processing – for example, the "hard edges" introduced with averaging across trials of data. In the top row of Figure 7.2, there are no hard edges – a hard-edged representation is not, without qualification and specification, "approximately true" of it. But at the same time, the way that the pixels are spottily arranged does suggest that a hard-edged pattern is an appropriate idealisation of it; the middle and bottom rows depict a regularity that is suggested by the top row. Second, we must remember the experimental artifice that occurs even before the production of the raw data. As mentioned in Section 1, the choice of visual stimuli affects the linearity of the responses of cortical neurons. Experiments are designed to generate regularity and enhance simple patterns (such as linear response). This is also the case with the experiment on the motor cortex whose results are presented in Figure 7.3. In the experiment, the monkey is constrained to keep the whole body as still as possible, except for the arm, and this limb is permitted only to move in eight radially symmetrical dimensions. This is a far more simple array of movements than a monkey would engage in if left free. Under free movement, the complexity of the neural responses in the cortex increases (Gao et al. 2017), and data recorded in such conditions would not even suggest the sinusoidal pattern presented in Figure 7.3 (Shenoy et al. 2013). Thus, the response to the factivist is that any talk of the "approximate truth" of ideal patterns is vague and consistently neglects the effort and experimental technique that goes into generating the ideal pattern. Further, the idealisation is effective because of its artifice and not in spite of it.

I will freely admit that there is a close connection – in visualisable cases, often a resemblance – between the features in the raw data and the features of the ideal patterns. For example, all the rows of Figure 7.2 show periodic increases and decreases in pixel density. In the example presented in Figure 7.3, the raw spike data (see Georgopoulos et al. 1982, Figure 7.4) and the averages fit with the cosine curve; both show a systematic increase and decrease in activity levels. However, characteristics that are central to the representational content of the model, such as the linearity and symmetry of neuronal responses in the examples above – are not, strictly speaking, there in the natural systems, especially when considered outside of the laboratory. In claiming that the linearity represented in the traditional neural models is an "ideal pattern," the point is that it is not there in the brain independently of the scientist, but that it arises in the course of an interactive process in which the scientific activities of experimental manipulation, data gathering and processing, grapple with the complexity of the brain.

3.3 Phenomena as Ideal Patterns

A useful way to characterise ideal patterns is to relate them to the concept of scientific phenomena. I begin with the influential account of Bogen and Woodward (1988). The point of their paper is to argue against the empiricist idea that raw data or observations are the target of model building, theorising, and explanation. They show that because of the irregularities introduced by instrument noise, and trial-to-trial variability in the target system, it is important to draw a distinction between data and phenomena.[10] Before proceeding with modelling etc., scientists first process the irregular data, performing statistical operations on it – which can be as simple as averaging across trials – in order to infer what the phenomenon is. The phenomenon is the actual target of modelling, theory, and explanation. Phenomena are *discovered*, according to Bogen and Woodward, through the performance of statistical inference on raw datasets. They are robustly real, in that they are thought of as existing in nature, independently of scientific activity (hence they are revealed and in no sense manufactured). The irregularities present in the data are taken as due to processes, not of interest to the scientist, that interfere with the manifestation of the phenomenon that would be, without such interference, much more regular. One can draw an analogy between signal processing and the task of inferring a phenomenon from the data: by subtracting the noise from the signal, it is possible to recover the signal (more or less) as it was before interference generated in transmission; likewise, the scientist attempts to characterise the phenomenon (more or less) as it was before the interfering and noise-inducing effects that occur with any material process in the laboratory.[11]

It should be obvious that my proposal to treat phenomena as ideal patterns comes just with the rejection of the assumption that phenomena exist independently of experimentally generated effects and data processing. This is in the spirit of Massimi's (2011) Kantian account. As she proposes,

> what sort of phenomena we infer depends on the way we have carved and 'massaged' those data. Of course, there are constraints, and in no way do I want to suggest that anything goes.
>
> (p. 104)

As in Bogen and Woodward's account, scientists are said to learn about phenomena by following careful experimental design, statistical analysis of data, data modelling, and so on. The difference is that now phenomena are conceived as the outcome of regularisations introduced by scientists, which operate on an inordinately complex natural process. It is supposed that the complexity and disorder that nature presents to us in our observations is *not* a mask of an underlying simplicity and regularity.[12] It cannot be assumed that what the scientist is doing is revealing a simplicity that is latent in nature, but instead the proposal is that the workings of experimental manipulations and data processing serve to generate an order that is simpler and hence more comprehensible to the scientist's mind, than what is to be found before such operations have taken place.

One reason to reject Bogen and Woodward's realistic approach to phenomena, at least for my examples from neuroscience, is to consider that in complex, biological systems like the brain (and unlike their canonical example of the melting point of lead) what is signal and what is noise is far from clear cut. There are innumerable processes in any organism or organ, some more of interest to the scientist than others. The partitioning of signal and noise is often driven by preliminary theory, and by pragmatics. Some trial-to-trial variability within datasets is due to experimentally introduced interference, such as instrument noise, but some is due to the endogenous variability of the system, ubiquitous in biology, such as circadian rhythm. Variation in neural response due to other things going on in the brain – behavioural and attentional context – is *not* "noise" from the perspective of the system, but to the neuroscientist it is an inconvenience to be discarded.

4 Conclusions

I have argued in this chapter that the trade-off between prediction and understanding, encountered in various branches of science, motivates a non-factivist account of scientific understanding, and that non-factivism is most tenable if we construe models as representing ideal rather than real patterns. One might be worrying, *if understanding is brought about, in such cases, by the modelling of patterns not straightforwardly there in nature, why should we want it? Should we not resist this version of understanding that comes to us via human-generated "illusions?"* This concern is misguided just because the "ideal pattern" is not "imaginary," something projected onto the natural system and unconstrained by its behaviour. Rather, it is the result of an interactive experimental process and is, therefore, quite tightly but not fully constrained. The point of calling it an "ideal pattern" is just to highlight the fact that it is not entirely human-independent; but it is not nature-independent either – it is both scientist- and nature-dependent. The intuition behind the non-factivist account is that understanding is a compromise between the unruly complexity to be found in nature, especially beyond the laboratory, and the limitations of the ability of scientists (who are human after all) to represent complex events. The notion of an ideal pattern is here intended to encapsulate this compromise.

A final point is that my approach attacks an assumption that Potochnik shares with opponents like Khalifa, namely, that a phenomenon or fact is real only if independent of the scientist. According to my broadly Kantian view, scientific facts are the product of an interaction between the subject and object of knowledge, and understanding occurs when the object is made to conform to the scientist's cognition (Chirimuuta under contract, chap. 2). As such, I am presenting an alternative notion of facts, and my account of scientific understanding could rightly be called factivist in this different sense.

Acknowledgements

Gratitude is owed to Kareem Khalifa for many helpful comments and Angela Potochnik for a fruitful discussion of this project, and also to the audience at the inaugural SURe workshop at Bordeaux, 2019, where this work was first presented.

Notes

1 A question in the background of this paper is that of the demarcation of understanding, taken as a separate topic, from explanation. Both understanding and explanation are epistemic aims of science, and the two terms tend to be used interchangeably by scientists. The point to keep in mind is that *understanding* is an inherently human-relative notion, in that a natural system may or may not be understood *by a group of human scientists*

(de Regt 2017: 17–18), whereas explanation is sometimes conceived in the human-independent, "ontic" sense (Craver 2014). However, "epistemic" or "pragmatic" explanation, which is to say, explanation as a product of human activity, and something valued by scientists because of the usefulness and/or the intellectual satisfaction it provides, obviously overlaps with understanding. Nothing in this paper turns on there being a neat separation between understanding and explanation in the epistemic/pragmatic sense.

2 See Gao et al. (2017) for the argument that one kind of simplicity – the low dimensionality of neural datasets – is introduced through experimental design; also Rust and Movshon (2005); Chirimuuta (under contract, chap. 5).

3 A difference between physics and neuroscience may be that neuroscience, as a branch of biomedical science, feels more pressure to be able to predict and manipulate neural systems beyond the ideal conditions of the laboratory, in order to achieve therapeutic interventions in uncontrolled contexts. As such, the ability to predict or decode neural responses in naturalistic contexts has often been accepted as the gold standard for modelling projects. Note also that in common with other success cases of scientific understanding (de Regt 2017), the linear models of the cortex enabled scientists to achieve certain practical outcomes. For example, the motor cortex models supported functioning brain machine interfaces and the visual cortex models led to the design of the first convolutional neural networks for object recognition, the ancestors networks used today (Fukushima 1980).

4 Elsewhere, I question the assumption that any of these models represents computations actually implemented in neural systems (Chirimuuta 2021). For the purpose of this paper, I take this assumption for granted.

5 See note 10.

6 Though bc-algorithm is not, in principle, unobservable, it could be observed by witnessing the process through which the figure was generated.

7 Since such models would not originate through the relaxing of assumptions made in an idealised model.

8 There is a point of connection here with the "impossible" idealisations frequently discussed by philosophers, such as frictionless planes, point masses, and perfectly rational men. These are often described by philosophers of science as "fictions," as if figments of the scientist's imagination. Instead, we should take them as exaggeration, to an extreme and impossible point, of regularities or features that are present in the observed systems. One finds this interpretation of idealisation in the much earlier and neglected account of Cassirer (1910/1923: 128).

9 Neural data were taken from Rodieck and Stone (1965: Figures 1 and 2); Dreher and Sanderson (1973). See Chirimuuta (2021) for a further discussion of this example.

10 An important note on terminology: Potochnik uses "phenomenon" to refer to what is observed. This is like Ian Hacking's notion of "surface phenomenon," in contrast with Bogen and Woodward's use of "phenomenon" to refer to the "hidden regularity" (Feest 2011).

11 See McAllister (1997) for a critical discussion of Bogen and Woodward on this point.

12 Cf. Cartwright (1983: 19), quoted at the end of Section 1.

References

Abbott, Laurence F., Stefano Fusi, and Kenneth D. Miller. 2013. "Appendix F, Theoretical Approaches to Neuroscience: Examples from Single Neurons to Networks." In *Principles of Neural Science*, Fifth Edition, edited by Sarah Mack, Eric R. Kandel, Thomas M. Jessell, James H. Schwartz, Steven A. Siegelbaum, and A.J. Hudspeth, 1601–1617. New York: McGraw Hill Professional.

Bogen, J., and J.F. Woodward. 1988. "Saving the Phenomena." *The Philosophical Review* 97 (3): 303–352. https://doi.org/10.2307/2185445.

Cartwright, Nancy. 1983. *How the Laws of Physics Lie*. Oxford: Oxford University Press. https://doi.org/10.1093/0198247044.001.0001.

Cassirer, Ernst. 1910/1923. *Substance and Function, and Einstein's Theory of Relativity*. Translated by William Curtis Swabey and Marie Collins Swabey. Chicago, IL: Open Court.

Chirimuuta, M. 2020. "Prediction versus Understanding in Computationally Enhanced Neuroscience." *Synthese* https://doi.org/10.1007/s11229-020-02713-0.

———. 2021. "Your Brain Is Like a Computer: Function, Analogy, Simplification." In *Neural Mechanisms: New Challenges in the Philosophy of Neuroscience*, edited by Fabrizio Calzavarini and Marco Viola, 235–261. Berlin: Springer.

———. under contract. *The Brain Abstracted: Simplification in the History and Philosophy of Neuroscience*. Cambridge, MA: MIT Press.

Craver, C.F. 2014. "The Ontic Account of Scientific Explanation." In *Explanation in the Special Sciences: The Case of Biology and History*, edited by M.I. Kaiser, O.R. Scholz, D. Plenge, and A. Hüttemann, 27–52. Dordrecht: Springer. https://doi.org/10.1007/978-94-007-7563-3_2.

Dennett, Daniel, C. 1991. "Real Patterns." *Journal of Philosophy* 88 (1): 27–51. https://doi.org/10.2307/2027085.

de Regt, Henk. 2017. *Understanding Scientific Understanding*. Oxford: Oxford University Press. https://doi.org/10.1093/oso/9780190652913.001.0001.

Dreher, B., and K.J. Sanderson. 1973. "Receptive Field Analysis: Responses to Moving Visual Contours by Single Lateral Geniculate Neurones in the Cat." *Journal of Physiology* 234: 95–118. https://doi.org/10.1113/jphysiol.1973.sp010336.

Elgin, Catherine Z. 2017. *True Enough*. Cambridge MA: MIT Press. https://doi.org/10.7551/mitpress/9780262036535.001.0001.

Feest, Uljana. 2011. "What Exactly Is Stabilized when Phenomena are Stabilized?" *Synthese* 182: 57–71. https://doi.org/10.1007/s11229-009-9616-7.

Fukushima, K. 1980. "Neocognitron: A Self-Organizing Neural Network Model for a Mechanism of Pattern Recognition Unaffected by Shift in Position." *Biological Cybernetics* 36 (4): 193–202. https://doi.org/10.1007/BF00344251.

Gao, Peiran, Eric Trautmann, Byron Yu, Gopal Santhanam, Stephen Ryu, Krishna Shenoy, and Surya Ganguli. 2017. "A Theory of Multineuronal Dimensionality, Dynamics and Measurement." *bioRxiv*. https://doi.org/https://doi.org/10.1101/214262.

Georgopoulos, A.P., J.F. Kalaska, R. Caminiti, and J.T. Massey. 1982. "On the Relations between the Direction of Two-dimensional Arm Movements and Cell Discharge in Primate Motor Cortex." *The Journal of Neuroscience* 2 (11): 1527–1537. https://doi.org/10.1523/JNEUROSCI.02-11-01527.1982.

Hooker, Giles, and Cliff Hooker. 2018. "Machine Learning and the Future of Realism." *Spontaneous Generations* 9 (1): 174–182. https://doi.org/10.4245/sponge.v9i1.27047.

Koyama, Shinsuke, Steven M. Chase, Andrew S. Whitford, Meel Velliste, Andrew B. Schwartz, and Robert E. Kass. 2010. "Comparison of Brain–computer Interface Decoding Algorithms in Open-loop and Closed-loop Control." *Journal of Computational Neuroscience* 29 (1): 73–87. https://doi.org/10.1007/s10827-009-0196-9.

Levins, R. 1966. "The Strategy of Model Building in Population Biology." In *Conceptual Issues in Evolutionary Biology*, edited by E. Sober, 18–27. Cambridge, MA: MIT Press.

Marr, David. 1982. *Vision*. San Francisco, CA: W. H. Freeman.

Marr, David, and Shimon Ullman. 1981. "Directional Selectivity and Its Use in early Visual Processing." *Proceedings of the Royal Society of London, B* 211: 151–180. https://doi.org/10.1098/rspb.1981.0001.

Massimi, Michela. 2011. "From Data to Phenomena: A Kantian Stance." *Synthese* 182: 101–116. https://doi.org/10.1007/s11229-009-9611-z.

McAllister, J. 1997. "Phenomena and Patterns in Data Sets." *Erkenntnis* 47 (1): 217–228. https://doi.org/10.1023/A:1005387021520.

Potochnik, Angela. 2017. *Idealization and the Aims of Science*. Chicago, IL: Chicago University Press. https://doi.org/10.7208/chicago/9780226507194.001.0001.

Rodieck, R. W., and J. Stone. 1965. "Response of Cat Retinal Ganglion Cells to Moving Visual Patterns." *Journal of Neurophysiology* 28 (5): 819–832. https://doi.org/10.1152/jn.1965.28.5.819.

Rust, N. C., and J. A. Movshon. 2005. "In Praise of Artifice." *Nature Neuroscience* 8 (12): 1647–1650. https://doi.org/10.1038/nn1606.

Shenoy, K. V., M. Sahani, and M. M. Churchland. 2013. "Cortical Control of Arm Movements: A Dynamical Systems Perspective." *Annual Review of Neuroscience* 36 (337–59). https://doi.org/10.1146/annurev-neuro-062111-150509.

Stringer, Carsen, Marius Pachitariu, Nicholas Steinmetz, Charu Bai Reddy, Matteo Carandini, and Kenneth D. Harris. 2019. "Spontaneous Behaviors Drive Multidimensional, Brainwide Activity." *Science* 364: eaav7893. https://doi.org/10.1126/science.aav7893.

Sullivan, Emily, and Kareem Khalifa. 2019. "Idealizations and Understanding: Much Ado About Nothing?" *Australasian Journal of Philosophy* 97 (4): 673–689. https://doi.org/10.1080/00048402.2018.1564337.

8 Topological *Explanations*
An Opinionated Appraisal

Daniel Kostić

1 Introduction

In recent years, there has been a growing interest in the nature of topological explanations in the sciences (Bechtel 2020; Craver 2016; Darrason 2018; Green et al. 2018; Huneman 2017, 2010; Jones 2014; Kostić 2020a, 2020b, 2018; Kostić and Khalifa 2021; Levy and Bechtel 2013; Matthiessen 2017; Rathkopf 2018; Ross 2021). Most of the accounts agree that a topological explanation appeals to topological properties (for instance, graph-theoretical properties of networks representing complex systems) when explaining certain properties or behaviors of real-world systems. This point has been put forward in a series of papers, starting with Huneman (2010), in which he argued that topological explanations provide a genuine alternative to mechanistic explanations as the mainstream paradigm of explanation in life sciences (Bechtel and Richardson 2010; Craver 2007; Machamer, Darden, and Craver 2000). Similar points have been made in Darrason (2018), Kostić (2018), and Rathkopf (2018), who by discussing different aspects of topological explanations in different areas of science, all argue that topological explanations provide a distinct explanatory strategy in which topological properties play a key role. I call this view "autonomism about topological explanations" and its proponents "autonomists." According to this view, there are cases in which topological properties explain independently of any causal or mechanistic considerations.

In subsequent literature, a few philosophical proposals about what makes these network models explanatory have emerged. The goal of this chapter is to assess them by providing an opinionated appraisal. To that effect, I will argue that my view compares favorably with these other conceptions of topological explanations. I will proceed as follows. Section 2 presents my account of topological explanations. In Section 3, I will compare it with other autonomists' accounts of topological explanations. In Section 4, I will defend my account from key criticisms raised by "mechanistic imperialists," who allege that all topological explanations, if explanations at all, are mechanistic.

DOI: 10.4324/9781003202905-9

2 A General Account of Topological Explanations

In my autonomist view, topological explanations describe counterfactual dependencies between topological properties and empirical properties. A network is a set of vertices and edges. Vertices in different areas of science may represent different things. For example, in neuroscience, vertices might represent neurons or brain regions, whereas the edges are synapses or functional connections between the regions. In computer science, they represent computers, routers, or web pages, while edges represent cables between computers and routers or hyperlinks between web pages. In ecological and food networks, vertices might be species and edges are predation relations. The topological properties are obtained by quantifying network properties of systems by using the graph theory.

For example, an explanation of "why is the rate of the spread of infection in a population as it is, rather than slower or faster?" is "because the population is a small-world network." Such an explanation supports counterfactuals such as: had the population been a regular or random network, the spread of infection in it would have been lower.

With these concepts in hand, I can now provide a general account of topological explanations:

a's being *F* topologically explains why *b* is *G* (*a* and *b* are often identical) if and only if:

(T1): *a* is *F* (where *F* is a topological property);
(T2): *b* is *G* (where *G* is an empirical property);
(T3): Had *a* been *F'* (rather than *F*), then *b* would have been *G'* (rather than *G*);
(T4): That *a is F* is an answer to the question why is *a G*.

The first condition, T1, distinguishes what kinds of properties are taken to be explanatory, and in that way determines whether an explanation is topological or of some other kind. For example, instead of topological properties, *F* could describe statistical properties, or geometrical properties, and the resulting explanations would be statistical or geometrical respectively. T2 ensures that *G* is a proper scientific *explanandum* (i.e., it is a description of an empirical phenomenon), for example, why a disease spreads at a certain rate in a population (Watts and Strogatz 1998). The third condition T3, secures explanatoriness, that is, T3 captures counterfactual dependency relations. In my earlier example, such a counterfactual would have the following form:

> had the topological properties of contagion relations network in a human population been different, the infection wouldn't have spread as quickly.

Finally, the fourth condition provides contextual criteria for using the counterfactual in two explanatory modes, that is, a horizontal or a vertical explanatory mode. The horizontal and vertical modes emerge from different question asking perspectives. Thus, the counterfactual in the horizontal mode takes the following form:

(T4H) *a* describes a counterfactual dependency between topological properties and empirical properties that are at the same level,

whereas in the vertical mode it takes this form:

(T4V) *a* describes a counterfactual dependency between topological properties and empirical properties that are at different levels.

The distinction between vertical and horizontal modes could be illustrated by using local and global topological properties. Local topological properties concern only a part of a network, whereas global ones concern topological properties of the whole network. An example of a local topological property is a node degree, which measures the number of edges maintained by a single node. An example of a global property is average path length, which measures how many edges on average must be traversed to reach any node in a network.

The idea about vertical and horizontal explanatory modes could be extended to more levels than the two general ones that I already mentioned. For example, one could conceive of three levels in ecological networks. In such networks, the nodes represent trophic compartments and edges are the carbon "flows" between trophic compartments. Trophic compartments represent organisms with the same ecological roles (e.g., same prey, predators, or the same metabolic rates). The edges represent the carbon flows or carbon transfers between trophic compartments. The flows can be defined as predator-prey relations, respiration, excretion, that is, the flow quantifies the exchange of carbon biomass (grams of organic carbon per square kilometer per year). In this case, the levels are defined relative to edges as opposed to nodes. And so here we could have a flow level between individual trophic compartments, a cycle level in which flows are embedded into multiple loops within a network (such a loop may exist among several trophic compartments), and the global flow at the whole network level (Niquil et al. 2020).

To illustrate this point, consider an example of an ecological network in which scientists are modeling the effect of offshore wind farms (OWF) on marine ecosystems (Nogues et al. 2021). Building a new OWF in a sandy area provides a new habitat for species like mussels, which is called the reef effect. The reef effect can be measured by looking at the topological recycling property at the node and network level. The node-level recycling is defined by the amount of carbon that is produced by one node and that will circulate into closed loops of flows. The global level is defined by the topological recycling of a whole network, that is, the amount of carbon that is produced by the network that will circulate into closed loops of flows.

The explanation-seeking question in the horizontal mode will be:

> How will the reef effect change the recycling property at the individual trophic compartment level?

Here, the explanation takes the horizontal mode, that is, it answers why the introduction of wind turbines changes the biomass of mussels. Due to wind turbines, the mussels will have an additional habitat, which increases the recycling property of their trophic compartment.

On the other hand, the explanation-seeking question in the vertical mode is:

> How will reef effect change the recycling property of the whole network?

As opposed to the previous example, the explanation here takes the vertical mode, that is, it answers why the introduction of wind turbines changes the biomass of the whole network (due to wind turbines, many species in different trophic comportments will have an additional habitat, which increases the recycling property of the whole network).

Distinguishing vertical and horizontal explanatory modes in topological explanations is important for understanding how different organizational levels of a system are functionally related as well as how exogeneous changes affect each of the levels. For example, vertical and horizontal explanatory modes provide complementary multilevel explanations of the effect of offshore wind farms on the aquatic ecological community.

Now with my account of topological explanations laid out in detail, in the next section I discuss other autonomists' accounts of topological explanations and highlight some of the major points of agreement between them and some of the fundamental problems in one of them.

3 Other Autonomists

Others have written on topological explanation too, and their views can be roughly categorized in two groups. First, *autonomists* about topological explanation, such as myself, maintain that topological explanations are a new and distinct kind of explanation (Darrason 2018; Huneman 2010, 2017; Jones 2014; Kostić 2020a, 2020b, 2018; Kostić and Khalifa 2021, 2022; Rathkopf 2018). Opposed to autonomists are *mechanistic imperialists*, who deny that topological explanations are a distinct kind of explanations and try to subsume topological under mechanistic explanations (Bechtel 2020; Craver 2016; DiFrisco and Jaeger 2019; Glennan 2017; Levy and Bechtel 2013; Matthiessen 2017) or deny their explanatoriness altogether (Craver 2016; Craver and Povich 2017; Povich 2018, 2021). This and the next section focus on autonomists, and subsequent ones discuss mechanistic imperialists.

Most autonomists focus on the distinguishing features of topological explanation. However, apart from Huneman (2017), a few develop any account of topological explanations' structure or source of explanatory power. Since I consider myself an autonomist too, I agree with almost all the points that my fellow autonomists make. These points are not covered in my own work, and hence I see the work of all autonomists as largely complementary.

For example, Rathkopf (2018) argues that mechanistic and topological explanations can be distinguished by their unique *explananda*, that is, topological explanations are normally used in nearly non-decomposable systems, whereas mechanistic explanations are typically used in decomposable systems. Further, as opposed to mechanistic explanations, topological explanations are actually fueled by the complexity in a system, that is, the very properties that make a system complex are the ones that are explanatory relevant.

Darrason (2018) also argues, just like Rathkopf, that topological explanations have unique *explananda*, for example, in network medicine they are often used to understand robustness and functional redundancy in certain diseases. She claims that this broader and more general perspective that the topological explanation brings is particularly well suited for distinguishing between monogenetic and polygenetic diseases, which on its own should be a major methodological feat. The uniqueness of *explananda* in topological explanations naturally fits well with the idea that mechanistic and topological explanations are complementary rather than competitors, that is, if topological and mechanistic explanations have different *explananda*, they cannot compete in the first place.

Others also claim that topological explanations are frequently more abstract than mechanistic explanations (Darrason 2018; Huneman 2010, 2017). Huneman (2010) also argues that apart from topological explanations having different *explananda* in terms of different kinds of physical phenomena, they are almost exclusively used to explain certain general properties of a system, but never to capture its dynamics or some more specific properties and behaviors. But this claim is not warranted by the science literature. As I have shown in various papers (Kostić 2020a, 2020b, 2018; Kostić and Khalifa 2021, 2022), topological explanations are very often used to capture the brain's (temporal) dynamics, or often some local and specific properties of various systems.

Since autonomism claims that some topological explanations are distinct from other kinds of explanations, I also see as autonomists the authors who do not necessarily develop positive arguments for this view but whose views are consistent with it. These authors focus on other epistemic achievements of topological explanations. For example, Green and colleagues (Green et al. 2018) and Serban (2020) argue that topological models frequently provide useful heuristics for discovering mechanisms, but they do not argue that this is the sole function of topological models. As such, their view is compatible with both autonomism and its denial. De Boer and colleagues (2021) examine the applicability of topological explanations in understanding psychiatric disorders.

Ross (2021) makes a closely related but slightly different point. Ross argues that even though the mainstream examples of topological explanation are non-causal, there are some borderline cases that involve both causal and non-causal topological properties. As Ross argues, topologies can be causal when edges in a graph represent causal interactions between a system's entities. These edges are sometimes connected in sequences that are called causal pathways, for example, cell signaling pathways, metabolic pathways, or ecological pathways. These causal pathways can form a network that represents a map or a web of connections in some domain, sometimes called a "wiring topology," and it is similar to wiring diagrams of electric circuits. Ross' argument for why these topologies are causal turns on the fact that wiring topology captures causal properties that can be intervened upon, because these graphs are directed and weighted, and the graphs' edges are causal relations. However, unlike mechanistic imperialists who deny topological autonomism, Ross is more ecumenical and allows that there could be both causal and non-causal distinctively topological explanations. I am very much in agreement with this pluralist idea. Khalifa and I (Kostić and Khalifa 2022) have even elaborated conditions under which a network model can provide a mechanistic explanation.

Now, since these accounts outline only general features of topological explanations, I see them as not only compatible, but also complementary to my own account of topological explanations. The only other account that provides an analysis of the structure and explanatory power in topological explanations is Huneman's (2017) account of topological explanations as a subspecies of structural explanations. A discussion of his account requires a bit more space, and I turn to it in the next subsection.

3.1 Topological Explanation as a Subspecies of Structural Explanations

Huneman (2017) has also offered a philosophical analysis of topological explanation. In what follows, I will present his view, and then I shall highlight several advantages of my own account.

Huneman's account of topological explanation is a limiting case of his broader analysis of "structural" explanations. Unlike my treatment of topological explanations, Huneman does not provide necessary and sufficient conditions for structural explanations. His clearest formulation is:

> . . . to the extent that [mathematical properties] are explanatory, [i] such properties do not concern variables that would be directly involved in the representation of the system (e.g., time, length, any measurable magnitude of the system)—rather, they constrain any possible causes/mechanisms to have/display a specific behavior/property/ outcome range, provided that the system satisfies a small set of conditions (e.g., being describable by a function in the functional class C*, etc.). . . . Whatever the causes in the system, because of one of such properties (T) their mathematical description must display a specific property P, which is mapped onto the *explanandum* P* we are interested in. In other words, [ii] were this mathematical proposition untrue, the system would not exhibit the property P*, since notwithstanding the actual mechanisms going on here, there [iii] would be no necessity for its mathematical description to display P. Hence those properties account for the *explanandum* since they provide us with the reason why any possible mechanism proper to the system will have to behave in a specific way that precisely includes having P*; to this extent, no particular mechanism could by itself give us this reason.
>
> (Huneman 2018, 686–687)

Since I am only concerned with topological explanations, I restrict myself to cases in which "*T*" is a topological property. For example, Huneman (2017, 693) offers the following topological explanation as one such example of his structural account of explanation:

- *Explanans:* Small-world networks are stable when random vertices are removed.

  Ecosystems are small-world networks, with species as their vertices.
- *Explanandum:* Ecosystems are stable when random species go extinct.

Here, the major premise in the *explanans* is a mathematical theorem with small-worldliness serving as the topological property *T*, and stability when random vertices are removed serving as Huneman's *P*. Ecosystems' stability when random species go extinct is Huneman's *P*.

Having presented Huneman's position, I will now argue that my view avoids problems that each of the three conditions numbered above— concerning representation, counterfactuals, and necessity, respectively— raise. Further, as the quotation makes clear, these three conditions underwrite Huneman's claim that topological explanations are not mechanistic. This section concludes by showing how the problems with

Huneman's account of structural explanation undercut his argument that topological explanations are non-mechanistic.

However, before proceeding, a small caveat is in order. Since Huneman does not state the precise scope of his account of topological explanations, my discussion of his view can be interpreted in two ways. If Huneman takes all topological explanations to be structural, then the arguments below are direct challenges to his view. If he does not, then my view provides a more comprehensive treatment of topological explanations, including those that are not structural in his intended sense. I remain neutral on which of these two interpretations is correct.

Turning to my criticisms of Huneman's view, consider condition [i]: that topological explanations "do not concern variables that would be directly involved in the representation of the system." Specifically, Huneman's account of structural explanation requires a "mathematical hierarchy," which he characterizes as follows:

> Calling "representative variables" those variables that stand in the model for the properties of the system—e.g., charge, mass, species abundance, activation/nonactivation (of a gate or a neuron), etc.—one can construe a hierarchy of variables: "metavariables" denoting operations on representative variables, "metametavariables" denoting operations on metavariables, etc. Any operation on variables can in turn be part of a set of operations, and as such be represented by a metavariable when this set is described, and so on.
>
> (Huneman 2018, 688)

Huneman takes structural explanations to require metavariables. However, this requirement is difficult to evaluate. For instance, consider a "topological" variable that assumes the value 0 if the ecosystem is small-world, 1 if it is random, and 2 if it is regular. Since this simply encapsulates information about predator-prey relations, it would appear to be a representative variable on Huneman's account. Moreover, if this topological variable did not represent predator-prey relations, then it is hard to see how the graph theory applies to ecosystems in this example, or, more generally, how the graph theory applies to any empirical system. Indeed, some of Huneman's other remarks exacerbate this worry:

> . . . metavariables do not refer to properties of the system (since they are not representative variables). "Reference" here concerns the question whether mathematical properties are directly mapped onto the systems properties—like representative variables—or not.
>
> (Huneman 2018, 688; emphasis in original)

By contrast, my account requires no special class of metavariables and thereby avoids these problems.

Turn now to condition [ii]: that structural explanations work chiefly by supporting mathematical counterpossibles, that is, counterfactuals with antecedents in which the relevant mathematical propositions are false (Huneman 2018, 675). Thus, in the explanation of ecosystems' stability, Huneman requires the following counterfactual to be true:

> Had small-world networks been unstable when random vertices are removed, then the ecosystem would have been unstable if random species were to go extinct.

Since this counterfactual's antecedent requires parts of graph theory to be false, it is a counterpossible. As is well known, such counterpossibles are semantically fraught (Nolan 2013). Indeed, some hold that since no possible worlds exist in which mathematical statements are false, counterpossibles are vacuously true. I consider it a virtue of my view that it does not require counterpossibles. Rather, my account only requires the ecosystem to be capable of having a different topological structure than it does, for example,

> Had the ecosystem not been a small-world network, then it would have been unstable if random species were to go extinct.

This is a garden-variety counterfactual. Assume, as Huneman does, that edges represent predation relations. Then, the counterfactual ranges over possible worlds in which, for example, the ecosystem has different predator-prey relations but the graph theory remains unchanged. Thus, my account requires no mathematical counterpossibles. Indeed, the chief paper from which Huneman draws this example (Montoya, Pimm, and Solé 2006) accords better with my account, for it considers the effects of removing one or more species from the ecosystem, but never considers what would happen if the graph theory were untrue. Thus, *contra* Huneman, the explanation here does not involve counterpossibles. Rather, it involves an empirically testable counterfactual, as my account proposes.

Next, consider condition [iii]: that topological explanations confer an especially strong form of necessity—"in a stronger modal way than the laws of physics" (Huneman 2018, 690)—on their *explananda*. However, not all topological explanations exhibit extraordinary modal strength. In the ecological example, it is possible for the topological recycling property to be higher or lower, but those possibilities are not results of some mathematical modal constraint (e.g., certain axioms in the graph theory being false); they simply depend on different rates of carbon flows between trophic compartments. Thus, in this case, the explanation works because of the counterfactual dependency between topological recycling and reef effect, and not because of graph-theoretic axioms. It might be thought that my account thereby fails to capture explanations involving high-grade necessity. This is a mistake. Let us follow Huneman in assuming that the explanation of the ecosystem's stability involves this "high-grade necessity." Then we can characterize the *explanandum* as follows:

> Why is it mathematically impossible that the ecosystem is unstable when new OWFs are built?

Admittedly, because this describes a mathematical impossibility, it is not obvious that this is an empirical statement. Hence, this might imperil satisfaction of T2. However, I assume that because it also describes an ecosystem, treating it as empirical in a broad sense is defensible. Thus construed, T3 thereby suggests that the topological explanation supports the following counterfactual:

> Had the ecosystem not been a network with topological recycling property F, then it would have been mathematically possible that the ecosystem is unstable when new OWFs are built.

Since anything physically possible is also mathematically possible, this counterfactual is true. Hence, my account covers topological explanations that exhibit mathematical necessity.

Thus, my view avoids three problems plaguing Huneman's account. First, my view does not appeal to the problematic notion of meta-variables. Second, unlike Huneman's view, mine avoids the notorious semantic puzzles that counterpossibles raise. Third, it covers topological explanations that confer mathematical necessity upon their *explananda*, as well as those that do not.

Moreover, these three problems undercut Huneman's argument that topological explanations are not a species of causal-mechanical explanation. If no strong distinction between representative variables and metavariables exists, then topological explanations might well be highly abstract causal-mechanical explanations. If topological explanations do not support counterpossibles, then causal-mechanical structure might underwrite the garden-variety counterfactuals they support. Finally, if topological explanations are not restricted to explaining necessities, then causal-mechanical differences—even if represented topologically—might account for when a contingent state of affairs does and does not obtain. Hence, closer scrutiny of Huneman's position undercuts his reasons to think that topological explanations are autonomous.

In the following section, I will provide independent reasons for thinking that topological explanations are autonomous from mechanisms.

4 Critiques of Autonomism

Autonomists' views have been challenged in the literature by the proponents of mechanistic explanations in two important ways. First, Craver (2016), Craver and Povich (Craver and Povich 2017), and Povich (Povich 2018, 2021) have argued that topological explanations do not model the right kind of stuff, and that without some ontic backing they are not really explanatory. The second mechanist objection is that networks are insufficiently distinct from mechanisms, especially in terms of organization, and thus if anything they are not a distinct kind of topological model, but merely a very abstract kind of mechanistic model (Bechtel 2020; DiFrisco and Jaeger 2019; Glennan 2017; Levy and Bechtel 2013; Matthiessen 2017). Given this, if they do provide any kind of explanation, it is a mechanistic one.

Khalifa and I (2022) have addressed these objections in much more detail, and hence I will only outline them here. Let me start with the mechanists' interpretation of network topology as mechanistic organization. To equivocate the topological properties of a network with the "organization" in the sense that the organization itself explains (or that the organization of a "mechanism" is responsible for the occurrence of the *explanandum*) means that topological properties can explain when all other aspects of the mechanism remain fixed. Hence, if topology alone is explanatory, then it is obviously a distinctively topological explanation. This either concedes precisely what is at stake or it trivializes the mechanistic conception of "organization." And second, to say that topology as the organization of a mechanism does not explain is to beg the question against the autonomist.

To develop these arguments in more detail, Khalifa and I proposed an analysis that we call a Mechanistic Interpretation of Topological Explanations or a MITE. According to MITE, a network model provides a mechanistic explanation, if and only if network nodes represent a mechanism's entities and activities; edges represent the interactions between entities and activities; the network model specifies how these entities and activities are organized to be responsible for a target phenomenon; and finally, the target phenomenon is at a higher level than the mechanism's entities and activities. If only one of these conditions is violated, the MITE fails, and the resulting explanation is not mechanistic. This, in itself, does not imply that the resulting explanation is distinctively topological either. We provided examples in which at least one of the MITE's requirements have been violated, but the explanation satisfied T1–T4, which means that the resulting explanation is topological and non-mechanistic (Kostić and Khalifa 2022).

In addition, mechanists who subscribe to ontic conceptions of explanation have provided an additional challenge to autonomism (Craver and Povich 2017). They claim that without ontic backing, topological explanations fail to account for directionality, that is, they incorrectly deem instances of logical contraposition as correct explanations. According to the directionality problem, topological models without ontic backing are bidirectional, yet no bidirectional model is explanatory. So, purveyors of this objection conclude that no topological models without ontic backing are explanatory. In response, Khalifa and I (2021) provide a more precise definition of the directionality problem, distinguish it from the asymmetry problem, and formulate what we call the "ontic irrelevance lesson." The ontic irrelevance lesson uses the ideas from my previous work (Kostić 2020a) to show that topological explanations can preserve directionality without any kind of ontic backing that Craver (2016), Craver and Povich (2017), and Povich (2018, 2021) require. Non-ontic bases of directionality of topological explanations that we proposed are the property (which is tied to T1 and T2), counterfactual (T3), and perspectival ones (T4) (Kostić and Khalifa 2021).

Khalifa and I argue (2021) that a topological explanation will be property directional when its contraposition or "reversal" does not involve topological properties in their "*explanantia*." For example, Helling and colleagues (Helling, Petkov, and Kalitzin 2019) offer a topological explanation in which mean functional connectivity[1] explains the dynamics of the onset of epileptic seizure (also called *ictogenicity*). If the reversal of their original case is an explanation, then ictogenicity must be a topological property. However, it is not. Hence, Helling et al.'s actual explanation satisfies the property requirement, but its reversal does not.

In counterfactual directionality, the original case satisfies a counterfactual, but its reversal does not. Using the study by Adachi and colleagues (Adachi et al. 2012), which shows that the functional connectivity between anatomically unconnected areas in the macaque cortex counterfactually depends on the anatomical connectivity network's overall frequency of three-node network motifs, Khalifa and I showed that a reversal of this counterfactual violates the very dependency that the original case postulates. Thus, if the relevant counterfactual is true, its reversal is false, and this difference alone suffices to capture the directionality of the explanation. Thus, it exhibits *counterfactual directionality*.

Finally, we argued that topological explanations must be answers to the relevant explanation-seeking questions, which is embodied in the T4-Perspectival condition. Thus, an explanation is *perspectivally directional* whenever its original satisfies T4 condition, but its reversal does not. Explanation-seeking questions allow that in the same context *at least one* answer will appeal to a topological property, but its reversal will likely *prohibit* any topological answers.

This is another aspect in which my account of topological explanations outperforms Huneman's. Namely, Huneman's example states:

1 *Explanans:* Small-world networks are stable when random vertices are removed.

2 Ecosystems are small-world networks, with species as their vertices.
3 *Explanandum:* Ecosystems are stable when random species go extinct.

However, this is bidirectional, because it can be reversed in the following way, and Huneman's account would incorrectly classify it as a correct explanation:

1 *Explanans:* Small-world networks are stable when random vertices are removed.
2 Some ecosystems are *unstable* when random species go extinct.
3 *Explanandum:* Some ecosystems are either *not* small-world networks or do *not* have species as their vertices.

So, this poses an additional problem with Huneman's account: its vulnerability to the directionality problem.

Two points remain to be made in defense of topological autonomism. Autonomism argues that some topological explanations are non-mechanistic. Moreover, I have claimed that T1–T4 provide sufficient conditions for genuinely autonomous topological explanations. However, Craver (2016, 704–706) argues that functional connectivity models are examples of topological models that do not provide explanations, chiefly because they do not represent mechanisms. Functional connectivity measures temporal dependency between time series of anatomically separated brain regions (van den Heuvel and Hulshoff Pol 2010, 519). In functional connectivity models, the nodes represent blood-oxygen-level-dependent (BOLD) signals (in fMRI) or EEG channels (in EEG recordings) and the edges represent synchronization correlations between BOLD signals or EEG channels. The idea is that if two BOLD signals (or EEG channels) are synchronized, the populations of neurons that the BOLD/EEG signals represent are functionally connected. Craver observes that functional connectivity models' nodes "need not . . . stand for working parts," that is, for the entities that constitute a mechanism. Rather, many functional connectivity models' nodes are conventionally determined spatiotemporal regions that are adopted mostly because they are "conveniently measurable units of brain tissue rather than known functional parts."

The spearpoint of his argument is that models representing correlations are merely evidential, but not explanatory (Craver 2016, 705). This argument turns on the idea that if nodes are not working parts of a mechanism and the edges are merely correlations, then functional connectivity networks do not model the right kinds of stuff, and because of that they cannot be explanatory. Just like the barometer reading is correlational evidence of a storm, but it does not explain the storm, the synchronization likelihoods in functional connectivity networks are evidence of the actual anatomical connections and BOLD/EEG signals are evidence of neuronal populations. To model the right kinds of stuff and be explanatory, synchronization likelihoods and BOLD/EEG signals would have to be working parts of a mechanism that is responsible for the phenomenon we want to explain. Since they are not, then functional connectivity models cannot be explanatory, according to Craver. This is the core of Craver's challenge.

Responding to Craver's challenge allows me to sketch what I will call "ambitious autonomism." According to this view, I can grant that in general, most models trafficking only in correlations between conventionally determined spatiotemporal regions are evidential but not explanatory, but I argue that if some functional connectivity models, such as the one discussed in this chapter, satisfy T1–T4 conditions and also violate some or all of the MITEs, then they are exceptions of the Craver's challenge, in that they fail to be mechanistic, but are still explanatory. Ambitious autonomism might draw inspiration from the fact that some functional connectivity networks capture pathological brain dynamics that structural connectivity (i.e., mechanistic structure) cannot. This is because variability in functional connectivity often occurs in the absence of direct anatomical connections (Helling, Petkov, and Kalitzin 2019; Honey et al. 2009; Moon et al. 2017; Suárez et al. 2020). This is a *prima facie* methodological reason to countenance functional connectivity explanations that are autonomous of any mechanistic explanation. Recall that functional connectivity edges are synchronization likelihoods. Thus, many of the functional connectivity models' topological properties are different ways of describing how well synchronized a brain is. For instance, the fact that the likelihood of an epileptic seizure (or ictogenicity) varies in proportion to mean functional connectivity suggests that "oversynchronization" of the brain explains seizures (Kalitzin et al. 2019, 7). Thus, if the brain dynamics counterfactually depends on variability in functional connectivity, whereas the direct anatomical connections remain fixed, then such a model satisfies T1–T4. This shows that functional connectivity models can be explanatory, and they are not always merely evidential.

In addition, recent philosophical work at the intersection of modeling and explanation further dispels this intuition in the case of explanatory functional connectivity models. Ambitious autonomists may interpret functional connectivity models as showing that the specific mechanistic entities and activities simply *don't matter* when it comes to explaining certain phenomena. In the case of ictogenicity, oversynchronization (i.e., high mean functional connectivity) suffices on its own. As such, functional connectivity models' representation of brain regions as arbitrary spatiotemporal regions is a kind of explanatory abstraction (Jansson and Saatsi 2017) or idealization in which mechanistic entities are caricatured as arbitrary spatiotemporal regions to highlight their explanatory irrelevance (Batterman and Rice 2014).

5 A Conclusion

Topological explanation is a recent scientific development, which by now has received significant philosophical attention. It already has some devoted proponents, autonomists, but it also has staunch critics from the ranks of new mechanistic philosophers. Among the autonomists, there is a broad agreement about some general features of topological explanation, which make them a unique and a novel kind of explanation. Such features are that they are often invariant relative to causal or microphysical details of real-world systems, they are fueled rather than hindered by the complexity of a system, they are particularly well suited for explaining nearly non-decomposable systems, they are often used to understand robustness and functional redundancy in certain systems, they are also more abstract than their mechanistic counterparts, and they are often non-causal. When it comes to their structure, there is also a broad agreement about certain features, for example, in contrast to traditional accounts of realization relation, the topological realization base is normally at a higher or the same level as the realized properties (Huneman 2018; Kostić 2018). However, when it comes to the precise analyses of topological explanation, there are two developed accounts: mine and Huneman's. As I have argued, I see my account as enjoying several advantages, such as avoiding problems with metavariables, counterpossibles, the problems with mathematical necessity, directionality problems, as well as being in closer dialogue with the scientific literature.

The critics of topological explanation have pointed out some really important issues that directly challenge the autonomy of topological explanations, for example, the ontic backing, the directionality, or the representational power of network models. However, the autonomists have ample resources to meet these challenges head on and disarm them in many different ways (Kostić 2020a, 2020b, 2018; Kostić and Khalifa 2021, 2022). Finally, autonomists are more ecumenical. For example, we argue that topological explanations can come in many different flavors, e.g., both causal (Ross 2021) and non-causal or distinctively topological and mechanistic (Kostić and Khalifa 2022); or that topological models can have many important non-explanatory uses, such as providing heuristics for discovering mechanisms or parts of mechanisms (de Boer et al. 2021; Green et al. 2018; Serban 2020).

This diversity of perspectives and views illustrates the richness and a broad potential that topological explanations bring to the table. Current accounts and debates about them are only the beginning of a fruitful development in the philosophy of scientific explanation.

Acknowledgments

I would like to thank Kareem Khalifa for his invaluable comments and insights that helped bring my work on topological explanations in general, but also specifically in this chapter, in much greater focus. I would also like to thank Kareem for the wonderful friendship.

Funding

I would like to acknowledge the funding by the Radboud Excellence Initiative.

Note

1 Which measures the average strength of the correlations that exist between any two nodes in a functional connectivity network.

References

Adachi, Yusuke, Takahiro Osada, Olaf Sporns, Takamitsu Watanabe, Teppei Matsui, Kentaro Miyamoto, and Yasushi Miyashita. 2012. "Functional Connectivity between Anatomically Unconnected Areas Is Shaped by Collective Network-Level Effects in the Macaque Cortex." *Cerebral Cortex* 22 (7): 1586–1592. https://doi.org/10.1093/cercor/bhr234.

Batterman, Robert W., and Collin C. Rice. 2014. "Minimal Model Explanations." *Philosophy of Science* 81(3): 349–376. https://doi.org/10.1086/676677

Bechtel, William. 2020. "Hierarchy and Levels: Analysing Networks to Study Mechanisms in Molecular Biology." *Philosophical Transactions of the Royal Society B: Biological Sciences* 375 (1796): 20190320. https://doi.org/10.1098/rstb.2019.0320.

Bechtel, William, and Robert C. Richardson. 2010. *Discovering Complexity: Decomposition and Localization as Strategies in Scientific Research.* MIT Press ed. Cambridge, MA: MIT Press.

Craver, Carl F. 2007. *Explaining the Brain: Mechanisms and the Mosaic Unity of Neuroscience.* Oxford: Clarendon Press.

Craver, Carl F. 2016. "The Explanatory Power of Network Models." *Philosophy of Science* 83 (5): 698–709. https://doi.org/10.1086/687856.

Craver, Carl F., and Mark Povich. 2017. "The Directionality of Distinctively Mathematical Explanations." *Studies in History and Philosophy of Science Part A* 63: 31–38. https://doi.org/10.1016/j.shpsa.2017.04.005.

Darrason, Marie. 2018. "Mechanistic and Topological Explanations in Medicine: The Case of Medical Genetics and Network Medicine." *Synthese* 195 (1): 147–173. https://doi.org/10.1007/s11229-015-0983-y.

de Boer, Nina S., Leon C. de Bruin, Jeroen J. G. Geurts, and Gerrit Glas. 2021. "The Network Theory of Psychiatric Disorders: A Critical Assessment of the Inclusion of Environmental Factors." *Frontiers in Psychology* 12 (February): 623970. https://doi.org/10.3389/fpsyg.2021.623970.

DiFrisco, James, and Johannes Jaeger. 2019. "Beyond Networks: Mechanism and Process in Evo-Devo." *Biology & Philosophy* 34 (6): 54. https://doi.org/10.1007/s10539-019-9716-9.

Glennan, Stuart. 2017. *The New Mechanical Philosophy*. First edition. Oxford: Oxford University Press.

Green, Sara, Maria Şerban, Raphael Scholl, Nicholaos Jones, Ingo Brigandt, and William Bechtel. 2018. "Network Analyses in Systems Biology: New Strategies for Dealing with Biological Complexity." *Synthese* 195 (4): 1751–1777. https://doi.org/10.1007/s11229-016-1307-6.

Helling, Robert M., George H. Petkov, and Stiliyan N. Kalitzin. 2019. "Expert System for Pharmacological Epilepsy Treatment Prognosis and Optimal Medication Dose Prescription: Computational Model and Clinical Application." In *Proceedings of the 2nd International Conference on Applications of Intelligent Systems*, 1–6. Las Palmas de Gran Canaria Spain: ACM. https://doi.org/10.1145/3309772.3309775.

Honey, Christopher J., Jean-Philippe Thivierge, and Olaf Sporns. 2010. "Can Structure Predict Function in the Human Brain?" *Neuroimage* 52 (3): 766–776. https://doi.org/10.1016/j.neuroimage.2010.01.071

Huneman, Philippe. 2010. "Topological Explanations and Robustness in Biological Sciences." *Synthese* 177 (2): 213–45. https://doi.org/10.1007/s11229-010-9842-z.

———. 2017. "Outlines of a Theory of Structural Explanations." *Philosophical Studies*, no. 1984: 1–38. https://doi.org/10.1007/s11098-017-0887-4.

———. 2018. "Realizability and the Varieties of Explanation." *Studies in History and Philosophy of Science Part A* 68 (April): 37–50. https://doi.org/10.1016/j.shpsa.2018.01.004.

Jansson, Lina, and Saatsi, Juha. 2017. "Explanatory Abstractions". *The British Journal for the Philosophy of Science*, 70 (3): 817–844. https://doi.org/10.1093/bjps/axx016

Jones, Nicholaos. 2014. "Bowtie Structures, Pathway Diagrams, and Topological Explanation." *Erkenntnis* 79 (5): 1135–1155. https://doi.org/10.1007/s10670-014-9598-9.

Kalitzin, Stiliyan, George Petkov, Piotr Suffczynski, Vasily Grigorovsky, Berj L. Bardakjian, Fernando Lopes da Silva, and Peter L. Carlen. 2019. "Epilepsy as a Manifestation of a Multistate Network of Oscillatory Systems." *Neurobiology of Disease* 130: 104488. https://doi.org/10.1016/j.nbd.2019.104488

Kostić, Daniel. 2018. "The Topological Realization." *Synthese* 195 (1): 79–98. https://doi.org/10.1007/s11229-016-1248-0.

———. 2020a. "General Theory of Topological Explanations and Explanatory Asymmetry." *Philosophical Transactions of the Royal Society B: Biological Sciences* 375 (20190314): 1–8. http://dx.doi.org/10.1098/rstb.2019.0321.

———. 2020b. "Minimal Structure Explanations, Scientific Understanding and Explanatory Depth." *Perspectives on Science 2019* 27 (1): 48–67. https://doi.org/10.1162/posc.

Kostić, Daniel, and Kareem Khalifa. 2021. "The Directionality of Topological Explanations." *Synthese* 199: 14143–14165. https://doi.org/10.1007/s11229-021-03414-y.

Kostić, Daniel and Kareem Khalifa. 2022. "Decoupling Topological Explanations from Mechanisms."

Philosophy of Science, 1-39. doi:10.1017/psa.2022.29

Levy, Arnon, and William Bechtel. 2013. "Abstraction and the Organization of Mechanisms." *Philosophy of Science* 80 (2): 241–261. https://doi.org/10.1086/670300.

Machamer, Peter, Lindley Darden, and Carl F. Craver. 2000. "Thinking about Mechanisms." *Philosophy of Science* 67 (1): 1–25. https://doi.org/10.1086/392759.

Matthiessen, Dana. 2017. "Mechanistic Explanation in Systems Biology: Cellular Networks." *British Journal for the Philosophy of Science* 68 (1): 1–25. https://doi.org/10.1093/bjps/axv011.

Montoya, José M., Stuart L. Pimm, and Ricard V. Solé. 2006. "Ecological Networks and Their Fragility." *Nature* 442 (7100): 259–264. https://doi.org/10.1038/nature04927

Moon, Joon-Young, Junhyeok Kim, Tae-Wook Ko, Minkyung Kim, Yasser Iturria-Medina, Jee-Hyun Choi, Joseph Lee, George A. Mashour, and UnCheol Lee. 2017. "Structure Shapes Dynamics and Directionality in Diverse Brain Networks: Mathematical Principles and Empirical Confirmation in Three Species." *Scientific Reports* 7 (1): 1–11. https://doi.org/10.1038/srep46606

Niquil, Nathalie, Matilda Haraldsson, Télesphore Sime-Ngando, Philippe Huneman, and Stuart R. Borrett. 2020. "Shifting Levels of Ecological Network's Analysis Reveals Different System Properties." *Philosophical Transactions of the Royal Society B: Biological Sciences* 375 (1796): 20190326. https://doi.org/10.1098/rstb.2019.0326.

Nogues, Quentin, Aurore Raoux, Emma Araignous, Aurélie Chaalali, Tarek Hattab, Boris Leroy, Frida Ben Rais Lasram, et al. 2021. "Cumulative Effects of Marine Renewable Energy and Climate Change on Ecosystem Properties: Sensitivity of Ecological Network Analysis." *Ecological Indicators* 121 (February): 107128. https://doi.org/10.1016/j.ecolind.2020.107128.

Nolan, Daniel P. 2013. "Impossible Worlds." *Philosophy Compass* 8 (4): 360–372. https://doi.org/10.1111/phc3.12027

Povich, Mark. 2018. "Minimal Models and the Generalized Ontic Conception of Scientific Explanation." *The British Journal for the Philosophy of Science* 69 (1): 117–137. https://doi.org/10.1093/bjps/axw019.

———. 2021. "The Narrow Ontic Counterfactual Account of Distinctively Mathematical Explanation." *The British Journal for the Philosophy of Science* 72 (2): 511–543. https://doi.org/10.1093/bjps/axz008.

Rathkopf, Charles. 2018. "Network Representation and Complex Systems." *Synthese* 195 (1): 55–78. https://doi.org/10.1007/s11229-015-0726-0.

Ross, Lauren N. 2021. "Distinguishing Topological and Causal Explanation." *Synthese* 198 (10): 9803–9820. https://doi.org/10.1007/s11229-020-02685-1.

Serban, Maria. 2020. "Exploring Modularity in Biological Networks." *Philosophical Transactions of the Royal Society B: Biological Sciences* 375 (1796): 20190316. https://doi.org/10.1098/rstb.2019.0316.

Suárez, Laura E., Ross D. Markello, Richard F. Betzel, and Bratislav Misic. 2020. "Linking Structure and Function in Macroscale Brain Networks." *Trends in Cognitive Sciences* 24 (4): 302–315. https://doi.org/10.1016/j.tics.2020.01.008

van den Heuvel, Martijn P., and Hilleke E. Hulshoff Pol. 2010. "Exploring the Brain Network: A Review on Resting-State FMRI Functional Connectivity." *European Neuropsychopharmacology* 20 (8): 519–534. https://doi.org/10.1016/j.euroneuro.2010.03.008.

Watts, Duncan J., and Steven H. Strogatz. 1998. "Collectivedynamics of 'Small-World' Networks." *Nature* 393 (6684): 440–442. https://doi.org/10.1038/30918.

9 Explanatory Power

Factive vs. Pragmatic Dimension

Juha Saatsi

1 Introduction

As science advances, it provides better explanations of natural phenomena. Today, we have good explanatory understanding of various phenomena that were understood only vaguely or not at all just a couple of decades ago. So, as science improves, its explanatory capacity increases: it provides better understanding through theories and models that have more *explanatory power*. I will explore how this notion of explanatory power cuts across philosophical issues of current interest concerning the explanatory role of idealisations and mathematics; the factivity of understanding; explanations' realist commitments; and how "explanatory power" is best construed.

I will begin by identifying in abstract terms two central dimensions of explanatory power. One is "factive," having to do with how explanations relate to mind-independent reality. The other is "pragmatic," having to do with how explanations relate to us as epistemic agents who use explanations to gain understanding. Although more fine-grained distinctions between different dimensions of explanatory power can be made in the context of specific philosophical accounts of explanation, the two-pronged coarse-grained distinction will prove useful in thinking about various issues at the intersection of explanation, understanding, and representation.

In the first half of this chapter, I will frame my discussion around two natural intuitions regarding explanatory power. Each is prima facie plausible in its own right, but together they seemingly give rise to a tension (Section 2). We can alleviate this tension by drawing a broad conceptual distinction between factive and pragmatic dimensions of explanatory power (Section 3) that can be made more precise in terms of specific ideas regarding scientific explanation and understanding (Section 4). In the second half of the chapter, I will first look at a number of benefits of demarcation between these two dimensions of explanatory power, focusing on understanding's factivity (Section 6) and its growth (Section 7).

DOI: 10.4324/9781003202905-10

2 Two Intuitions Concerning Explanatory Power

The following two intuitions concerning explanatory power are prima facie very plausible. (1) *Increasing explanatory power is due to representing the world better.* (2) *More powerful explanations provide better understanding.*

The first is supported by intuitions regarding the factivity of actual (as opposed to merely potential) explanations. If an explanatory law statement or a model is changed so that it represents the world less well, how could that yield a better explanation? It could, of course, yield a *lovelier* explanation in Peter Lipton's (2004) sense: an explanation that would be better *if it were true*. But, as Lipton argues, when science aims at providing deeper or more powerful explanations, loveliness is at best a means to achieving that aim. (Conspiracy theories tend to furnish rather lovely explanations, for example.) It seems intuitively epistemically misguided to *mis*represent the world in an effort to boost an explanation's explanatory virtues; for example, its scope, precision, mechanistic detail, unifying power, and simplicity. For instance, if an explanation of some phenomenon P is "improved" by incorporating an abstract mathematical framework that unifies it with a disparate phenomenon $P*$, it is not clear what has been explanatorily gained if the unification is entirely spurious and only pertains to our formal representation of P and $P*$. Or, if an explanatory model of P is "improved" so that it represents P in a more fine-grained way by incorporating mechanistic details that misrepresent the actual mechanism behind P, it is natural to think that we have ended up with a merely potential explanation. Similarly, radically idealised explanatory models, such as the Schelling model of social segregation, can furnish great potential or "how-possibly" explanations (Reutlinger et al., 2018), but it seems wrong to regard them as powerful explanations of why actually there is social segregation (Sullivan and Khalifa, 2019). These intuitions are contestable, of course, but they provide prima facie support for (1).

Turning now to (2), it is straightforwardly supported with reference to the epistemic role of explanations: what explanations are *for*. We care about explanations because they provide understanding. An explanation that does not provide any understanding seems to lack the epistemic value that we place on explanations. We value science in part because as it advances it provides better and better understanding of the world. This central aspect of scientific progress is naturally captured by saying that science provides better, more powerful, explanations. It seems oxymoronic to say that explanation E is more powerful than explanation $E*$, but that E provides no more understanding than $E*$ does. This intuition provides prima facie support for (2).

Within a broadly realist framework, these two intuitions about explanatory power could even be regarded as platitudes. More powerful explanations provide better understanding, since that is what explanations are *for*. And increasing explanatory power is due to representing the world better, since *actual* explanations are factive.[1] So far, so good, but (1) and (2) may also seem to be in tension. Taking them at face value, in conjunction, may suggest that *increasing scientific understanding is a matter of scientific theories and models representing the world better and better*. This seemingly intuitive implication is often implicitly endorsed in philosophy, lying behind various explanation-centred arguments.[2] But we know that this implication is false! We know that increasing scientific understanding is not just a matter of representing the world better. This can be clearly seen by paying attention to the explanatory use of idealisations, "theories emeritus," nonfactive "explanatory perspectives," and explanatory approximation schemes. The relevant aspects of science are well-known and discussed at length in the literature.[3]

Given the plentiful evidence against the idea that increasing scientific understanding is just a matter of representational improvement, we can ask how the two prima facie plausible intuitions (1) and (2) are best understood, so that they do not suggest something we know to be false.

3 Two Dimensions of Explanatory Power

In thinking about (1) and (2), it is worth distinguishing between two dimensions of explanatory power. This enables us to account for the extent to which (1) and (2) *pull in different directions*, while nevertheless being platitudes in their own right.

On the one hand, we can continue regarding (actual) explanations as factive by associating explanatory power in connection with (1) with the amount of true explanatory information contained in or captured by an explanation. Exactly what kind of information is regarded as explanatory depends on our theory of explanation, of course. According to some philosophers, explanatory information is information about difference-makers (Strevens, 2008); according to others, it is information answering what-if-things-had-been different (*what-if*) questions (Woodward, 2003); and so on. Whatever it is, we can maintain that explanatory information is factive – namely information that can be expressed with true propositions – and that the more such information an explanation provides, the better the explanation, ceteris paribus. This is the *factive* dimension of explanatory power. It is a feature that an explanation has in relation to its explanandum.

On the other hand, we also need to recognise that explanatory *understanding* is not a feature of an explanation in and of itself, but rather something that *we* stand to gain from it. We can take into account this relationship between explanations and ourselves (as cognitive agents whose understanding is at stake) by associating explanatory power with an explanation's capacity to provide us understanding. It is a feature that an explanation has in relation to its users. An explanation's capacity to provide understanding is not just a matter of how much explanatory information is contained or captured by an explanation; what matters also is how that explanatory information is (re)presented, so that it has the capacity to provide understanding to us. Exactly what we take this capacity to amount to depends on our theory of understanding, of course, but whatever it is, it is a feature of an explanation that is dependent not only on its informational content, but critically also on how we can use it to gain understanding. This is the *pragmatic* dimension of explanatory power.

With these two dimensions of explanatory power in mind, we can disambiguate between the instances of "explanatory power" in the two platitudes above: (1) Increasing explanatory power is due to representing the world better. (2) More powerful explanations provide better understanding. The first instance of "explanatory power" refers to a quality of explanations that is independent from us: the amount of factive explanatory information contained in an explanation. The second refers to a quality of explanations that depends on us: how much understanding we can gain from an explanation. This disambiguation allows us to see why (1) and (2) do not imply that increasing scientific understanding is (just) a matter of scientific theories representing the world better and better. The two instances of "explanatory power" refer to the two different dimensions, each of which can be to an extent increased (or decreased) independently from the other. On the one hand, a theory or model that represents the world better can contain or capture more explanatory information, thus increasing explanatory power along the factive dimensions, without increasing its capacity to provide us understanding. On the other hand, by misrepresenting the world in suitable respects, and thus losing out on the factive dimension of explanatory power, a model can nevertheless enhance its capacity to provide us understanding if it renders its remaining factive explanatory information more usable to us.

Distinguishing between the two dimensions of explanatory power enables us to schematically see why (1) and (2) do not give rise to a problematic tension. I will next flesh this out by indicating how prominent conceptions of scientific explanation and understanding coherently fit this two-dimensional conception of explanatory power.

4 Explanation and Understanding

What could factive explanatory information be, first of all? I am sympathetic to the idea that explanations provide information about how an explanandum (variable) counterfactually depends on an explanans (variable). This idea is at the heart of a prominent counterfactual-dependence account of causal explanation (Woodward, 2003), which can arguably be extended to also cover various non-causal explanations (French and Saatsi, 2018; Jansson and Saatsi, 2019; Reutlinger, 2018; Woodward, 2018). With this conception of explanation in mind, the factive dimension of explanatory power can be simply identified with the amount of appropriate counterfactual information that an explanation contains. An explanation that provides (approximately) true answers to a broader range of *what-if* questions regarding the explanans variable is more powerful, ceteris paribus (Hitchcock and Woodward, 2003).

Extant analyses of explanatory power in the context of the counterfactual-dependence account fill in the details. These analyses are more fine-grained than the two-fold distinction that I made above. For example, Hitchcock and Woodward (2003) identify six different ways in which the degree of invariance of an explanatory generalisation can vary – that is, there are six ways in which explanations can differ in the range of *what-if* questions they cover. Relatedly, but carving the matter at different joints, Ylikoski and Kuorikoski (2010) count five dimensions of explanatory power. These analyses assume that explanatory information is factive, and also that there are always pragmatic and contextual factors in play in the assessments of explanatory power. My two-fold distinction focuses on highlighting the difference between explanatory power as a feature of explanations due to their representational relationship to mind-independent reality, and explanatory power as a feature of explanations due to their relationship to us.

Moving on to understanding, what could this be? It is commonplace to associate understanding with some kind of "grasping" of an explanation. This notion of grasping is often left somewhat unarticulated (see Belkoniene, 2021). A prominent thought – one that I am very sympathetic with – is that "grasping" is best construed as an ability. In the context of the counterfactual-dependence account of explanation, it is natural to identify it with an agent's ability to correctly perform counterfactual *what-if* inferences to extract counterfactual-dependence information contained in an explanatory theory or model (Kuorikoski and Ylikoski, 2015; Ylikoski and Kuorikoski, 2010). Obtaining understanding in this sense obviously requires that there actually is true modal information contained in the explanation to begin with, that is, that the explanation is sufficiently powerful in the factive sense. At the same time understanding in this sense is also unavoidably dependent on the cognitive salience and extractability of the relevant modal information relative to an agent equipped with appropriate reasoning skills. An explanation can render relevant modal information more or less readily available to an agent, depending on the latter's training, skills, cognitive limitations, and other such contextual factors (de Regt, 2017; Kuorikoski and Ylikoski, 2015; Ylikoski and Kuorikoski, 2010).

These prominent conceptions of explanation and understanding serve to flesh out the two dimensions of explanatory power in a coherent way.

5 Benefits

Distinguishing between pragmatic and factive dimensions of explanatory power can help throw light on various issues at the confluence of explanation, understanding, and representation. I will briefly review some potential benefits in this section before delving deeper into the question of understanding's factivity (Section 6) and its growth (Section 7).

Explanatory role of idealisations and fictional presuppositions. Let us first look at the role of different intentional misrepresentations in the context of explanatory modelling. According to some philosophers, such falsehoods sometimes play an essential role in explanations (e.g., Batterman, 2001, 2005a, 2009; Bokulich, 2008a, 2008b, 2009, 2012). This contrasts with the more commonly held view that the false parts or aspects of explanatory models always correspond to explanatorily irrelevant features of the systems being represented (e.g., Strevens 2008). Let us focus on Bokulich (2017), who acknowledges as "counterintuitive" the notion that "the idealizations or [fictional] falsehoods themselves do real explanatory work," while nevertheless maintaining that "sometimes it is in part because of their falsehoods – not despite them – that models explain" (111). According to Bokulich (2009), "some fictions can give us genuine insight into the way the world is, and hence be genuinely explanatory and yield real understanding" (94).

Lumping together "genuine physical insight," "genuine explanatoriness," and "real understanding" in this way cries out for unpacking and articulating what each of these notions refers to. In fairness to Bokulich, she has explicitly associated the explanatoriness of what she calls "structural model explanations" with how they correctly capture patterns of structural dependencies in the world (Bokulich, 2008a). This chimes well with the counterfactual-dependence account of explanation. But, as indicated above, the latter account can be regarded as factive in the sense that explanatory power can be identified with true counterfactual information contained in an explanation, so it is not very clear how the non-veridical aspects of explanatory models can then do "real explanatory work" or yield "physical insight."

Distinguishing between factive and pragmatic dimensions of explanatory power helps by allowing us to precisify Bokulich's position by locating the contribution of idealisations and fictional presuppositions on the pragmatic side of the equation: such falsehoods improve explanatory models simply by rendering them more user-friendly, allowing cognitive agents to extract (more easily) factive explanatory information. If understanding is further construed as an ability to make correct counterfactual inferences on the basis of a model, we can more clearly see how falsehoods can be conducive to increasing explanatory understanding by contributing solely to the pragmatic dimension of explanatory power. And although Bokulich's references to "real physical insight" and "real explanatory work" are ambiguous and potentially misleading, I am very sympathetic overall to her analysis of, for example, semiclassical mechanics as a discipline in which "reasoning with fictional classical structures [. . .] can yield explanatory insight and *deepen our understanding*" (Bokulich, 2008a, 20, my emphasis). In the light of the factive/pragmatic distinction, I would, however, insist on explicitly associating the explanatory role of idealisations and fictions firmly with the pragmatic dimension of explanatory power.[4]

The explanatory role of mathematics. The distinction between pragmatic and factive dimensions of explanatory power can similarly help in analysing mathematical explanations in science.[5] The indispensability of mathematics to scientific explanations is undeniable: mathematics-laden models often furnish better, more powerful explanations than those expressed by purely nominalistic means. Although mathematics arguably often has a merely representational function in explanatory models, there is a lively debate around examples in which mathematics is purportedly "distinctively" explanatory (e.g., Baker, 2009, 2017; Baron 2020; Barrantes, 2020; Lange, 2013).[6] This debate is multifaceted and I cannot do justice to its nuances here, but let us focus on a contrary viewpoint according to which there actually are *no* mathematical explanations of empirical phenomena (Kuorikoski, 2021). How can such a claim be maintained in the face of the abundant evidence that by appealing to mathematical theories we end up with more powerful explanations?

Our two-fold distinction can help answering this question. When one looks more closely at the explanatory virtues of mathematical explanations in contrast to their nominalistic counterparts, it can be argued that their explanatory virtues often pertains specifically to explanations' *degree of integration* and *cognitive salience* (Knowles and Saatsi, 2021). The former is a matter of an explanation's connectedness to other theoretical frameworks, and its importance lies in enhancing one's inferential ability with respect to counterfactual *what-if* questions (Ylikoski and Kuorikoski, 2010). The latter relates to an explanation's pragmatic virtue of making explanatory reasoning more easily followed, rendering explanatory implications and scope cognitively transparent and more easily evaluated, and so on (*ibid.*, 215). All these explanatory virtues to which mathematics can contribute point along the pragmatic dimension of explanatory power in a way that is entirely compatible with the denial that mathematical truths/facts/entities in and of themselves are explanatory (Saatsi, 2016b); for example and more specifically, the study by Knowles and Saatsi (2021) of the contribution of mathematics to the virtue of explanatory generality. They argue in the context of the counterfactual-dependence account that even with respect to the so-called "distinctly" mathematical explanations mathematics increases explanatory generality only relative to our limited cognitive powers, maximising explanations' cognitive salience in particular.

Kuorikoski (2021) goes further to articulate a notion of "formal understanding" that pertains to understanding, not of empirical phenomena, but of our systems of reasoning and representation:

> Such understanding is constituted by the abilities in making correct what-if inferences about systems of reasoning. It is formal in that the relevant abilities concern the form of the inferences. What would follow from special relativity if the symmetry group of space-time was de Sitter rather than Poincaré? What would the properties of a competitive equilibrium be if preferences were nonhomothetic? These kinds of questions concern primarily our systems of reasoning and representation, not directly what we are representing and reasoning about (i.e., space-time or specific markets). The answers are nevertheless fully objective and ontic in that they are grounded on the objective features of our representational systems – what could justifiably be inferred from given assumptions if the properties of the system of reasoning were different.

Kuorikoski is explicit in drawing "a sharp distinction between this formal understanding and explanation proper," where the latter involves "explanatory relationships [that] do not hold between representations, concepts, or descriptions but between the things in the world being represented, conceptualized, and described" (191). Information about such worldly explanatory relationships is factive in the sense associated with the factive dimension of explanatory power. By contrast, "formal understanding" directly involves our systems of reasoning, not the phenomenon being reasoned about – hence, it is akin to "the distinction between understanding a model and understanding with a model" (*ibid.*, 207). So, although there are objective (e.g., mathematical) facts about these systems of reasoning, in so far as an increase in formal understanding is conducive to increasing understanding of some *empirical phenomenon*, this is still measured by explanatory power purely along the pragmatic dimension.

Explanatory autonomy of non-fundamental theories. Can non-fundamental sciences provide explanations that are more powerful than those provided by fundamental physics? Are there contexts in which the ideal gas law, for example, can furnish an explanation that is more powerful than those furnished by statistical mechanics? Many philosophers have argued that the answer to both of these questions is yes, thus defending the (intuitively plausible) explanatory autonomy of non-fundamental sciences (e.g., Garfinkel, 1981; Jackson and Pettit, 1992; Woodward, 2003). Weslake (2010) criticises various extant analyses of such explanatory autonomy, and he puts forward his own account of "explanatory depth" that is meant to track the explanatory value of non-fundamental theories and explain their explanatory autonomy. According to Weslake, it is the *abstractness* of a higher-level explanation, in the sense of it applying to a wider range of counter-nomically possible systems, that can render it "deeper" (viz. more powerful) in some contexts. For example, Weslake notes that "the ideal gas law is independent of whether the underlying mechanics is Newtonian or quantum mechanical," which "means that there are physically impossible systems to which the macroscopic explanation applies but to which the microscopic explanation does not" (287).

The distinction between factive and pragmatic dimensions of explanatory power provides an alternative (and in my view preferable) perspective on explanatory autonomy. The latter can be simply explained in terms of higher-level theories offering more powerful explanations (ceteris paribus) along the pragmatic dimension. There is no reason to think that the explanatory autonomy is a feature of explanatory theories or models independent of cognitive agents who use scientific theories to explain and understand.

For a quick illustration, consider the well-worn example of Putnam's peg. Why does not a square peg of 1 inch side go through a hole of 1 inch diameter? A mundane geometrical explanation answers this question in a way that not only provides information about the factive explanatory dependencies (e.g., what if the relative sizes/shapes were different thus and so?), but also does that in a way that renders those explanatory dependencies cognitively salient so as to maximise (in a typical context) the explanatory power along the pragmatic dimension. Although a detailed microphysical account of the two bodies' solid physical composition, shapes, and motions would contain all the same information, it would render this information cognitive opaque and hidden in the mass of other, explanatorily irrelevant information. This points to a simple analysis of the explanatory power of high-level, non-fundamental theories. This account can be contrasted with Weslake's analysis, according to which the geometrical explanation is "deeper" by virtue of being multiply realisable with respect to microphysical laws alternative to the actual world. This analysis of explanatory depth is problematic in that it is not at all clear as to how such nomological possibilities are connected to the way we appeal to and appreciate the explanatory power of the geometrical explanation.

6 Is Understanding Factive?

A lot has been written about the issue of factivity of understanding (Elgin, 2017; Khalifa, 2017; Lawler, 2021; Mizrahi, 2012; Potochnik, 2017; Strevens et al., 2017). The ongoing debate around this issue is motivated by a clash between a prima facie intuition that understanding is factive, and the fact that scientific understanding can be facilitated by falsehoods. Insa Lawler (2021) explains:

> Prima facie, understanding is factive: its content can only contain true propositions (or at least approximations to the truth). If I seem to understand a phenomenon but it turns out that my account of the phenomenon contains false propositions, one would say that it looked as if I understood it, but I actually did not. Yet, science is replete with falsehoods that are considered to foster or facilitate understanding rather than prevent it.
>
> (6860)

One of the benefits of drawing the distinction between factive and pragmatic dimensions of explanatory power is that it goes a long way to resolve this tension. The intuition that understanding is factive reflects the factive dimension of explanatory power. If I claim to have explanatory understanding of some phenomenon, the level of my understanding is bound by the amount of true explanatory information contained in my account of the phenomenon. In other words, it is the truth-content of my account that underwrites the maximum degree of understanding that I can have on the basis of this account.[7] This basic factivity intuition notwithstanding we should also recognise that it is not enough for an account of a phenomenon to contain true explanatory information, for that information also needs to be suitably accessible to human beings who are trying to understand it. Therefore, if you have two accounts both of which contain the same explanatory information, one can be more or less powerful than the other in the pragmatic sense of providing us more or less understanding, and the falsehoods that science is replete with sometimes have the function of enhancing theories' capacity to provide us more understanding, as required by our cognitive needs and limitations.

This schematic resolution of the tension can be fleshed out with reference to more specific ideas regarding explanation and understanding. Let us again identify the former with information that correctly answers counterfactual change-relating *what-if* questions, and the latter with an *ability* to reliably make the relevant counterfactual inferences (see Section 4). With these philosophical theories in mind, understanding can be regarded as "factive" in the sense that it can only be had through an explanatory theory or model that contains true counterfactual information, but at the same time understanding is clearly not "factive" in the sense that this ability to make correct counterfactual inferences could well be increased by non-veridical aspects of our theory or model. There can, thus, be a tradeoff between (i) representing explanatory dependences in a less veridical way that increases our ability to make correct counterfactual inferences (the pragmatic dimensions of explanatory power), and (ii) representing these dependencies in a more veridical way (along the factive dimension) that reduces our inferential ability. (See also Kuorikoski and Ylikoski, 2015; Ylikoski and Kuorikoski, 2010.)

This perspective saves the basic factivist intuition about explanation. It complements existing defences of factivism regarding understanding. Consider Lawler (2021), for instance, whose strategy is to distinguish between the "content" of understanding as the *product* of scientific reasoning or modelling, and the *process* of obtaining it. According to Lawler, idealisations and other such "felicitous legitimate falsehoods" merely play an epistemic enabling role for "extracting" from theories and models (approximate) truths that comprise the content of understanding. "These falsehoods can play an epistemic enabling role in the process of obtaining understanding but are not elements of the explanations or analyses that constitute the content of understanding" (*ibid.*, 6860). This allows Lawler to defend the view that "understanding is factive: its content can only contain true propositions (or at least approximations to the truth)" (*ibid.*).

This view is kindred to mine, but Lawler does not take sides on the debate on the nature of understanding or explanation, only suggesting that "one might say that a subject understands a phenomenon only if they grasp an explanation or analysis of it" (*ibid.*, 6859). What is being grasped, the explanation, is the "content" of understanding (*ibid.*). That being so, she is committed to the claim that (genuine) explanations are (approximately) true. Whatever role the falsehoods play, these are somehow connected to the "production" of explanations, while not being a "part" of the explanations themselves.

When it comes to figuring out whether a given mathematical feature or idealisation (e.g., the thermodynamic limit or renormalisation group fixed points) is a "part" of a scientific explanation, I prefer to operate in the context of a *specific theory of explanation*. With a theory of explanation in hand, one can look at the explanatory role of different theoretical posits, and argue that mathematics, idealisations, and fictional posits play a broadly representational or instrumental role in procuring factive information that is doing the explanatory heavy lifting (e.g., Reutlinger and Saatsi, 2018; Saatsi, 2016b). Demarcating between different kinds of explanatory roles in this fashion can further be connected to the two dimensions of explanatory power, as indicated in Section 4. In my analysis, in comparison to Lawler, such a demarcation is dependent on a more substantive view on what it is to explain and what understanding consists of.

7 The Growth of Scientific Understanding

Another benefit of the distinction between factive and pragmatic dimensions of explanatory power comes in relation to capturing the growth of scientific understanding. This aspect of scientific progress involves an epistemic achievement of steadily improving scientific explanations. Our account of explanatory power should make sense of this progress in terms of the interrelationships of the theories and models that contribute to it. Paradigmatic examples of explanatory progress include, for instance, increasing understanding of tidal phenomena, and increasing understanding of the rainbow, glory, and other related optical phenomena. Arguably, a naturalistic philosopher ought to regard these as examples of lasting epistemic achievements regardless of her stance on scientific anti-realism, for there is an accepted scientific narrative of the development of relevant theoretical explanations from early modern science to the current day. Importantly, despite radical shifts in the relevant theories' conceptual frameworks and ontological commitments – shifting from Newtonian gravitational force to curved space-time, for example, and from Newtonian light corpuscles to ether waves and electromagnetic waves – there is also a cumulative growth of scientific understanding.

In Saatsi (2019), I explore in some detail the interplay between the accumulation of understanding as a lasting epistemic achievement, on the one hand, and the varying theoretical perspectives on the phenomena being understood (e.g., gravity, or light), on the other. My analysis sees value in these shifting "explanatory perspectives" even when they do not form a lasting epistemic achievement by themselves, but rather function as a cognitive scaffolding for scientists operating in a particular theoretical context that involves specific ways of thinking about and representing the phenomenon at stake mathematically and metaphysically. Such explanatory perspectives can contribute to understanding broadly along the lines of de Regt's (2017) account, by enhancing the intelligibility of theories and models. According to de Regt, "a phenomenon P is understood scientifically if and only if there is an explanation of P that is based on an intelligible theory T and conforms to the basic epistemic values of empirical adequacy and internal consistency" (92).

The way I think of it, the various features of T that enhance its intelligibility increase T's explanatory power along the pragmatic dimension (in the relevant context). This is only half the story, however, as it does not account for the *epistemic achievement* of understanding that accumulates over potentially radical theory-shifts as science advances.[8] The key to capturing this epistemic progress is to also pay attention to the factive dimension of explanatory power that tracks the human-independent sense in which theories latch better and better onto explanatory features of reality. (For example, the electromagnetic theory of light contains more explanatory information than the ether theory, and the general theory of relativity contains more explanatory information than Newtonian gravity.) To the extent that these factive explanatory features are "grasped" by cognitive agents, we gain more and more understanding. For example, this allows us to make sense of the way in which Descartes and Newton managed to contribute to the accumulating scientific understanding of rainbows: despite operating in the context of explanatory perspectives that were idiosyncratic and misguided from our current theoretical vantage point, Descartes' and Newton's theorising provided them the ability to correctly answer critical *what-if* questions involving explanatory variables that carry over to contemporary explanatory models of the rainbow (see Saatsi, 2019, for a more detailed discussion).

In addition to helping us capture large-scale historical trends of epistemic progress in science, the factive-pragmatic distinction is also useful with respect to more fine-grained aspects of scientific understanding that involves mutually incompatible explanatory perspectives. In the context of contemporary theories of the rainbow, for example, one can consider the epistemic achievement of the semiclassical Complex Angular Momentum (CAM) method as furnishing a ray-theoretic perspective that clashes with what is regarded as a complete and fundamental wave theoretic account of rainbow scattering, namely, the Lorentz-Mie theory. The latter is a model of Maxwell's electromagnetic theory representing an "ideal" rainbow (comprising plane waves and perfectly spherical raindrops), all the optical properties of which it deductively entails. This model maximises the factive dimension of explanatory power, as it contains all the answers to different *what-if* questions about the (ideal) rainbow. The CAM method, which occupies the theoretical borderland between the geometrical ray theory and the wave theory, goes beyond this fundamental explanatory framework by increasing explanatory power along the pragmatic dimension: it allows us to represent key features of the dynamics of electromagnetic radiation in a way that makes them transparent to us. This is nicely expressed by the main architect behind CAM, Herch Nussenzveig (1992):

> A vast amount of information on the diffraction effects that we want to study lies buried within the Mie solution. In order to understand and to obtain physical insight into these effects . . . it is necessary to extract this information in a "sufficiently simple form."

(45)

This simplicity, Nussenzveig goes on to emphasise, lies "to some extent . . . in the eye of the beholder" (*ibid.*, 210). I think all this is nicely captured by the distinction between the factive and pragmatic dimensions of explanatory power, which thus helps to throw light on how studying an approximation scheme (CAM method), in relation to the complete and fundamental theory (Lorentz-Mie theory) that provides an exact solution, can make a substantial epistemic contribution to understanding. In essence, the issue with the fundamental theory is its lack of cognitive salience. As Nussenzveig (quoting Eugene Wigner) poetically puts it: "It is nice to know that the computer understands the problem, but I would like to understand it too" (*ibid.*).[9]

8 Conclusion

Many philosophers begin with the assumption that actual explanations are factive. If genuine understanding is just a matter of "grasping" an actual explanation, this basic factivity assumption can suggest that increasing scientific understanding is a matter of scientific theories and models representing the world better and better. This is problematic since genuine understanding can be increased in ways that involve idealisations, approximations, and other kinds of theoretical presuppositions that essentially incorporate falsehoods or are not regarded as representational by the scientists. We can alleviate this tension by drawing a distinction between the pragmatic and factive dimensions of explanatory power. This distinction allows us to hang on to the basic factivity assumption regarding the relationship between explanations and reality, as well as to the natural idea that the primary function of explanations is to give us understanding, and that better explanations provide more understanding.

Notes

1 Not everyone accepts such a realist framework. For example, these connections between explanation, understanding, and representation can be severed from a thoroughly pragmatist perspective like that of van Fraassen (1980).
2 For a relevant discussion of explanatory indispensability arguments and explanationist arguments in the scientific realism debate, see Saatsi (2016a, 2018).
3 For example, Potochnik (2020) and Rice (2021a, 2021b) discuss the indispensable role of idealisations in providing understanding. Bokulich (2016, 2008a, 2008b) and de Regt and Gijsbers (2017) discuss the role of "theories emeritus" (e.g., Newtonian gravity) in explanatory contexts, and Saatsi (2019) discusses the role of non-factive explanatory perspectives (e.g., ether-theoretic conception of light).
4 This is broadly aligned with Sullivan and Khalifa (2019, §4).
5 There is clearly a potential overlap between explanations' mathematical presuppositions and explanatory fictions; a mathematical fictionalist might, indeed, identify the two.
6 Prominent, well-worn examples involve Königsberg's bridges and cicada life cycles.
7 The way Lawler captures the factivity intuition in the quote above requires too much from understanding, I think. Do we want to say that Newton did not have understanding of the tides, for example, when it turned out that his account contained various false (not even approximately true) propositions?
8 In my view, this accumulation is not well captured by de Regt's account, but I will not argue for this here.

9 For a more detailed philosophical discussion of the explanatory use of semi-classical optics, see Pincock (2011) and Saatsi (2019), as well as Batterman (2005b) and Belot (2005). For a review of the CAM method, see Adam (2002) and Nussenzveig (1992).

References

Adam, J. A. 2002. "The Mathematical Physics of Rainbows and Glories." *Physics Reports* 356(4–5), 229–365. doi: 10.1016/S0370-1573(01)00076-X.

Baker, A. 2009. "Mathematical Explanation in Science." *The British Journal for the Philosophy of Science* 60(3), 611–633. doi: 10.1093/bjps/axp025.

Baker, A. 2017. "Mathematics and Explanatory Generality." *Philosophia Mathematica* 25, 194–209. doi: 10.1093/philmat/nkw021.

Baron, S. 2020. "Counterfactual Scheming." *Mind* 129, 535–562. doi: 10.1093/mind/fzz008.

Barrantes, M. 2020. "Explanatory Information in Mathematical Explanations of Physical Phenomena." *Australasian Journal of Philosophy* 98(3), 590–603. doi: 10.1080/00048402.2019.1675733.

Batterman, R. W. 2001. *The Devil in the Details: Asymptotic Reasoning in Explanation, Reduction, and Emergence.* New York: Oxford University Press.

Batterman, R. W. 2005a. "Critical Phenomena and Breaking Drops: Infinite Idealizations in Physics." *Studies in History and Philosophy of Science Part B: Studies in History and Philosophy of Modern Physics* 36(2), 225–244. doi: 10.1016/j.shpsb.2004.05.004.

Batterman, R. W. 2005b. "Response to Belot's 'Whose Devil? Which Details?'" *Philosophy of Science* 72, 154–163. doi: 10.1086/428073.

Batterman, R. W. 2009. "Idealization and Modeling." *Synthese* 169(3), 427–446. doi: 10.1007/s11229-008-9436-1.

Belkoniene, M. 2021. "Grasping in Understanding." *British Journal for the Philosophy of Science.* doi: 10.1086/714816.

Belot, G. 2005. "Whose Devil? Which Details?" *Philosophy of Science* 72(1), 128–153. doi: 10.1086/428072.

Bokulich, A. 2008a. "Can Classical Structures Explain Quantum Phenomena?" *The British Journal for the Philosophy of Science* 59(2), 217–235. doi 10.1093/bjps/axn004.

Bokulich, A. 2008b. *Reexamining the Quantum-Classical Relation: Beyond Reductionism and Pluralism.* Cambridge: Cambridge University Press.

Bokulich, A. 2009. "Explanatory Fictions." In M. Suárez (Ed.), *Fictions in Science: Philosophical Essays on Modeling and Idealization*, pp. 91–109. London: Routledge.

Bokulich, A. 2012. "Distinguishing Explanatory from Nonexplanatory Fictions." *Philosophy of Science* 79(5), 725–737. doi: 10.1086/667991.

Bokulich, A. 2016. "Fiction as a Vehicle for Truth: Moving Beyond the Ontic Conception." *The Monist* 99(3), 260–279. doi: 10.1093/monist/onw004.

Bokulich, A. 2017. "Models and Explanation." In L. Magnani and T. Bertolotti (Eds)., *Springer Handbook of Model-based Science*, pp. 103–118. New York: Springer.

de Regt, H. W. 2017. *Understanding Scientific Understanding.* Oxford: Oxford University Press.

de Regt, H. W. and V. Gijsbers. 2017. "How False Theories Can Yield Genuine Understanding?" In S. R. Grimm, C. Baumberger, and S. Ammon (Eds.), *Explaining Understanding: New Perspectives from Epistemology and Philosophy of Science*, pp. 50–75. New York: Routledge.

Elgin, C. Z. 2017. *True Enough*. Cambridge, MA: MIT Press. http://dx.doi.org/10.7551/mitpress/9780262036535.001.0001.

French, S. and J. Saatsi. 2018. "Symmetries and Explanatory Dependencies in Physics." In A. Reutlinger and J. Saatsi (Eds.), *Explanation Beyond Causation: Philosophical Perspectives on Non-Causal Explanations*, pp. 185–205. Oxford: Oxford University Press.

Garfinkel, A. 1981. *Forms of Explanation: Rethinking the Questions in Social Theory*. New Haven, CT: Yale University Press.

Hitchcock, C. and J. Woodward. 2003. "Explanatory Generalizations, Part II: Plumbing Explanatory Depth." *Noûs* 37, 181–199. doi: 10.1111/1468-0068.00435.

Jackson, F. and P. Pettit. 1992. "In Defense of Explanatory Ecumenism." *Economics & Philosophy* 8(1), 1–21. doi: 10.1017/S0266267100000468.

Jansson, L. and J. Saatsi. 2019. "Explanatory Abstractions." *British Journal for the Philosophy of Science* 70, 817–844. doi: 10.1093/bjps/axx016.

Khalifa, K. 2017. *Understanding, Explanation, and Scientific Knowledge*. Cambridge: Cambridge University Press.

Knowles, R. and J. Saatsi. 2021. "Mathematics and Explanatory Generality: Nothing but Cognitive Salience." *Erkenntnis* 86, 1119–1137. doi: 10.1007/s10670-019-00146-x.

Kuorikoski, J. 2021. "There Are No Mathematical Explanations." *Philosophy of Science* 88(2), 189–212. doi: 10.1086/711479.

Kuorikoski, J. and P. Ylikoski. 2015. "External Representations and Scientific Understanding." *Synthese* 192(12), 3817–3837. doi: 10.1007/s11229-014-0591-2.

Lange, M. 2013. "What Makes a Scientific Explanation Distinctively Mathematical?" *The British Journal for the Philosophy of Science* 64(3), 485–511. doi: 10.1093/bjps/axs012.

Lawler, I. 2021. "Scientific Understanding and Felicitous Legitimate Falsehoods." *Synthese* 198, 6859–6887. doi: 10.1007/s11229-019-02495-0.

Lipton, P. 2004. *Inference to the Best Explanation*. London: Routledge.

Mizrahi, M. 2012. "Idealizations and Scientific Understanding." *Philosophical Studies* 160(2), 237–252. doi: 10.1007/s11098-011-9716-3.

Nussenzveig, H. M. 1992. *Diffraction Effects in Semiclassical Scattering*. Cambridge: Cambridge University Press.

Pincock, C. 2011. "Mathematical Explanations of the Rainbow." *Studies in History and Philosophy of Science Part B: Studies in History and Philosophy of Modern Physics* 42(1), 13–22. doi: 10.1016/j.shpsb.2010.11.003.

Potochnik, A. 2017. *Idealization and the Aims of Science*. Chicago, IL: University of Chicago Press.

Potochnik, A. 2020. "Idealization and Many Aims." *Philosophy of Science* 87(5), 933–943. doi: 10.1086/710622.

Reutlinger, A. 2018. "Extending the Counterfactual Theory of Explanation." In A. Reutlinger and J. Saatsi (Eds.), *Explanation Beyond Causation*, pp. 74–95. Oxford: Oxford University Press.

Reutlinger, A., D. Hangleiter and S. Hartmann. 2018. "Understanding (with) Toy Models." *The British Journal for the Philosophy of Science* 69(4), 1069–1099. doi: 10.1093/bjps/axx005.

Reutlinger, A. and J. Saatsi. 2018. *Explanation Beyond Causation: Philosophical Perspectives on Non-Causal Explanations*. Oxford: Oxford University Press.

Rice, C. 2021a. *Leveraging Distortions: Explanation, Idealization, and Universality in Science*. Cambridge, MA: MIT Press.

Rice, C. 2021b. "Understanding Realism." *Synthese* 198, 4097–4121.

Saatsi, J. 2016a. "Explanation and Explanationism in Science and Metaphysics." In Z. Yudell and M. Slater (Eds.), *Metaphysics and the Philosophy of Science: New Essays*, pp. 163–191. New York: Oxford University Press.

Saatsi, J. 2016b. "On the 'Indispensable Explanatory Role' of Mathematics." *Mind* 125, 1045–1070. doi: 10.1093/mind/fzv175.

Saatsi, J. 2018. "Realism and the Limits of Explanatory Reasoning." In J. Saatsi (Ed.), *The Routledge Handbook of Scientific Realism*, pp. 200–211. London: Routledge.

Saatsi, J. 2019. "Realism and Explanatory Perspectives." In M. Massimi and C. McCoy (Eds.), *Understanding Perspectivism: Scientific Challenges and Methodological Prospects*, pp. 65–84. New York: Routledge.

Strevens, M. 2008. *Depth: An Account of Scientific Explanation*. Cambridge, MA: Harvard University Press.

Strevens, M. et al. 2017. "How Idealizations Provide Understanding." In S. R. Grimm, C. Baumberger, and Ammon (Eds.), *Explaining Understanding: New Perspectives from Epistemology and Philosophy of Science*, pp. 37–48. New York: Routledge.

Sullivan, E. and K. Khalifa. 2019. "Idealizations and Understanding: Much Ado About Nothing?" *Australasian Journal of Philosophy*, 673–689. doi: 10.1080/00048402.2018.1564337.

van Fraassen, B. C. 1980. *The Scientific Image*. Oxford: Clarendon Press.

Weslake, B. 2010. "Explanatory Depth." *Philosophy of Science* 77(2), 273–294. doi: 10.1086/651316.

Woodward, J. 2003. *Making Things Happen: A Causal Theory of Explanation*. Oxford: Oxford University Press.

Woodward, J. 2018. "Some Varieties of Non-Causal Explanation." In A. Reutlinger and J. Saatsi (Eds.), *Explanation Beyond Causation*, pp. 117–137. Oxford: Oxford University Press.

Ylikoski, P. and J. Kuorikoski. 2010. "Dissecting Explanatory Power. *Philosophical Studies* 148(2), 201–219. doi: 10.1007/s11098-008-9324-z.

Part II

Understanding and Scientific Realism

10 Understanding the Success of Science

Christopher Pincock

1 Introduction

This chapter sketches a new defense of scientific realism based on understanding the success of science (Sections 1 and 2) and then considers what features understanding must have for this defense to succeed (Sections 3 and 4). There are, thus, at least four contentious terms that are central to my discussion: scientific realism, its defense, scientific understanding, and scientific success. To make the discussion tractable, I will simply assume that someone is a scientific realist just in case they believe that they know of various unobservable entities and some of their characteristics, and that this knowledge is based on scientific investigations. In this chapter, I illustrate the scientific realist position using an example from electrostatics. Some of the entities that are central to this example are objects with positive and negative electric charges. These objects cannot be observed; however, the scientific realist claims to know of the existence of these charged objects, and also some aspects of their interactions. For example, like charges repel one another and opposite charges attract one another via a force that is determined by Coulomb's law: $F = k\,(q_1 q_2)/r^2$. Here, the q_i are the magnitudes of charges, r is the distance between them, and k is a constant fixed by the units.

One way that scientific knowledge is different from ordinary knowledge is that scientific knowledge arises by the deliberate activities of reflective agents. In an ordinary case, I may come to know where I left my glasses through a haphazard process of searching. This process ends when I perceive, and thereby come to know, that I left my glasses in the kitchen. When pressed to defend my claim to know that I left my glasses in the kitchen, I may not have much to say. Scientific knowledge, by contrast, aims at being much more articulate. An agent plans and carefully conducts an experiment so that the results can be communicated to other agents. One goal of this activity is to make whatever knowledge that arises from this experiment for that agent into sharable knowledge that is available to others. In our case, we will sketch an experiment that provides the experimenter with knowledge of the existence and character of unobservable charged objects. When this experimenter communicates

DOI: 10.4324/9781003202905-12

the results of this experiment, other agents may also have this knowledge. For this process to function, the experimenter must be able to defend their claim to know. This defense will involve much more than just asserting what they suppose to be the case. In addition, they must clarify how the experiment was conducted and how the experimental results bear on the existence and character of these unobservable objects. A defense of a claim to scientific knowledge requires responding to the sorts of doubts that other scientists are likely to raise. If such responses are not forthcoming, then the experimenter's claims to know will be rejected by their scientific peers.

In Section 2, I will sketch a defense of scientific realism that is inspired by this typical scientific procedure. I adopt a broadly naturalistic methodology that insists that philosophical arguments should resemble our best available scientific arguments and that philosophical doubts should be based on recognizably scientific considerations. This methodology constrains how one should conceive of a philosophical defense of scientific realism. The correctness of scientific realism requires knowledge of unobservable entities, and so a defense of scientific realism will involve a defense of some knowledge of unobservable entities. This defense takes the form of an argument whose conclusion is that there is knowledge of unobservable entities (by scientific means). The premises of this argument will provide reasons to accept the conclusion. These reasons are not a guarantee that the conclusion is true, and the premises will themselves be uncertain. The defense of scientific realism will be a good one, though, if it passes the level of scrutiny that is appropriately imposed on a novel scientific claim. This requires that the defense of scientific realism be capable of persuading a neutral party, but not that the defense persuade a resolute skeptic or an anti-realist.

The defense of scientific realism that I develop here involves scientific understanding. Scientific understanding is a distinctive kind of cognitive achievement. Here, I will focus on so-called objectual understanding: when an agent has objectual understanding, they stand in a demanding kind of cognitive relation to some object. One kind of object that is central to this defense is phenomena. A phenomenon is a repeatable type of event, state, or process (Bogen & Woodward 1988). Our experiment concerns a phenomenon known as electrostatic induction. The phenomenon of electrostatic induction occurs whenever a charged body is placed next to a conductor, and this placement brings about a new electrostatic force of attraction between the charged body and the conductor. As I will explain in Section 2, the core of my defense of scientific realism is an inference from an objectual understanding of a phenomenon (such as electrostatic induction) to knowledge of the existence and character of some unobservable objects (such as charged particles). This inference is legitimate to the extent that an understanding of a phenomenon involves knowledge of the existence and character of some unobservable entities.

A successful defense of scientific realism may then proceed by making it plausible to a neutral party that this sort of understanding exists.

It remains to sketch the kind of evidence that can be offered in favor of the claim that an agent has achieved this sort of scientific understanding. Scientific realists typically emphasize the success of science, but exactly what counts as a success that is adequate for realism is contentious. Here, I will focus on control of the phenomenon of interest, for example, electrostatic induction. In an experiment, an agent may be able to create and manipulate some phenomenon. When this creation and manipulation provides the proper kind of understanding of the phenomenon, then the agent may rightly claim to know of the existence and character of the unobservable entities involved in this creation and manipulation. The aim of Section 2 is to spell out this process.

2 Recasting the No Miracles Argument

The most well-known defense of scientific realism is the no miracles argument. Putnam is often credited with its most compact formulation: "The positive argument for realism is that it is the only philosophy that does not make the success of science a miracle" (given at Psillos 1999, 71). Putnam's formulation suggests a single, global argument whose conclusion is a claim about our best, current science as a whole. But more recently, scientific realists have formulated a more generic argument that is apt to be applied in a series of limited, local cases. I take Psillos' case for selective scientific realism to be the most compelling (Psillos 1999). Psillos restricts his focus to cases where some theory-laden investigation is able to generate a novel, predictive success. In some of these cases, the very theories that informed that investigation also offer potential explanations for why that prediction occurs by an appeal to the mechanisms involving unobservable entities. If those potential explanations make essential use of these unobservable entities, then that is often sufficient evidence to conclude that those entities exist and have the characteristics needed for that explanation to go through: "The best explanation of the instrumental reliability of scientific methodology is that: the theoretical statements which assert the specific causal connections or mechanisms by virtue of which scientific methods yield successful predictions are approximately true" (Psillos 1999, 78). The selective scientific realist will deploy this same generic argument when the right kind of predictive success is achieved, and thereby make the case for a limited range of knowledge of unobservable entities.

As Psillos has sometimes emphasized, one inspiration for this sort of defense of scientific realism is what Feigl called "the Copernican turn" (Feigl 1950, 41, given at Psillos 2011, 308). The Copernican turn requires a scientist to take whatever evidence they have assembled in favor of some theoretical claims, and to use their theories to explain that evidence and the scientist's knowledge of that evidence. Feigl illustrates this aim with an example from electrostatics: "The divergence of the goldleaves of the electroscope which epistemically serves as an indicator of the presence of charges is immediately deducible from the theoretical assumptions of electrostatics, i.e. primarily from the Coulomb law of attraction and repulsion" (Feigl 1950, 40). Feigl's discussion makes clear that the point of this deduction of the evidence is to explain the evidence along with the scientist's knowledge of the evidence:

> If knowledge (as behavior) is not to remain an utter mystery or miracle, it is clear that the knowing organism itself must find a place in the world it knows. Whatever object can be reached by empirical knowledge must, no matter how indirectly, be related, (yes, causally related) with the processes in the knowing organism.
>
> (1950, 41)

What is initially simply a report on one's observations is then treated, as it were, "cosmologically" by embedding what is observed in a causally integrated nexus that includes not only what is responsible for the objects observed, but also the scientists themselves and their actions. Through this embedding, indexical terms like "I", "here," and "now" are eliminated using terms that place the observations in an objective time and place: "The epistemic uniqueness of the base corresponds only to an objective specificity and focal character of a spatio-temporal region in the cosmological account" (1950, 41). For example, in the electrostatics case, the behavior of the electroscope is traced to the presence or absence of charged particles via Coulomb's law. In addition, the role of the scientist in building and manipulating this experimental setup is used to explain how the scientist knows what they do by indicating the agent-independent causal relationships.

Feigl's case of the experimental creation and manipulation of electrostatic induction fits perfectly into Psillos' generic argument for selective scientific realism. The experimenter builds their experimental apparatus using their electromagnetic theory, drawing in particular on Coulomb's law for electrostatic forces. The agent models the metals in this apparatus as conductors, with charged particles that are free to move on the surface of these metals in response to charged bodies being placed near the apparatus.[1] When the gold leaves move as the agent expects using their theory, the agent acquires evidence that Coulomb's law is correct, and that the hypothesized charged particles really do exist, even though they are too small to be observed. So, in such circumstances, the best explanation of the experimental manipulation of the gold leaves is that these theoretical claims are true, and that the mechanisms in question really are operating to produce the experimental effects.

Recall that my goal is to reformulate this kind of defense of scientific realism in terms of understanding. Through adopting their electrodynamic theory, the agent acquires an understanding of electrostatic induction. More specifically, the agent understands electrostatic induction as that phenomenon is being created and manipulated in their experiment. For this understanding to inform a defense of scientific realism, it must have implications for knowledge. What is it about the agent's understanding of electrostatic induction that indicates that the agent also knows of the existence and character of unobservable charged particles? As I have reconstructed the case, the agent believes their theory, and deploys their belief in these charged particles when they build and manipulate their experimental apparatus. The resulting success of these manipulations is the evidence that these beliefs are true. So in carrying out the experiment, and coming to understand this phenomenon, the agent also acquires evidence that some of their beliefs are true. This means that the experiment can afford knowledge of such unobservable entities.

One aspect of this process is crucial, although it is somewhat in the background of Psillos' discussion of his argument. However, this aspect seems to be the motivation for Feigl's talk of a "Copernican turn." Copernicus succeeded in understanding the trajectories of the planets by recognizing that the earth was a planet. This led him to embed his earth-based observations in a more cosmological coordinate system centered on the sun. The earth's own motion was, thus, a factor in the measurements that Copernicus had available, but this factor could be eliminated, and a more objective representation of the situation was thereby achieved. The same "cosmological" aim is essential to any instance of this sort of defense of selective scientific realism. The phenomenon in question and one's evidence that one is manipulating it successfully needs to be characterized in objective terms that dispense with indexical terms or other contextual restrictions. In putting forward a claim to know of the existence and character of these unobservable charged particles, the agent is claiming that this phenomenon is due to these charges quite generally. The claim to know is not qualified or restricted: everywhere and always, electrostatic induction arises due to the causal factors set out in the electrodynamic theory. The agent is not claiming just that this is what is going on in this experimental context or for this scientific community.

For this kind of knowledge of the objective world to result from an experimental success, the evidence acquired through the experiment must be made correspondingly objective or cosmological. To start, we can imagine an agent who conducts their experiment and records their personal observations in their "protocols." These observations only tell them what has happened here and now to them. But their theoretical understanding of their actions and the experiment indicates to the agent how these personal observations can be transformed into evidence of the character of some objective phenomenon. Their understanding of the phenomenon, then, must incorporate a grasp of the independence of that phenomenon from the particular circumstances of its creation in that experiment. By grasping that independence, the agent is thereby in a position to carry out the kind of objective rendering of what they have found. If this independence is not grasped, then the agent is not a position to know that some unobservable entities exist and are responsible for the phenomenon they have manipulated.

Here, then, is my proposed generic argument for a defense of selective scientific realism based on the special sort of scientific understanding just surveyed:

1 An agent conducts an experiment that successfully creates and manipulates an instance of some phenomenon, drawing in part on their theoretical beliefs concerning this phenomenon.
2 The agent grasps that the phenomenon in question obtains *independently* of their actions and scientific community.
3 The agent grasps that the observable features of this phenomenon *depend on* the existence and character of some unobservable entities that are posited by their theoretical beliefs.
4 The agent has a good understanding of the phenomenon.
5 Therefore, the agent knows of the existence and character of some unobservable entities.

Premise (1) is the basis for premise (2), and both premise (1) and premise (2) are the basis for premise (3). The good understanding required for premise (4) must incorporate at least the grasp of the propositions given in premises (2) and (3). When all these conditions are met, it is highly likely that (5) obtains.

One recent account of objectual understanding can inform this defense of scientific realism. In "Beyond Explanation: Understanding as Dependency Modelling," Dellsén defends what he calls a "dependency modelling account of understanding" according to which

> DMA: S understands a phenomenon, P, if and only if S grasps a sufficiently accurate and comprehensive dependency model of P (or its contextually relevant parts); S's degree of understanding of P is proportional to the accuracy and comprehensiveness of the dependency model (or its contextually relevant parts).
>
> (Dellsén 2020, 1268)

Premises (2) and (3) of my argument spell out two of the essential elements of the understanding required for the defense of scientific realism. In Dellsén's terms, we are to imagine that the agent has a model of their experiment that includes not just what occurred, but also how what occurred was dependent on and independent of various salient factors. Premise (2) mandates that this model represent the experiment's results as obtaining for any situation, no matter how the electrostatic induction is created. Premise (3) says that this model also represents what the results do depend on, namely the movements of these charges. Premise (1) adds that this dependency model of the experiment is built using the theoretical resources that the agent has provisionally adopted.

It is worth noting that Dellsén himself makes it very easy for an agent to understand a phenomenon. As the title of an earlier paper "Understanding Without Justification or Belief" (Dellsén 2017) suggests, Dellsén maintains that an agent may grasp the right kind of dependency model for some phenomenon, and thereby understand that phenomenon, without any justification at all, and also without anything like a belief in the claims that their model makes concerning that phenomenon. However, this possibility does not undermine the argument that I am proposing. I am considering a case when an agent *does* believe the theory in question, and deploys some of those beliefs in their experiments. When these beliefs and the experimental results are responsible for their grasp of the propositions given in premises (2) and (3), then, I am claiming, the resulting state of understanding will make it highly likely that they know of the existence and character of those unobservable entities. That is the contentious core of this understanding-based defense of scientific realism. Nothing about Dellsén's account of understanding is incompatible with this defense, although, of course, he may not wish to endorse it.[2]

If Dellsén is right about objectual understanding, then it is easy to argue that the truth of premises (1), (2), and (3) makes it highly likely that (4) obtains. The contentious step will then be from the truth of these premises to the truth of the conclusion (5). Other accounts of scientific understanding make understanding more demanding. For example, in different ways, Kelp (2017) and Khalifa (2017) require that a good state of understanding just be a special kind of knowledge. If this is right, then the inference from the truth of premises (1)–(4) to (5) is easy. But there will be many cases where premises (1), (2), and (3) obtain but (4) does not. Grasping a proposition is not knowing it, and even for an agent who knows the right propositions, the knowledge may not be integrated in the way that either Kelp or Khalifa requires.

The substance of an understanding-based defense of realism must, thus, clarify how to connect understanding to knowledge, and then respond to concerns that such a connection is implausible or unlikely. This is not my aim here. Instead, my goal is to consider what accounts of scientific understanding would block any development of this sort of defense. I first identify a necessary condition for this defense in Section 3 and then argue in Section 4 that Potochnik's recent proposal would undermine this necessary condition.

3 Scientific Understanding and Knowledge

The argument sketched in the previous section assumes that the knowledge of unobservable entities that arises through the understanding of the phenomenon proceeds in part through the grasp of the independence of that phenomenon from the experimenter and their scientific community. For example, in the case of electrostatic induction, the experimenter came to know of the existence of unobservable charged particles only because they grasped that the phenomenon of electrostatic induction depends on those particles in an objective way that is valid quite generally. Although that instance of the phenomenon is due to the experimenter's actions, the phenomenon has an independent and objective character that is described, at least in part, by the electrodynamic theory that includes Coulomb's law.

I will call this assumption an *independence condition* on understanding. An agent's understanding of some phenomenon X satisfies the independence condition just in case the agent grasps that X may obtain with the very same characteristics independently of the agent's actions or the operations of their scientific community. In this section, I consider the effects of denying that scientific understanding can meet this independence condition. I claim that if someone denies that understanding can meet the independence condition, then that person cannot defend scientific realism.

Suppose, then, that an agent conducts an experiment that involves the creation and manipulation of electrostatic induction. According to whatever proposed account of scientific understanding is being considered, let us suppose further that the agent acquires an understanding of electrostatic induction. However, we can stipulate that according to this account of scientific understanding there is some barrier to the agent grasping that the phenomenon occurs independently of their actions and their scientific community. If this independence condition is not met, the agent may still grasp that the phenomenon depends on the existence and character of some unobservable objects. But if they are unable to grasp that the phenomenon occurs independently of their actions and their scientific community, then they are not able to defend their claim to know of the existence and character of these unobservable objects. A natural and ordinary question for another agent to ask is "how objective are these unobservable entities that you are proposing?" If the agent cannot answer this question, then their claim to know will be rejected by other scientists.

To see how this exchange can develop in an ordinary scientific case, consider an experimental drug trial that is conducted without a control group. In the general population, 90% of people with a given disease recover in 7 days. When 500 people with the disease are treated with the new drug, suppose that 95% of these people recover in 7 days. If a scientist puts this result forward as evidence for the effectiveness of the drug through some unobservable mechanism, the first question they would be asked is how they had controlled for the so-called "placebo effect." This is the well-known phenomenon that sick people tend to recover more often when they believe that they are being treated, even if the treatment is an ineffective placebo. Doubts tied to the placebo effect are an instance of a legitimate scientific worry about our independence condition: to suggest that the higher rate of recovery is merely due to the placebo effect is one way of saying that the higher rate of recovery is dependent on the experimenter's actions. The experimenter's actions are responsible for the change, and so nothing about the effectiveness of the new drug can be inferred. The obvious way to remedy this kind of defect is to conduct the experiment with a control group whose treatment resembles the treatment of the drug trial group as much as possible.[3] This would allow an experimenter to assemble evidence that the independence condition is met, and so a necessary condition on their claim to know is satisfied.

I conclude that if an account of scientific understanding places principled restrictions on understanding that preclude satisfying our independence condition, then that account is unable to deliver a defense of scientific realism. There are two preliminary objections to this conclusion that are important to discuss right away. First, one might argue for the existence of genuine knowledge that cannot be defended in the way that I have required. As noted in Section 1, I may come to know that I left my glasses in the kitchen through perception, and be unable to respond to doubts about this knowledge. Other forms of basic knowledge have a similarly inarticulate character: my memory provides knowledge of where I was born, but a claim to know where I was born based only on my memory is hard to defend. Inspired by these cases, one might suppose that there are cases of scientific knowledge where an agent genuinely does know about some unobservable entities, but is unable to respond to legitimate doubts about their knowledge. The scientific community may not accept such claims to know, but that does not show that the agent lacks this kind of basic scientific knowledge.

My reply to this objection is that knowledge of the existence and character of unobservable entities is not basic knowledge, and it should not be approached on the model of basic kinds of knowledge like perception or memory. There may, of course, be basic knowledge that is involved in a defense of scientific realism. For example, any report on an experiment will rely on perception and memory. Scientific doubts about these elements can be addressed in ordinary ways such as by replicating the experiment. Philosophical doubts about perception and memory are not at issue in this defense of scientific realism, as I have assumed that only concerns that are recognizably scientific are salient to this defense.

The second objection to consider is that there may be a defense of scientific realism that does not involve scientific understanding. If so, then one's account of scientific understanding would have no implications for one's capacity to defend scientific realism. It is difficult to prove that any defense of scientific realism must go through claims about understanding, and I do not propose to defend that strong claim here. If a defender of scientific realism has a proposal that gets around the problems I consider here, then it would constitute a valuable contribution to the debate about scientific realism. Most accounts of scientific realism either give up the defense of knowledge or else defend their claim to know through an appeal to explanation or understanding. Explanationist defenses of scientific realism have immediate implications for understanding, at least if we assume that grasping an explanation can afford objectual understanding. Versions of scientific realism that do not require knowledge are certainly worth considering, but I have set them aside due to reasons of space.

To conclude this section, I would like to illustrate how one clear way of denying the independence condition goes along with an inability to defend scientific realism. Here, I focus on Giere's "perspectival realism." This proposal renders the achievements of science essentially contextual. That is, there is no way to detach that achievement from the context that enabled it, and present it in an unqualified or absolute way. As Giere describes the view,

> For a perspectival realist, the strongest claims a scientist can legitimately make are of a qualified, conditional form: "According to this highly confirmed theory (or reliable instrument), the world seems to be roughly such and such." There is no way legitimately to take the further objectivist step and declare unconditionally: "This theory (or instrument) provides us with a complete and literally correct picture of the world itself".
>
> (Giere 2006, 5–6)

Giere's objectivist target here is actually poorly framed. The scientific realism that results from the defense sketched in the previous section does not purport to offer "a complete and literally correct picture" of the phenomena under investigation (let alone the whole world). But the scientific realist does aspire to detach a claim to know from the context that enables it. This aim is encapsulated in premise (2) and our independence condition. In our electrostatics case, some preliminary conclusions could be presented in a way that makes reference to a given experimental or theoretical context: "According to our highly confirmed theory of electrostatics, the movement of the gold leaves is due to the changing distribution of electric charges." The realist goes further, though, and insists that agents can know more than this. A scientist can know, in an unqualified sense, that the movement of the gold leaves is due to the changing distribution of electric charges. What is known here is a genuine feature of the world. It is not qualified or conditioned by the context that enabled its discovery.

In his critical discussion of perspectivism, Chakravartty has helpfully identified this core issue: for the perspectivist, "Knowledge of scientific ontology is bound within specific contexts because our epistemic abilities do not extend as far as perspective-transcendent knowledge" (Chakravartty 2017, 177). Chakravartty examines the arguments offered for perspectivism and convincingly shows how they fail. Here, I emphasize a different point: adopting perspectivism blocks any defense of scientific realism. This is perhaps not a very surprising result for Giere's perspectivism: the position is flatly incompatible with a claim to outrightly know any features of reality. However, it is still instructive to see how Giere's perspectivism blocks any defense of scientific realism because this will provide a roadmap for how to proceed in more contentious cases.

A perspectivist like Giere offers principled reasons to reject our independence condition. That is, no matter how well an agent understands some phenomenon like electrostatic induction, the agent is not able to grasp how the phenomenon can occur independently of the agent's actions or the operations of their scientific community. That is, as far as they can tell, the features that they find the phenomenon to exhibit in their experiments could very well be dependent on the agent's actions or some of the operations of their scientific community.

Rejecting the independence condition leads a perspectivist to a highly qualified conception of scientific knowledge as essentially contextual or perspectival. There is, thus, a huge difference between the knowledge that the scientific realist ascribes to the agent and the knowledge that the perspectivist ascribes to the agent. The scientific realist moves from the understanding of some phenomenon to knowledge of the existence and character of some objects that that phenomenon depends on. In the case of electrostatic induction, the realist supposes that an agent who understands this phenomenon knows something about unobservable charges. The perspectivist cannot endorse this reasoning. Instead, they attribute knowledge that is about something else: what is occurring in that context, which is jointly produced by the agent's actions along with the target phenomenon. Knowledge of what is occurring in that context has no clear implications for what will occur in any other context. The knowledge that the perspectivalist opts for is like the knowledge of the results of the drug trial without a control group. In the drug trial case, the results include the effects of the drug (if any) along with the placebo effect, and there was no way initially to separate these two contributions to the experimental results. The perspectivalist supposes that this problem is a permanent one as we can never detach what we find from its experimental or theoretical context. And so even if the agent can know, with respect to some context, that Coulomb's law obtains, they cannot draw any inferences about what will happen in any other context. By contrast, the realist forms legitimate expectations for what would happen in a wide variety of new experimental contexts, from picking up bits of cloth to bending streams of tap water. That is the "cash value" of outrightly knowing rather than knowing in a way that is confined to some context.

4 A Challenge to Potochnik

Giere's perspectivism is motivated by a broad worry about the possibility of detaching scientific knowledge from its context of origin. By contrast, Potochnik has defended a more nuanced account of understanding that is more narrowly focused on the prevalence of idealization in scientific practice (2017). Still, it seems that Potochnik is committed to denying our independence condition. In a recent discussion of her account of scientific understanding, for example, she notes that "I do not think that science generates a unified understanding or explanatory store but rather different, crosscutting varieties of understanding, even of a given phenomenon" (2020, 940). These varieties arise based on the goals of the investigators. The worry is, then, that an agent's understanding of some phenomenon will be based on the goals of the agent's research community. If these goals partly constitute the understanding, then our independence condition will be violated, and the proposed defense of scientific realism will fail.

To develop this worry in more detail, I will briefly review the key elements of Potochnik's approach to understanding. For Potochnik, when an agent investigates a phenomenon, they may come to understand that phenomenon, but only in a qualified way. The investigation is invariably directed by an "epistemic aim" (2020, 937) or "research questions" (2020, 940), which, in turn, help to determine whether the resulting cognitive state constitutes genuine understanding. This allows an idealization to be an essential part of many states of understanding. In her apt formulation, "Idealizations are assumptions made without regard for whether they are true and often with full knowledge they are false" (2020, 934). These assumptions will be part of the state of understanding when they enable an agent to understand a feature of the phenomenon that is of interest to the agent.

Potochnik has developed a special vocabulary to characterize the features of a phenomenon that are salient to understanding that phenomenon. They are what she calls *causal patterns* that are *embodied* in the phenomenon. More fully, "scientific understanding of some phenomenon requires (a) grasping a causal pattern (b) that is embodied in the phenomenon and (c) focal to the cognizer(s)" (2020, 936). I will assume that a phenomenon is a repeatable type of state, process, or event. A causal pattern is a regularity of a causal kind that may be "limited in scope and that may permit exceptions" (2020, 935). Crucially, our representation of a causal pattern may not make these limitations or exceptions explicit. One of Potochnik's examples is the ideal gas law. It is a representation that "ignores molecular size and intermolecular forces" (2020, 935). In this respect, the representation is arrived at through assumptions that are believed to be false of actual gases. At the same time, such "idealizations can be used to facilitate representation of simple patterns to generate scientific understanding" (2020, 935). In this case, representing an actual gas with the ideal gas law can constitute genuine understanding, but only if the aims of the researcher focus their attention on specific aspects of the gas. These include "how temperature increasing in a sealed container of gas with a fixed volume increases the pressure" (2020, 935). This is an example of a causal pattern that is embodied in the phenomenon of some gas in a sealed container.

Although an agent's representation of the embodied pattern may require an idealization, whether or not the pattern is embodied by the phenomenon is fixed independently of the agent's representation. At the same time, an agent's understanding of the phenomenon is irredeemably representational, and so the conditions on that representation are essential to the state of understanding. This is the reason that Potochnik concludes that understanding is nonfactive: "an account that is less accurate of a phenomenon (i.e., more idealized) can generate better understanding of that phenomenon when it depicts the causal pattern focal to those who seek understanding" (2020, 941). By contrast, "knowledge is factive" (2020, 941). So even if the ideal gas law is central to the state of understanding some gas in a sealed container, an agent cannot know that the ideal gas law governs the gas for the simple reason that this law is not true of any real gas. For Potochnik, an agent can know "the pattern described by the ideal gas law is embodied in this system" (2020, 941). This knowledge is consistent with the phenomenon embodying other causal patterns, including patterns that would frustrate the pattern picked out by the ideal gas law.

Consider a case where an agent has an understanding of some phenomenon P. According to Potochnik's account of understanding, that agent does not grasp that the phenomenon obtains independently of their actions and scientific community. This is because the agent's understanding is constituted in part by their research interests. The research interests do not merely enable a grasp of some characteristic of the phenomenon that is detachable from those interests. Instead, the research interests permeate their understanding to such an extent that they cannot suppose that what they understand of the phenomenon obtains independently of those research interests. This has immediate implications for what the agent can claim to know about that phenomenon. As Potochnik recognizes,

> the object of our scientific knowledge is not technically the phenomena scientists investigate but the causal patterns those phenomena embody. Science generates understanding of phenomena, and it does so via knowledge of the causal patterns they embody. Knowledge and understanding go hand in hand, but there is a gap between their objects.
>
> (2020, 942)

The agent may come to know that a given causal pattern is embodied in their phenomena, but this knowledge is insufficient to defend scientific realism. The understanding achieved from within one research community involves knowledge that is restricted by the aims of that community. So, there is no way for an agent in one community to draw any conclusions about what would occur if their research interests changed or if they aimed at studying the phenomenon in a fully objective fashion.

Potochnik could reply that an agent can combine a series of qualified knowledge claims that some causal pattern is embodied in some phenomena (for some research program) in order to know a conclusion about how the phenomenon is (independently of any research program). For example, a common representation that incorporates Coulomb's law proves apt to generate understanding of the electroscope phenomenon along with others, such as picking up pieces of cloth, having balloons stick to walls, and bending thin streams of tap water. Perhaps this repeated use of Coulomb's law could afford knowledge that Coulomb's law holds absolutely or in a way that is independent of the research interests of scientists. It is unclear to me, though, how Potochnik can validate this inference for it looks to be an instance of the argument for scientific realism that I considered back in Section 2. I have argued that this argument must proceed through a state of understanding that involves a grasp of a dependency model. Crucially, an agent needs to grasp that the phenomenon in question obtains independently of the agent's community, including the interests of the agents doing the experiments. Even if we have a series of phenomena that are known to embody a causal pattern, the knowledge of this embodiment does nothing to establish what will happen for other phenomena. So, this approach to understanding and knowledge is unable to validate the core defense of scientific realism developed here. Unless some other defense is available, Potochnik's account of scientific understanding undercuts a defense of scientific realism.

5 Conclusion

Recent refinements of scientific realism emphasize its limited, piecemeal character. For such a realist, there is no simple or monolithic way to understand the success of science. Instead, there is only a series of successes, including novel, accurate predictions and technological applications, where each should be understood in their own, local terms. I have sketched a defense of this selective form of scientific realism that relies on scientific understanding and that incorporates the central aspects of Feigl's Copernican turn. This sort of understanding requires appreciating the objective character of the phenomenon. Any approach to scientific understanding that allows for this can make the right connections between understanding a phenomenon and knowing unqualified propositions concerning that phenomenon. This analysis should help the realist to see how best to deploy recent work on understanding in the service of their defense of realism. It also presents a dilemma for some accounts of understanding: either amend the account of understanding or give up on the prospects for a defense of scientific realism.

Acknowledgments

An earlier version of this chapter was presented remotely at the Scientific Understanding and Representation 3 Workshop at Radboud University, Netherlands in April 2021. I am grateful to the organizers and the audience for their helpful suggestions. This chapter has benefited enormously from a series of conversations with Angela Potochnik. I would also like to thank Collin Rice and the editors of this volume for their invaluable comments on earlier versions of this chapter.

Notes

1 See, for example, Griffiths (2013, sections 2.5.2, 3.2.2).
2 See also Dellsén (2016). My discussion has been influenced by Rice's "understanding realism," but I do not think he would endorse my defense of scientific realism. See Rice (2021).
3 Woodward (2003, section 3.1) motivates his interventionist account of causes using this kind of case.

References

Bogen, James, and James Woodward. 1988. "Saving the Phenomena." *Philosophical Review* 97:303–352. https://doi.org/10.2307/2185445.

Chakravartty, Anjan. 2017. *Scientific Ontology: Integrating Naturalized Metaphysics and Voluntarist Epistemology*. New York: Oxford University Press.

Dellsén, Finnur. 2016. "Scientific Progress: Knowledge versus Understanding." *Studies in the History and Philosophy of Science* 56:72–83. https://doi.org/10.1016/j.shpsa.2016.01.003.

Dellsén, Finnur. 2017. "Understanding without Justification or Belief." *Ratio* 30:239–254. https://doi.org/10.1111/rati.12134.

Dellsén, Finnur. 2020. "Beyond Explanation: Understanding as Dependency Modelling." *British Journal for the Philosophy of Science* 71:1261–1286. https://doi.org/10.1093/bjps/axy058.

Feigl, Herbert. 1950. "Existential Hypotheses: Realistic versus Phenomenalistic Interpretations." *Philosophy of Science* 17:35–62. https://doi.org/10.1086/287065.

Giere, Ronald. 2006. *Scientific Perspectivism*. Chicago: University of Chicago Press. https://doi.org/10.7208/chicago/9780226292144.001.0001.

Griffiths, David J. 2013. *Introduction to Electrodynamics*. Fourth edition. New York: Pearson.

Kelp, Christoph. 2017. "Towards a Knowledge-Based Account of Understanding." In *Explaining Understanding*, eds. S. Grimm, C. Baumberger, and S. Ammon, 251–271. New York: Routledge.

Khalifa, Kareem. 2017. *Understanding, Explanation, and Scientific Knowledge*. New York: Cambridge University Press. https://doi.org/10.1017/9781108164276.

Potochnik, Angela. 2017. *Idealization and the Aims of Science*. Chicago: University of Chicago Press. https://doi.org/10.7208/chicago/9780226507194.001.0001.

Potochnik, Angela. 2020. "Idealization and Many Aims." *Philosophy of Science* 87:933–943. https://doi.org/10.1086/710622.

Psillos, Stathis. 1999. *Scientific Realism: How Science Tracks Truth*. New York: Routledge.

Psillos, Stathis. 2011. "Choosing the Realist Framework." *Synthese* 180:301–316. https://doi.org/10.1007/s11229-009-9606-9.

Rice, Collin. 2021. *Leveraging Distortions: Explanation, Idealization, and Universality in Science*. Cambridge, MA: MIT Press. https://doi.org/10.7551/mitpress/13784.001.0001.

Woodward, James. 2003. *Making Things Happen: A Theory of Causal Explanation*. New York: Oxford University Press. https://doi.org/10.1093/0195155270.001.0001.

11 Truth and Reality

How to Be a Scientific Realist Without Believing Scientific Theories Should Be True

Angela Potochnik

Scientific realism is a thesis about the success of science. Most traditionally: science has been so successful at prediction and guiding action, because its theories are true (or approximately true or increasing in their degree of truth). Here is Anjan Chakravartty in the *Stanford Encyclopedia of Philosophy*:

> Scientific realism is a positive epistemic attitude toward the content of our best theories and models, recommending belief in both observable and unobservable aspects of the world described by the sciences . . . most people define scientific realism in terms of the truth or approximate truth of scientific theories or certain aspects of theories.

If science is in the business of generating true theories, then we should turn to those theories for explanatory knowledge, predictions, and guidance of our actions and decisions.

At least at first glance, views that are popular in contemporary philosophy of science create challenges for this traditional form of scientific realism. The embrace of model-based science introduces questions about whether theories are even the main epistemic currency in science. And though in the quote above Chakravartty includes models in his formulation of realism, it is not clear how traditional realism applies to them. Models are not definitive like theories may be: it is common to employ a variety of different models to the very same phenomena. Further, the occurrence of tradeoffs among modeling priorities, such as predictive accuracy versus generality, calls into question whether the same scientific accounts can even deliver explanations, predictions, and policy-guidance. Perhaps we need to employ a variety of models to accomplish all our goals. Finally, the widespread importance of idealization in both models and theories is taken by some to cast doubt on the idea that the best scientific accounts are true, approximately true, or increasing in degree of truth over time.

Nevertheless, the basic idea behind scientific realism that science has been and will continue to be epistemically successful is deeply appealing.

DOI: 10.4324/9781003202905-13

In this chapter, I use the challenges of modeling and widespread idealization to motivate the view that scientific realism should be fully divorced from the idea that science is in the business of generating true theories. In Section 1, I will say more about these challenges to a traditional realism as I understand them, trying to base this in views broadly endorsed in our field. Then, in Section 2, I will motivate the possibility of a realism that is consistent with a rejection of the idea that science is in pursuit of truth. Again here, my aim is to carve out a general place in philosophical space, but I will also say a little about the version of this view that I have advocated. Finally, in Section 3, I will address why this deserves to be called scientific realism and gesture toward what I think may be some advantages of this way of going about being a realist. Realism without a strict commitment to truth better accommodates the nature of our present-day scientific achievements, and it opens up an interpretation of cumulative scientific progress that resists the antirealist's pessimism.

1 The Trouble with Truth

I will begin by describing how views endorsed in philosophy of science put pressure on the idea, bound up with traditional realism, that science aims for truth. Three ideas developed in the literature on scientific modeling are in tension with that idea.[1]

First of all, models have reduced ontological commitments compared with theories. As a result, there is more of an air of convenience in the choice of modeling approach than in theory selection. According to Morrison (2016), for instance, scientific theories have truth conditions that indicate what must be the case in order for them to apply, whereas models incorporate idealized assumptions to aid in their application to specific types of systems, with the result that the models do not have straightforward truth conditions. In many fields, it is common practice to employ multiple different models of the same phenomena (Weisberg, 2013), and the same models are sometimes redeployed to very different kinds of phenomena. (See Potochnik, 2012, for example, regarding how game theory has been reinterpreted and redeployed across evolutionary biology and behavioral sciences.) Wimsatt (1987) outlines several ways in which even models recognized to be inaccurate (false, as he puts it) can, nonetheless, lead to truer theories, such as by suggesting refinements to existing theory, enabling scientists to localize errors to particular aspects of the model, or as a limiting model to define an extreme for the phenomenon of interest.

Second, following from the point that it is common to employ multiple models to investigate a single phenomenon, philosophers have also observed that there are often or always tradeoffs among the advantages of different modeling approaches. The biologist Richard Levins wrote the seminal paper on this point (1966), arguing that scientists must choose the extent to which their models prioritize accuracy versus generality and that it is often decided to sacrifice some accuracy in favor of greater generality of application. Odenbaugh (2003), Matthewson and Weisberg (2009), and others have more fully worked through these ideas. I want to make two points about this in the present context. First, this provides reason to think that a single model will not accomplish all epistemic ends for a given investigation; in Potochnik (2015a), I develop this idea with an eye to the diverse aims of science. Second, scientists regularly prioritize other values for their models other than accuracy. Compromises in truth or accuracy are regularly made, at least in model-based science, to further other scientific goals, including applicability across related phenomena. Such a sacrifice of accuracy seems, at least at first consideration, to contribute to epistemic success rather than to compromise it.

Third, as Bokulich (2009, 2011), among others, has emphasized, model-based explanations proceed with substantial assistance from idealizations or even fictions. If false posits or fictions facilitate at least some varieties of scientific explanations, this role is another form of sacrificing some truth to facilitate other aims. Philosophers working on scientific explanation have developed a variety of views about how to accommodate the contribution of idealizations to explanation, many of which aim at maintaining a commitment to the truth of successful explanations. But this contribution poses a potential challenge—even if the challenge is solvable—to the idea that scientific accounts should be (wholly) true. I will say more below to motivate the particular challenge that idealizations pose to a traditional truth-centric realism.

These points about scientific models are well appreciated in philosophy of science, and they make articulating epistemic success in terms of truth less obviously the right path forward for scientific modeling than for scientific theorizing. Insofar as model-based science is an important form of science, it seems that truth, approximate truth, or increasing truth may not always best describe the accomplishments—or even the aim—of science. Insofar as scientific models can explain, successful scientific explanations may not always be true or approximately true. Indeed, philosophers with influential early accounts of scientific models also suggested attenuated versions of realism on that basis, including Giere's (1988) constructive realism and Suppe's (1989) quasi-realism. Others have suggested that the discussion about scientific realism can proceed independently from considering such details of scientific modeling. Chakravartty (2001) argues that model-based science inherits the same traditional challenges facing scientific realism. Psillos (2011), in turn, argues that scientific realism is fully consistent with the use of models in science and, in particular, with models' apparent reliance on idealizations. And, in the *Stanford Encyclopedia of Philosophy* article quoted above (2017), Chakravartty formulates all the statements about scientific realism as claims regarding "theories and models," generically.

The article does not address any promise or challenges for realism specifically related to modeling.

In the rest of this section, I will focus particularly on the challenge posed by idealizations. By idealizations, I mean assumptions made without regard for whether they are true and often with full knowledge that they are false. Familiar examples include the assumptions that a gas is ideal, that a phenotypic trait reproduces to its degree of success, and that humans are perfectly rational agents. Though idealizations have perhaps received the most attention in discussions of scientific modeling, they are not used just in models but also in a host of scientific laws and theories. Idealizations are widely recognized to contribute to scientific understanding; many of our most heralded scientific explanations involve idealizations. This goes beyond the model-based explanations that are Bokulich's focus to include theory- and law-based explanations—consider, for instance, Cartwright's (1983) decades-earlier discussion of how the laws of physics rely on idealizations. Though Chakravartty (2001) mostly places the question of the significance of idealization for scientific realism to one side, he does say in passing, "Ultimately, the question of how to make sense of idealization may pose the greatest challenge to the realist" (329).[2]

Here is how that challenge might be developed. Idealizations are common in our scientific accounts—including not just models but also theories and laws. As several philosophers have emphasized, idealizations persist in at least some of our best scientific accounts and contribute positively to the explanatory value of those accounts. Examples include Cartwright's (1983) seminal treatment of idealizations in the laws of physics, Elgin's (2004, 2017) articulation of how scientific understanding benefits from compromising truth, and my (Potochnik, 2015a, 2017) argument for the widespread use of idealizations to further diverse scientific aims, including understanding different features of the same phenomena. From the broadly recognized role of idealizations in scientific explanations, these philosophers have drawn the conclusion that explanations need not be true—that the understanding they produce is non-factive. That is, these philosophers reject *veritism* about scientific explanation. Of course, none of these views hold that we are indifferent to the truth value of our scientific accounts or prefer falsehoods to truths. The claims made are subtler, such as that idealizations contribute directly to the epistemic success of our scientific explanations, and that less accurate explanations are sometimes better than more accurate alternatives.

Put in terms of (successful) scientific explanations, these are my grounds for rejecting veritism:

- An explanation Y can explain the target phenomenon without being entirely true of it.
- An explanation Y can better explain the target phenomenon than does an explanation Z that is more accurate of it.

- A posit P in explanation Y can be crucial for Y explaining X without being remotely true.

Here are very brief illustrations of each idea, using the tried-and-true example of the ideal gas law. An application of the ideal gas law can explain the approximate behavior of a real gas with molecules exhibiting attractive or repulsive interactions, despite the real gas's properties deviating from the ideal gas law's predictions. And, in many circumstances, the ideal gas law explanation is preferable to a more accurate explanation with the van der Waals equation. Finally, the posit that a gas is ideal, that is, composed of non-interacting point particles, is crucial for an ideal gas law explanation but may nonetheless deviate significantly from the properties of the real gas. (A note about example choice: I assume a toy example suffices insofar as disagreement about veritism does not hinge on whether there are examples with these features, but rather whether these are sufficient reasons to reject veritism.)

Other philosophers accept the idea that idealizations contribute positively to scientific explanation but seek to retain veritism about explanation. Such views have been developed by Lawler (2021), Sullivan and Khalifa (2019), and Pincock (2021), among others. One motivation for developing a veritism about explanation that accommodates idealization is the perception that this will enable scientific realism to be retained (e.g., Rice, 2019). Nonetheless, for the sake of argument in the current investigation, I will assume that these features of idealization warrant reconsidering whether or to what extent truth is a requirement or possibly even an aim of our scientific accounts. There may be a way to rehabilitate a central role for truth. But on the face of it, the important roles for idealizations—false assumptions—in our scientific accounts of the world, as well as the features of model-based science discussed earlier in this section, seem to support looking for a scientific realism that does not require a commitment to the truth, approximate truth, or increasing truth of our scientific accounts. At the very least, such a realism can remain noncommittal on these questions of the extent to which truth is a benchmark for epistemic success in science, thus sidestepping the complications of scientific modeling and idealization that have received significant attention of late in philosophy of science.

2 Realism about What?

In Section 1, I used scientific modeling and idealizations to motivate the idea that it might be beneficial to develop a form of scientific realism that does not require strict truth, increasing truth, or potentially even the pursuit of true scientific accounts. Now I want to outline the form that such a realism might take. As in the previous section, my aim is to carve out a general place in philosophical space, but I also develop a specific version of the view that I would like to advocate, influenced especially by the use of idealizations in the science described above.

The first task is to find a generic formulation of scientific realism that does not already presume a commitment to truth; then, we can work out the specifics of how to make good on a commitment to such a realism. Here is Anjan Chakravartty again, a bit later in the *Stanford Encyclopedia of Philosophy* entry drawn from above:

> What all of these approaches [of defining realism in terms of epistemic achievement] have in common is a commitment to the idea that our best theories have a certain epistemic status: they yield knowledge of aspects of the world, including unobservable aspects.

Without too much of a stretch, we can broaden this formulation to apply to any scientific accounts used to explain our world, which might include models and laws as well as theories, and perhaps still other varieties of accounts. Then, at its most basic, realism is the idea that *our best scientific accounts qualify as epistemic achievements and yield knowledge of the world*. This idea strikes me as at the heart of a commitment to realism. This formulation also creates room to maneuver: if there is a natural way to interpret scientific accounts as epistemic achievements that yield knowledge, despite any sacrifices of truth, then that can enable a form of realism to be preserved even if veritism must be rejected. This approach is akin to Asay (2013), who argues against "truth-mongering" approaches to scientific realism that locate the debate in specifics about the theory of truth embraced. Asay instead advocates what he calls a metaphysical approach to realism, locating the true debate in questions of ontological commitments. Though Asay does not identify a need for distancing scientific realism from the commitment to the truth of our scientific accounts, his articulation of what should be at issue for realists versus antirealists accommodates such a distancing.

So, in what way—if any—do our scientific accounts yield knowledge bearing on observable and unobservable entities (despite potential sacrifices of truth)? On this initial question, insight can be gained from a convergence of views among recent discussions of idealization in explanation. Bokulich (2011), Batterman and Rice (2014), and I (Potochnik, 2015, 2017), among others, emphasize patterns. Here is Bokulich: "The model explains the explanandum by showing how elements of the model correctly capture the pattern of counterfactual dependence of the target system" (2011, 39). Batterman and Rice articulate the role of models in "[explaining] universal patterns across diverse real systems" (2014, 350). And, my take:

> scientific knowledge consists of truths about causal patterns. Grasping those truths about causal patterns comprises understanding of

the phenomena embodying the patterns. ... [T]ruths of causal patterns are by and large partial untruths about phenomena, accomplished with the use of idealizations.

(2017, 119)

Though Elgin (2004, 2017) focuses less explicitly on patterns, her emphasis on the aim of exemplification suggests implicit reliance on patterns or a similar concept as well—instantiation of a property or relation across some range of circumstances. This suggests a potential way forward: such patterns might be the objects of our scientific knowledge. This would be a form of selective realism, perhaps akin to a kind of structural realism: realism about the relations depicted in our scientific accounts, not taken as descriptions of the targets of investigation directly but of the patterns that those targets of investigation embody.[3]

The key move here is allowing non-identity between the target of investigation and the object of knowledge. You may study/learn about one thing, while the object of knowledge is technically something else. This non-identity creates two issues to address. The first issue regards the difference between the target of investigation and the object of knowledge. If the latter is not the same as the former, then what is it? Is the object of knowledge structure, patterns, or something else? Implicit in this question is the need to justify why this, whatever we settle on, should be taken to be the object of scientific knowledge. The second issue regards the connection between the target of investigation and the object of knowledge if we accept that this is not a relation of identity.[4] What makes this knowledge qualify as an epistemic achievement *about* the target of investigation? There must be some special relationship—if not identity—for this to be the case. So, to summarize, the two questions are:

1 **The Difference:** What is the object of knowledge if not the phenomena targeted in scientific investigation?
2 **The Connection:** What makes this knowledge qualify as an epistemic achievement about the target of investigation?

In the remainder of this section, I will describe my suggestions for how to address these two questions. Along the way, I will gesture toward allied positions.

2.1 The Difference

Above, I suggested on the basis of some recent views about the explanatory value of idealizations that the object of scientific knowledge might be *patterns*. In particular, following Bokulich's and my formulations, I propose as the object of scientific knowledge patterns in counterfactual dependence—by and large, in the manipulability relations constituting causal dependence (Woodward, 2003). In (Potochnik, 2015, 2017), termed these / *causal patterns*/. Postulating causal patterns rather than causal relationships or processes as the object of scientific knowledge reflects the scientific practice of strategically isolating difference-makers to indicate the scope of the dependence, that is, the scope of what Woodward calls the invariance of the causal relationship.

Thus, what I have in mind for what I dubbed "the difference" is causal patterns. Idealized scientific accounts cannot generate knowledge of the targets of investigation directly insofar as they are not true of those targets. But they can and do generate knowledge of causal patterns embodied by those targets. Beyond according with the insights of several recent accounts of idealization, this also captures well science's frequent focus on general laws, patterns, and tendencies. Many causal patterns, like the pattern depicted by the ideal gas law, are broad regularities that are limited in scope and can have exceptions. On the other hand, sometimes scientists are interested in accounting for the interplay of multiple causes, which motivates a focus on a more specific causal pattern with a narrower scope, that is, present in a smaller set of phenomena. The van der Waals equation, with additional parameters specifying a particular molecular size and attractive force, depicts a more specific pattern with a narrower scope. As this illustrates, any target of investigation—such as the behavior of some real gas—embodies multiple, perhaps countless, causal patterns. The strategic isolation of a set (small or large) of causal factors gives rise to a focus on just one causal pattern among many embodied in the target of investigation. Just as it is a matter of choice which phenomena are targeted in scientific investigation, it is a matter of choice which causal pattern is targeted in the investigation of a given phenomenon.

This many-one relationship between causal patterns and target phenomena is, in my view, of central importance. Insofar as our scientific knowledge regards causal patterns, the resulting knowledge is inevitably an incomplete perspective on the target of investigation—and, when idealizations facilitate the knowledge, a perspective enabled by false posits of features of the target that are incidental to the focal pattern. Positing causal patterns as the object of scientific knowledge introduces a variability in our scientific knowledge that goes beyond what phenomena our investigations target. The nature of our scientific knowledge is, on this view, relational to the sets of causes deemed interesting by investigators or their audience.[5]

This may be considered a form of perspectival realism. The view is in some respects similar to Massimi's (2019) perspectival realism, particularly her emphasis on realism regarding robust phenomena. On the other hand, for my view, the status of qualifying as knowledge is not indexed to research context, which is often thought to be entailed by perspectivism. Rather, the role for perspective merely is in what knowledge is prioritized, that is, knowledge of which causal pattern(s).

So, in my view, the difference between the objects of scientific knowledge and targets of investigation is that of causal patterns versus the phenomena embodying those (and other) patterns. The key difference regarding truth and knowledge is that idealizations (such as the posit that a gas is ideal) can contribute directly to knowledge of a causal pattern, even as they introduce falsehoods of phenomena. Why create this gap, what I have called the difference, at all? First, based on considerations raised in Section 1, to accommodate or even take our lead from scientific practices that appear to be driven by considerations other than maximizing the truth of our scientific accounts about the phenomena they target. Second, the landing point—a form of realism that accommodates how science has developed many different accounts of the very same phenomena—is, in my view, a welcome one. And third, I will offer a promissory note: I argue for some further philosophical advantages to embracing this kind of approach to realism in Section 3.

2.2 The Connection

Now that I have outlined a view for the difference between the targets of investigation and the objects of scientific knowledge, what is the connection? That is, why take knowledge of causal patterns to constitute an epistemic achievement *about* the phenomena targeted in scientific investigations? The brief answer is that, in my view, when a causal pattern is embodied in a phenomenon, knowledge of the causal pattern can explain the phenomenon. But, given the many-one relationship between causal patterns and target phenomena, there is more that must be said about the conditions in which knowledge of the former suffices to explain the latter. Here, I will sketch an approach, but it requires moves that I will not be able to make fully precise or adequately defend in this chapter. For a fuller articulation and defense in a different philosophical context, see Potochnik (2015b,2017).

Let us presume that a given scientific account qualifies as knowledge of some causal pattern, since what is at issue here is how that knowledge relates to the target of investigation such that it constitutes an epistemic achievement about that target. In my view, to explain the target of investigation, such an account must relate to the target phenomenon, how that phenomenon is characterized in the explanandum, and the audience in the proper ways. First, the pattern must be embodied by the target phenomenon. This entails that the causal claim(s) are approximately true—importantly, they need not be a comprehensive account of the phenomenon but merely depict the influence of at least one causal factor. Even the same causal factor(s) may be depicted differently in characterizations of distinct causal patterns, but the claims themselves still must be approximately true (tout court, not merely within a specific research context). Second, the causal pattern must account for the explanandum, that is, the features of the target phenomenon specified in the call for explanation. Roughly put, this means that what the depiction of the causal pattern, including any idealizations, would entail is specified by the (approximately true) explanandum as having occurred. Any given causal pattern embodied in a phenomenon can only explain features conforming to that pattern, not unrelated features of the phenomenon or deviations from the pattern. Third and finally, the depicted causal pattern must address the cognitive needs of those seeking explanation. Given the many-one relationship between causal patterns and target phenomena and the inevitable relationality of depicting one causal pattern rather than another, as described in Section 2.1, there is an ambiguity regarding which causal pattern is explanatory that needs to be resolved with reference to the agenda of those seeking explanation. This is influenced by their interests and background knowledge, as well as by incidental circumstances, potential biases, and blind spots. Specifying the explanandum cannot resolve the ambiguity (Potochnik, 2016).

Our simple ideal gas law example can show what these requirements amount to. The ideal gas law succeeds as an explanation of the pressure in a balloon doubling when the volume was halved (the explanandum) when: (1) the depicted relationship between pressure and volume is roughly true (relationship to target phenomenon); (2) this application of the ideal gas law entails the pressure roughly doubling (relationship to the explanandum); and (3) the ideal gas law application addresses the explanation-seekers' cognitive needs (relationship to the audience). This last criterion would be satisfied, for example, for someone with a basic understanding of the relevant physics but who has forgotten the exact relationship between pressure and volume or what other variables, like temperature, that relationship depends on. This would not be the case for someone who well remembers the ideal gas law and how it applies here but mistakenly thought the temperature had also increased to an extent that the balloon's pressure would have caused it to pop. That requires a different causal pattern, featuring different causal influences, to explain. A background assumption here is that all scientific explanations are relative to the audience's cognitive needs in this way. The shared background knowledge of a scientific community may constrain the variety of causal patterns that are explanatory, but variability will persist due to which specific questions are posed, which other phenomena are of interest, which methods are employed, and so on.

This, briefly, is what I take to be necessary for knowledge of a causal pattern to constitute an epistemic achievement—in the form of an explanation—of the target of investigation. The best explanation of a given explanandum features a causal pattern (1) embodied by the phenomenon, (2) accounting for the explanandum, that (3) best addresses the audience's cognitive needs. If increased accuracy of the phenomenon contributes to this, it does so only incidentally. More development of this view is needed, which you can find in Potochnik (2015b, 2017). Here, I am simply sketching a view with the potential to connect the target of investigation to the object of scientific knowledge to flesh out a candidate for the approach to selective realism I motivate in this chapter. This view creates the desired space between the objects of scientific knowledge and the targets of investigation. The former are multiple, whereas the latter—empirical reality—is single; for this reason, knowledge of the former inevitably yields only partial and perspectival insights into the latter.

3 A Realism Worth Having

Let us take stock. I have suggested that science achieving or even aiming for true accounts (of target phenomena) should not be a requirement for scientific realism. There may be a deep relationship between the objects of scientific understanding and the objects of scientific knowledge that is not identity. In particular, perhaps science explains phenomena by providing knowledge of (some of) the laws, patterns, and tendencies that they embody. This kind of view requires specifying what the objects of scientific knowledge are, what I called the difference, and how that qualifies as an epistemic achievement *about* the targets of investigation, what I called the connection. I have sketched one way of working out those components of this form of realism: We come to understand phenomena by generating knowledge of causal patterns that they embody.

This approach suggests that the question of realism might be reformulated from "Does science uncover legitimate knowledge (even about unobservables)?" to "About what, if anything, does science uncover legitimate knowledge?" In other words, on my suggested reframing, the question of realism regards what, if anything, we should be realists about on the basis of our scientific findings. I have proposed causal patterns as a candidate—in the epistemic sense of claiming that we have scientific knowledge of them and thus also in the ontic sense of positing their existence. This is one version of a selective realism.

Causal patterns might well not be the *only* kind of thing we should be realists about on the basis of scientific findings; that is not a requirement of the view I have outlined. For instance, I think that some existence claims about individual events also qualify as legitimate scientific knowledge, such as the claim that life on Earth evolved from one or a few common ancestors and the claim, of some iodine solution, that adding the starch will turn it blue. In some cases, scientific explanations may also provide knowledge of target phenomena themselves. I have pointed to idealization and features of scientific modeling as reasons to doubt that, by and large, this is so; explanations that do not involve models or idealizations may not face the attendant complications.

The selective realism introduced here is meant to replace a realism predicated on the requirement that scientific accounts aim for or achieve truth about the targets of investigation directly. It is meant to replace realism about phenomena investigated in science with a realism about some of the causal patterns embodied by those phenomena. In this last section, I want to explore some advantages of this approach to realism.

To start, this realism about causal patterns better accommodates some features of our present-day scientific achievements. The motivation for this approach is accommodating scientific practices bound up with the production of scientific knowledge that do not seem to be in pursuit of truth (of target phenomena) or at least not truth alone, as detailed in Section 1. Beyond that, this approach predicts a continuing plurality of scientific accounts of the very same phenomena whenever scientists are interested in different features of those phenomena rather than movement toward a single, integrated account. I take it that this well describes the present situation in many fields, as evidenced by the philosophical popularity of various pluralisms. This approach reconciles this continuing plurality of accounts with scientific realism taken as a commitment to the epistemic success of science and with the expectation of a unitary empirical reality—both of which I, at least, find to be compelling ideas worth maintaining if possible.

I wonder whether a causal pattern realism such as I have sketched here might not also open up an interpretation of cumulative scientific progress that resists the antirealist's pessimism. If we are traditional realists, then we must conclude that science has fallen short of knowledge again and again, and that theory change does not, by and large, seem to be getting us closer to the truth. Many superseded scientific theories posit radically different views of the world, so they must have been radically incorrect. Thus, the pessimistic meta-induction: Why think our current accounts are true, when their many predecessors were not? However, if we are realists about knowledge of causal patterns, the outlook is perhaps rosier. Taking our scientific achievements to constitute objective understanding of phenomena based in knowledge of some among many causal patterns enables a subtler accounting of what past scientific achievements have had right. The relationality of knowledge of causal patterns to the cognitive needs of the investigators and their audiences enables a possible explanation for radically different accounts. Though all cannot be true, they might all achieve or at least pursue knowledge of different causal patterns. Further, the sensitivity of our stores of scientific knowledge to the cognitive circumstances of scientists and their audiences can help legitimize the social influences on past scientific findings that other realisms may take to be threats.

Of course, some scientific conjectures are simply wrong: they posit causal patterns that do not exist. But science arguably *has* latched onto some of the many causal patterns embodied in phenomena of interest.

Many previously accepted theories captured causal patterns that are embodied in phenomena, even if they were later replaced by other theories that depicted other patterns, including sometimes more refined relatives of the original pattern. Causal pattern realism may be able to account for theory change resembling the reduction of prior theories to their replacements in terms of related causal patterns and account for radical theory change in terms of transforming priorities, questions, and background assumptions. Of course, one would need to motivate the plausibility of this approach for specific instances of theory change, which is a project that is, unfortunately, not only beyond what I can accomplish in this chapter but also more or less beyond my expertise. So, my articulation of how a causal pattern realism can accommodate theory change will, by necessity, remain suggestive.

Stanford (2003) and Tulodziecki (2017), among others, have argued against selective realism's ability to rebut the pessimistic meta-induction. Both highlight how jettisoned theories fare poorly with regard to successful reference and approximate truth from the perspective of our current theories, which are presumed by the realist to be true. A causal pattern realism such as I have described, however, does not posit the truth of our current best theories—at least not their truth of the targets of investigation. Because different accounts of the same phenomena may target distinct and even unrelated causal patterns, misconstruing other features of the target phenomena along the way, these accounts may emphasize different aspects and even posit different entities. This does not give us grounds for claiming that a past theory's successes will be conserved from the perspective of current theory but rather grounds for suspecting that past scientific accounts succeeded at generating scientific knowledge of causal patterns apt for the epistemic and social locations in which they were formulated. There can be different and even incompatible ways to define causal variables that give rise to patterns of counterfactual dependence that are invariant over some range of conditions. Note that this does leave intact the ability to diagnose shortcomings—empirical and programmatic—in past research.

I propose that the same approach I have suggested for accommodating past scientific change also suffices as a basis for confidence in the epistemic success of our present scientific accounts. Our current best theories are probably not true. Some of our current best theories even contradict one another. All of those theories, along with scientific models and perhaps other varieties of scientific accounts, stand a good chance of being jettisoned or significantly altered sometime in the future. But even if that comes to pass, it need not undermine their contribution to human understanding. If humans now use these accounts to yield knowledge of real causal patterns that satisfy our cognitive needs, then it does not matter if they are not strictly true, or turn out not to be true, of the phenomena they target. Some postulated causal patterns may turn out not to obtain at all, but many more are real—revealing important aspects of the world, even as they are incomplete and shaped by our present epistemic and social locations.

Whatever knowledge science generates is in response to our cognitive needs. That much is uncontroversial, I think. And our cognitive needs regularly motivate simplified accounts of our complex world. Simplified accounts privilege some aspects of phenomena to draw connections with other related phenomena, while neglecting other aspects. This is one reason for the widespread value of idealizations in scientific accounts. In this chapter, I have argued that such simplified accounts can qualify as knowledge even if they are not strictly true of phenomena insofar as they are knowledge of something else. It has often been said that science focuses on laws, regularities, and repeat phenomena rather than one-off events. It is not a far stretch to suggest that science achieves an understanding of phenomena—the events and occurrences of our world—by generating knowledge of some of the laws, patterns, and tendencies that these phenomena embody. On the view I have developed, the resulting scientific knowledge is not technically of the targets of scientific investigation but of some of the causal patterns that they embody. This causal pattern realism enables scientific realism to be maintained regardless of scientists' habit of sacrificing truth of phenomena for other epistemic priorities, and it holds promise for a realism-friendly interpretation of past scientific change and present scientific knowledge. Science is always in revision, but along the way, it is amassing ever more understanding via knowledge of causal patterns.

Acknowledgments

Thanks to Kareem Khalifa, Insa Lawler, and Elay Shech for organizing this volume; to Chris Pincock for fruitful exchanges and helpful feedback on this chapter; and to Insa Lawler, Kareem Khalifa, Michela Massimi, Aaron Novick, Darrell Rowbottom, the Rotman Institute of Philosophy at Western University, and the Caltech philosophy of physics reading group for their feedback and suggestions.

Notes

1 Models and theories are two forms of what we might generically call "scientific accounts." I will skirt the issue of how models and theories relate to one another, other than rejecting the idea associated with the semantic view of theories that models are simply interpretations of scientific theories.
2 Levy (2017) explores the prospects for realism in light of challenges from modeling and idealization. The approach I develop here is most like his characterization of potential responses to perspectivism.
3 This is particularly akin to Saatsi's (2019) proposed selective realism about counterfactual dependence, as these relationships are at the heart of causal patterns.

4 Or, at least, if we accept that this is not always a relation of identity; this approach only requires that the objects of knowledge, at least sometimes, be distinct from the targets of investigation.

5 This relationality is, I gather, a distinctive feature of my view (Potochnik, 2017) in comparison with the other discussions of idealization that advocate a focus on patterns. See Potochnik (2020) for a discussion of this difference from Elgin's (2017) view, in particular.

References

Asay, J. 2013. "Three Paradigms of Scientific Realism: A Truthmaking Account." *International Studies in the Philosophy of Science* 27: 1–21. https://doi.org/1 0.1080/02698595.2013.783971.

Batterman, Robert W. and Collin Rice. 2014. "Minimal model explanations." *Philosophy of Science* 81(3): 349–376. https://doi.org/10.1086/676677

Bokulich, Alyssa. 2009. "Explanatory Fictions," in *Fictions in Science: Philosophical Essays on Modeling and Idealization*, Mauricio Suárez (Ed.). Routledge Studies in the Philosophy of Science, Chapter 6, 91–109. New York: Routledge.

Bokulich, Alyssa. 2011. "How Scientific Models Can Explain." *Synthese* 180: 33–45. https://doi.org/10.1007/s11229-009-9565-1.

Cartwright, Nancy. 1983. *How the Laws of Physics Lie*. Oxford: Oxford University Press. https://doi.org/10.1093/0198247044.001.0001.

Chakravartty, Anjan. 2001. "The Semantic or Model-Theoretic View of Theories and Scientific Realism." *Synthese* 127: 325–345.

Elgin, Catherine. 2004. "True Enough." *Philosophical Issues* 14: 113–131. https:// doi.org/10.1111/j.1533-6077.2004.00023.x.

Elgin, Catherine. 2017. *True Enough*. Boston: MIT Press. https://doi. org/10.7551/mitpress/9780262036535.001.0001.

Giere, Ronald N. 1988. *Explaining Science - A Cognitive Approach*. Chicago: University of Chicago Press. https://doi.org/10.7208/chicago/9780226292038. 001.0001.

Lawler, Insa. 2021. "Scientific Understanding and Felicitous Legitimate Falsehoods." *Synthese* 198 (7): 6859–6887. https://doi.org/10.1007/s11229-019-02495-0.

Levy, Arnon. 2017. "Modeling and Realism: Strange Bedfellows?" in *Routledge Handbook of Scientific Realism*, Juha Saatsi (Ed.), Chapter 19, 237–249. New York: Routledge. https://doi.org/10.4324/9780203712498-20.

Massimi, Michela. 2019. "Realism, Perspectivism, and Disagreement in Science." *Synthese* https://doi.org/10.1007/s11229-019-02500-6.

Matthewson, John and Michael Weisberg. 2009. "The Structure of Tradeoffs in Model Building." *Synthese* 170: 169–190. https://doi.org/10.1007/s11229-008-9366-y.

Morrison, Margaret. 2016. "Models and Theories," in *Oxford Handbook of Philosophy of Science*, P. Humphreys (Ed.). Chapter 18, 378–396. Oxford: Oxford University Press. https://doi.org/10.1093/oxfordhb/9780199368815.013.32.

Odenbaugh, Jay. 2003. "Complex Systems, Trade-offs, and Theoretical Population Biology: Richard Levin's 'Strategy of Model Building in Population Biology' revisited." *Philosophy of Science* 70: 1496–1507. https://doi. org/10.1086/377425.

Pincock, Christopher. 2021. "A Defense of Truth as a Necessary Condition on Scientific Explanation." *Erkenntnis*. https://doi.org/10.1007/s10670-020-00371-9.

Potochnik, Angela. 2012. "Modeling Social and Evolutionary Games." *Studies in History and Philosophy of Science Part C: Studies in History and Philosophy of Biological and Biomedical Sciences* 43: 202–208. https://doi.org/10.1016/j.shpsc.2011.10.035.

Potochnik, Angela. 2015a. "The Diverse Aims of Science." *Studies in History and Philosophy of Science* 53: 71–80. https://doi.org/10.1016/j.shpsa.2015.05.008.

Potochnik, Angela. 2015b. "Causal Patterns and Adequate Explanations." *Philosophical Studies* 172: 1163–1182. https://doi.org/10.1007/s11098-014-0342-8.

Potochnik, Angela. 2016. "Scientific Explanation: Putting Communication First." *Philosophy of Science* 83: 721–732. https://doi.org/10.1086/687858.

Potochnik, Angela. 2017. *Idealization and the Aims of Science*. Chicago: University of Chicago Press. https://doi.org/10.7208/chicago/9780226507194.001.0001.

Potochnik, Angela. 2020. "Review of Catherine Elgin's *True Enough*." *Philosophical Review* 128: 363–366. https://doi.org/10.1215/00318108-7537361.

Psillos, Stathis. 2011. "Living with Models." *Synthese* 180: 3–17. https://doi.org/10.1007/s11229-009-9563-3.

Rice, Collin. 2019. "Understanding Realism." *Synthese* 198 (5): 4097–4121. https://doi.org/10.1007/s11229-019-02331-5.

Saatsi, Juha. 2019. "Realism and Explanatory Perspectivism," in *Understanding Perspectivism: Scientific Challenges and Methodological Prospects*, M. Massimi & C.D. McCoy (Eds.), Chapter 4. New York: Routledge.

Stanford, Kyle. 2003. "Pyrrhic Victories for Scientific Realism." *The Journal of Philosophy* 100: 553–572.

Sullivan, Emily and Kareem Khalifa. 2019. "Idealizations and Understanding: Much Ado About Nothing?" *Australasian Journal of Philosophy* 97: 673–689. https://doi.org/10.1080/00048402.2018.1564337.

Suppe, Frederick. 1989. *The Semantic Conception of Theories and Scientific Realism*. Chicago: University of Illinois Press.

Tulodziecki, Dana. 2017. "Against Selective Realism(s)." *Philosophy of Science* 84. https://doi.org/10.1086/694004.

Weisberg, Michael. 2013. *Simulation and Similarity: Using Models to Understand the World*. Oxford: Oxford University Press. https://doi.org/10.1093/acprof:oso/9780199933662.001.0001.

Wimsatt, William. 1987. "False Models as Means to Truer Theories," in *Neutral Models in Biology*, M.H. Nitecki & [A. Hoffman (Eds.), Chapter 2, 23–55. Oxford: Oxford University Press.

Woodward, J. 2003. *Making Things Happen: A Theory of Causal Explanation*. Oxford: Oxford University Press. https://doi.org/10.1093/0195155270.001.0001.

12 Defensible Scientific Realism
A Reply to Potochnik

Christopher Pincock

Marxists often argue about how best to achieve a communist revolution. They agree on the ultimate aim, but not on the means to achieve that aim. I take my debate with Angela Potochnik to have a similar structure: both of us agree that science provides knowledge, and that this feature of science should be maintained against doubts motivated by the practice or history of science. But we disagree about the most effective means of maintaining scientific realism. In my chapter, I began my examination of scientific realism by supposing that it is subject to a certain kind of defense. This contributes to the somewhat schematic character of my contribution, where I argue that we can defend a realist's claims to know in a special type of situation. But my contribution does not engage with the complicated practice of actual science. By contrast, Potochnik draws on a rich account of scientific practice. This informs her account of what scientific knowledge amounts to. But this focus on practice leaves open how defensible these claims to knowledge will turn out to be.

In this brief reply, I want to examine in more detail what is involved in Potochnik's causal pattern realism and consider again how the defense I sketched in my contribution can be adapted to this form of realism. It might seem that any sort of selective scientific realism can be defended using the kind of argument that I sketched. In my contribution, I argued that this is not the case. Any form of selective scientific realism that involves contextual limitations on knowledge, explanation, or understanding is not defensible. This argument needs to be fleshed out in more detail if it is to convince Potochnik (and others) to avoid these sorts of contextual limitations. I doubt I can persuade her to make these changes, but I believe that the exchange can still be fruitful as it clarifies how various forms of selective scientific realism fit with different styles of defense.

In her contribution, Potochnik makes clear that she is a causal pattern realist. Scientists investigate phenomena, which are repeatable types of events, states, or processes. Some phenomena involve unobservable entities such as molecules or charged particles. When these investigations go well, scientists come to know that a causal pattern is embodied in

DOI: 10.4324/9781003202905-14

such a phenomenon. Causal patterns exist and are embodied by some phenomenon in a completely objective and mind-independent way. And so, if scientists succeed in knowing of the existence and embodiment of causal patterns, it is perfectly appropriate to say that they have acquired knowledge of a feature of the objective world.

In my contribution, I supposed that a scientific realist must insist on knowledge of the existence and character of unobservable entities like molecules or charged particles. From Potochnik's discussion, it seems that a scientist may know that a causal pattern is embodied in a phenomenon without knowing of the existence and character of any unobservable entities. For example, if all the causal patterns that a scientist knows about are observable, then this is knowledge that the traditional anti-realist about science would be happy to accept. So, I think it is fair to consider a case where there is something unobservable about the embodiment of a causal pattern in a given phenomenon. In my electrostatics case, the theory claims that there are charged particles that come in one of two kinds. Like charges repel one another and unlike charges attract one another based on their distance. What sort of causal pattern is thereby claimed to be embodied in an electrostatic phenomenon? I suppose that such patterns involve the right kind of counterfactual relationship between variables, or what Potochnik calls "factors." These, in turn, can be treated as families of properties. On this reading, one salient causal pattern is how varying the distance between like charges varies the force of repulsion between them. In an experimental manipulation of electrostatic induction, the experimenter decreases the distance between large numbers of like charges and thereby brings about the divergence of the gold leaves on their electroscope. The observable causal pattern that relates their movements to the divergence of the gold leaves is said to be due to the unobservable causal pattern that relates their movements to the movements of unobservable like charges, which, in turn, are related to the divergence of the gold leaves.

The causal pattern realist claims that these sorts of theory-informed experimental manipulations are apt to provide knowledge of some unobservable causal patterns. More specifically, the scientist can come to know that this unobservable causal pattern is embodied in that very phenomenon. This is knowledge that something has these properties, but the character of the property bearer does not seem relevant to the embodiment of the causal pattern. Potochnik's causal pattern realism, thus, has some similarities to what Woodward once called "instrumental realism": for successful theories, "what is worth taking most literally and realistically . . . [are] claims about relational structures and patterns and particularly their claims about how changing one quantity, property, or feature will change some other quantity" (Woodward 2003, 114). Causal pattern realism is, thus, a highly restrictive form of scientific realism. This should make it easier to defend. In what follows,

I argue that causal pattern realism cannot be defended if contextual limitations are placed on knowledge, explanation, or understanding.

Can Potochnik adapt my defense of selective scientific realism to her causal pattern realism? Let us see how the steps of that defense are changed when tailored to causal pattern realism. Here, the changes are italicized:

1 An agent conducts an experiment that successfully creates and manipulates an instance of some phenomenon, drawing in part on their theoretical beliefs concerning this phenomenon, *including a belief that this phenomenon embodies an unobservable causal pattern.*

2 The agent grasps that *the embodiment of the unobservable causal pattern* in question obtains *independently* of their actions and scientific community.

3 The agent grasps that the observable features of this phenomenon *depend on* the *embodiment of an unobservable causal pattern* that is posited by their theoretical beliefs.

4 The agent has a good understanding of the phenomenon.

5 Therefore, the agent knows of the *embodiment of an unobservable causal pattern.*

Here, I have substituted "embodiment of an unobservable causal pattern" for "existence and character of some unobservable entities." In my contribution, I argued that if explanation or understanding is constituted in part by an agent's research interests, this defense will break down at step (2). If the agent is unable to grasp that the causal pattern is embodied independently from their research interests, then whatever understanding they achieve in step (4) will be unlikely to go along with the knowledge claimed in step (5). This is because the knowledge claimed in step (5) requires a strong sort of independence, including independence from the agent's research interests. Potochnik and I agree that scientific knowledge is the knowledge of an objective matter of fact. So, if the understanding fails to include a grasp of this independence, then the agent will fail to know of this feature of the objective world.

What should Potochnik's attitude toward step (2) be? One option that she could take is that it is not necessary. On this reading, the argument goes through even without step (2) because step (2) is redundant. Taking this option involves arguing that meeting step (1) and step (3) is sufficient to obtain the right kind of good understanding described in step (4). And when step (4) is arrived at through this route, it is highly likely that the agent acquires knowledge of the embodiment of that unobservable causal pattern. The worry I have about this defense of causal pattern realism is that the state of understanding is partly constituted by the agent's research interests. In her contribution, Potochnik emphasizes the contextual aspects of an explanation. A genuine explanation "addresses the explanation-seekers cognitive needs" and the best explanation "best addresses the audience's cognitive needs" (this volume, 160–161). As I note in my original contribution, Potochnik has the same contextual account of scientific understanding. So, it would be natural to insist that step (2) is irrelevant to the sort of understanding her account allows. But then it is hard to see how achieving such a state of understanding would make it likely that the agent knew, in an interest-independent way, of the embodiment of the pattern. Potochnik is here clear that she does not wish to make the knowledge of patterns itself context-specific, for she says that "the status of qualifying as knowledge is not indexed to research context" (this volume, 158). So, there is a gap between the context-specific understanding that her account requires for step (4) and the context-free knowledge in step (5).

Another option is that step (2) plays an important role in the argument, but that Potochnik can provide sufficient motivation for this step in the cases where the knowledge of unobservable causal patterns is acquired. On this option, even if the understanding in step (4) is context-specific, when it is informed by a grasp of independence in step (2) it is highly likely to go along with context-free knowledge of the embodiment of that causal pattern. How, though, can an agent grasp that the pattern is embodied independently of their actions and research community? Here, I would appeal to my discussion of the placebo effect in my original contribution. Notice that according to Potochnik every phenomenon embodies many causal patterns, including those that involve an experimenter's actions. In any situation where an agent is inferring the existence of a new causal pattern through their experimentation, they must be able to sort out when an observable feature genuinely depends on a new causal pattern and when an observable feature arises as an artifact of their experimentation. There is a common realist strategy to address this sort of worry: the best explanation of the observable features found in the experiment is that they depend on this objectively existing unobservable causal pattern. But this strategy is not available to Potochnik. As we have just seen, she makes cognitive needs central to the goodness of a proposed explanation. Elsewhere, she emphasizes that our best explanations are permeated with idealized representations, including idealized representations of the very causal patterns at issue. So, if Potochnik's accounts of explanation and understanding are correct, it is not clear how an agent can satisfy step (2) in this defense.

In conclusion, it is important to clarify what I think these considerations establish. They are meant to establish that this defense of scientific realism is not available to those who index knowledge, explanation, or understanding to an agent-relative context such as a goal or a community. I am not arguing, though, that Potochnik is wrong when she says that we have knowledge of the embodiment of unobservable causal patterns. There very well may be the kind of knowledge that she insists on. But if a defense of these claims to know is critical to maintaining one's realism in the face of doubts, then this kind of realism is in trouble precisely because it cannot be defended.

References

Potochnik, Angela. this volume. "Truth and Reality: How to Be a Scientific Realist without Believing Scientific Theories Should Be True."

Woodward, James. 2003. "Experimentation, Causal Inference, and Instrumental Realism." In *The Philosophy of Scientific Experimentation*, ed. H. Radder, 87–118. Pittsburgh: University of Pittsburgh Press.

13 Different Ways to Be a Realist

A Reply to Pincock

Angela Potochnik

In his chapter in this volume, Christopher Pincock develops an argument for scientific realism that incorporates what is required for scientific understanding and, then, what is required for that understanding to suffice as a basis for scientific knowledge. From there, he argues that Giere's (2006) and my (2017, 2020) commitment to the context-dependence of scientific understanding or knowledge renders our views unable to account for an essential step in how scientists come to know, namely, the extrapolation of findings from specific experimental and observational contexts.

Meanwhile, in my chapter in this volume, I focus on the apparent challenges to scientific realism introduced by scientific modeling and, especially, idealization. I use these challenges to motivate a scientific realism according to which the objects of scientific knowledge are causal patterns. Knowledge of causal patterns, in turn, provides understanding of the phenomena embodying these patterns. I dubbed this "causal pattern realism." In this response, I will sketch a revised version of Pincock's proposed argument for realism that is consistent with causal pattern realism. Then, I will respond to Pincock's concern that the context-dependence of understanding I endorse would, if true, interfere with the scientific community's ability to extrapolate from specific experimental and observational contexts as needed to develop knowledge. My goal is not to convince anyone to be a causal pattern realist but rather to create the space for such a view, taking into account the concerns motivating Pincock.

1 Different Kinds of Realism

Both Christopher Pincock and I are scientific realists. Further, I accept something like Pincock's proposed argument for realism. A good starting point for determining the alignment and differences between our versions of realism is, I suggest, to ask *about what* we are realists. In my chapter in this volume, I urge a careful consideration of what is taken to be the target of scientific realism, that is, the objects of scientific knowledge.

DOI: 10.4324/9781003202905-15

I will begin by characterizing what Pincock's chapter is advocating realism *about* (on the basis of our best scientific findings). In several places, Pincock indicates that this is a question of the reality of unobservable objects. For instance, in the conclusion of his enumerated argument from understanding to knowledge (this volume, 140): "Therefore, the agent knows of the existence and character of some unobservable entities." This is a classic focus of debates about scientific realism: whether scientific findings suffice for us to know of the existence of, for example, subatomic particles. However, earlier in the chapter, when introducing the experiment demonstrating Coloumb's law, Pincock offers what I take to be a different target for realism. He says, "the best explanation of the experimental manipulation of the gold leaves is that these theoretical claims are true, and that the mechanisms in question really are operating to produce the experimental effects" (this volume, 138).

So, here are four candidates for what Pincock thinks we should be realists about:

1 Existence of unobservable entities
2 Character of unobservable entities
3 Truthmakers of theoretical claims
4 Mechanisms under investigation

Of course, one might hold a realism that combines these: perhaps what it is to have knowledge of the existence and character of unobservable entities just is to know theoretical claims bearing on those entities, and perhaps knowledge of these theoretical claims just is knowledge of the mechanisms governing their behavior.

However, in my view, these four candidates for the objects of scientific knowledge do not align so neatly, and three are not apt targets for a general scientific realism. Regarding (1), the existence of unobservable entities is only very occasionally what is at issue in scientific discovery. Consider the extraordinary amount of scientific research conducted on Covid-19 in the early 2020s. Scientists determined quickly that SARS-Cov-2 was the novel coronavirus responsible for Covid-19. At that point, research turned toward downstream questions bearing (variously) on understanding and controlling the spread and effects of this virus.

Regarding (2) and (3), scientific discoveries do, arguably, more often bear on knowledge of the character of unobservable entities, and such knowledge plausibly consists of theoretical claims. But knowledge of the character of some entity is open-ended in a way that knowledge of its existence is not. How many theoretical claims must be known to suffice for realism about the character of an entity? Merely requiring knowledge of one theoretical claim in which the entity factors is surely too weak, but knowledge of all true theoretical claims in which the entity factors seems too formidable a standard. For this reason, (3) might be a preferable candidate. We can speak of knowing theoretical claims, even if we are uncertain about whether these claims suffice as a basis for knowing the character of some (unobservable) entities. (Of course, there is also the well-worn issue of how to meaningfully distinguish between observable and unobservable entities. Sidestepping this issue is another advantage of focusing on the objects of theoretical claims as a target for scientific realism.)

Regarding (4), many theoretical claims in science arguably do not regard, or at least do not directly regard, mechanisms. Above, I indicated that the existence of entities is a narrow segment of the targets of scientific investigation, and the same is so for repeat processes carried out by the coordinated activity of some entities (the sense of "mechanism" emphasized by the new mechanists, e.g., Machamer et al., 2000). Perhaps by "mechanisms under investigation," Pincock instead means something more generic like "how this entity behaves." In that case, though, a realism targeting mechanisms under investigation has the same difficulty as a realism targeting the character of entities: how much knowledge of a mechanism suffices for us to be realists about that mechanism? Surely the knowledge of mere existence or a sliver of knowledge of its workings is not sufficient but complete knowledge is too high a standard. A related downside to (2), the character of unobservable entities, and (4), mechanisms under investigation, as targets for scientific realism is that whether something is posited as real does not seem like it should be a vague category, yet each of these targets requires judgment calls regarding how much knowledge suffices for realism.

Therefore, (1)–(4) identified above are not, I think, interchangeable targets for scientific knowledge. And, in my opinion, (3)—theoretical claims—is the most promising candidate on the list.

I will now briefly explore how this relates to the causal pattern realism I outlined in my chapter in this volume. In that chapter, I emphasized the divide between knowledge of theoretical claims and knowledge of the character of entities (or, we could add, mechanisms) under investigation. I think we should be realists about—that is, posit that we have scientific knowledge of—the objects of well-corroborated theoretical claims. But those objects are not unobservable entities, mechanisms, or even the phenomena under investigation. Rather, science's theoretical claims (when successful) by and large yield knowledge of *causal patterns*. Causal patterns—real patterns à la Dennett (1991) involving manipulability relations à la Woodward (2003)—are embodied by specific phenomena. However, claims about causal patterns are usually not (strictly) true of the phenomena embodying them due to widespread idealization, and they bear on only some limited aspects of the phenomena (Potochnik, 2017). So, with my focus on causal pattern realism, I endorse Pincock's focus on theoretical claims as a target for realism, but have given reasons to think that this target diverges from the entities and mechanisms under investigation.

Only minor revisions to Pincock's argument for realism from understanding are required to bring it in line with the causal pattern realism I have motivated. As Pincock summarizes his interpretation of the electrostatic case study on which he focuses, "the agent believes their theory, and deploys their belief in these charged particles when they build and manipulate their experimental apparatus" (this volume, 139). In this statement, the beliefs in question regard both theoretical claims and the existence of unobservable objects. Alternatively, from my perspective, it may be that an agent takes their theory or model to adequately capture the target phenomenon for the purposes at hand, that is, the agent believes that the theory or model is true *of a causal pattern* embodied by the phenomenon. I have argued in previous work (2017, 2020) that this can suffice for scientific understanding (of the target phenomenon) and scientific knowledge (of the causal pattern).

2 Different Roles for Context

I have suggested that minor edits make something like Pincock's understanding-based argument for realism available for my causal pattern realism as well. But a primary aim of Pincock's chapter is to show that an argument for realism such as he deploys is not available to perspectivalists, a camp in which he includes me. Indeed, Pincock argues that an element of my view is inconsistent with essential requirements for the scientific community to gain knowledge. To have any hope of maintaining my causal pattern realism, I must address that criticism.

The trouble, as Pincock sees it, regards the context-dependence of understanding on my account. This is a version of what he takes to be a general difficulty with perspectival realism, including Giere's (2006), that any scientific knowledge qualifies as such merely from the perspective of a particular scientific context. Pincock summarizes how this falls short of realism: "What is known here is a genuine feature of the world. It is not qualified or conditioned by the context that enabled its discovery" (this volume, 145). He accurately points out that I have argued for a similarly perspectival character to scientific explanations (and understanding). In my view, the character of an explanation depends not just on the explanandum but also on the occasioning research interests—the context in which the explanation is formulated. The problem with this, according to Pincock, is as follows (this volume, 148):

> The agent may come to know that a given causal pattern is embodied in their phenomena, but this knowledge is insufficient to defend scientific realism. The understanding achieved from within one research community involves knowledge that is restricted by the aims of that community. So there is no way for an agent in one community to draw any conclusions about what would occur if their

research interests changed or if they aimed to study the phenomenon in a fully objective fashion.

I agree with Pincock's interpretation of my view that a causal pattern's ability to explain (or, equivalently, to engender understanding) depends on the research interests of those seeking explanation; see Potochnik (2016) for my fullest defense of this idea. One scientist's or lab's or scientific field's explanation may well not be an explanation for a scientist, lab, or field with different questions, even regarding the very same phenomenon.

However, there is a distinction available that can protect my view from the conclusion that scientific knowledge cannot transcend the specific research context of its discovery. In my view, if the knowledge of some causal pattern discovered in a particular research context is unenlightening in a different research context, this is not because it ceases to obtain but simply because it no longer engenders understanding. Other researchers may well have different questions about the phenomena under investigation, questions that are answered with information about different causal patterns. In my view, and in agreement with Pincock's line quoted earlier, causal patterns are "genuine [features] of the world" (this volume, 145). Their existence "is not qualified or conditioned by the context that enabled its discovery" (this volume, 145). Yet the content of our scientific knowledge, nonetheless, does depend on the research interests that occasion this knowledge. This combination is possible because the phenomena investigated in science embody many, even countless, causal patterns—more than scientists will ever seek to know. One can have some (objective, context-independent) knowledge of a phenomenon without having all knowledge of that phenomenon, and even without having the right knowledge for one's purposes.

To summarize, for my causal pattern realism, the context-dependence of scientific knowledge consists not in whether a causal pattern exists but merely in whether a causal pattern is of interest—and thus properly explanatory. Different scientific communities will disagree about the importance of some causal patterns, whereas other causal patterns will never come to be investigated by any practicing scientists. Our scientific knowledge is, thus, always partial and bears the indelible mark of our interests. However, it is still full-fledged knowledge. This form of context-dependent explanation does not result in a merely contextual definition of scientific knowledge, for the context-relativity does not regard the truth (of causal patterns) but the cognitive value to the explainers or knowers.

References

Dennett, Daniel. 1991. "Real Patterns." *The Journal of Philosophy* 88: 27–51. http://dx.doi.org/10.2307/2027085.

Giere, Ronald. 2006. *Scientific Perspectivism.* University of Chicago Press. http://dx.doi.org/10.7208/chicago/9780226292144.001.0001.

Machamer, Peter, Lindley Darden, and Carl Craver. 2000. "Thinking about Mechanisms." *Philosophy of Science* 67: 1–25. http://dx.doi.org/10.1086/392759.

Pincock, Christopher. this volume. "Understanding the Success of Science."

Potochnik, Angela. 2016. "Scientific Explanation: Putting Communication First." *Philosophy of Science* 83: 721–732. https://doi.org/10.1086/687858.

Potochnik, Angela. 2017. *Idealizations and the Aims of Science.* University of Chicago Press. https://doi.org/10.7208/chicago/9780226507194.001.0001.

Potochnik, Angela. 2020. "Idealization and Many Aims." *Philosophy of Science* 87: 933–943. http://dx.doi.org/10.1086/710622.

Potochnik, Angela. this volume. "Truth and Reality: How to Be a Scientific Realist without Believing Scientific Theories Should Be True."

Woodward, James. 2003. *Making Things Happen: A Theory of Causal Explanation.* Oxford: Oxford University Press. https://doi.org/10.1093/0195155270.001.0001.

14 Realism about Molecular Structures

Amanda J. Nichols and Myron A. Penner

1 Introduction

Advances in natural sciences over the last 150 years have resulted in increased understanding of the constituent parts of matter. This understanding is remarkable in that on the one hand, one cannot make a direct observation of the atomic and subatomic building blocks of medium-sized objects. On the other hand, current understanding of the natures of these "blocks" is supported by precise, well-confirmed measurements of various kinds, and this enables technological advances that touch every area of modern life. The tension that emerges from the inability to "see" things like molecules, atoms, and electrons while at the same time depending on accurate models of their properties is the main fuel that has fired the realism/anti-realism debate in philosophy of science throughout the last century. Realists tend to see the success of a scientific theory—success defined in terms of empirical adequacy, generating novel predictions, and explanatory power—as evidence of the theory's accuracy in describing a mind-independent, objective reality, even if that theory describes entities and structures that are not directly observed. Some philosophers resist the realist tendency to connect a theory's empirical success with its truth, pointing out that many theories that have been successful in the past were rejected as new data came to light, and a new successor theory comes into view. Why think that our current models will fare any better?

We think that the development of, and empirical support for, the current molecular theory provides strong evidence for realism about molecular structures, even though these structures are not observed in direct, standard ways. By "realism about molecular structures," we mean that our current models of molecular structures are correct with respect to (i) the number and kind of atoms they depict, (ii) the bond connections between atoms (i.e. what is bonded to what), and (iii) the orientation and approximate shape of the molecule in space (i.e. its stereochemistry). We are not making claims about the precise shape of individual atoms, the subatomic properties of bond connections,[1] nor are we weighing in

DOI: 10.4324/9781003202905-16

on debates concerning the nature, if any there be, of orbitals or on the ontology of quantum fields.[2]

Our argument proceeds as follows. First, we specify further what we mean by "realism about molecular structures." Second, we look at two lines of evidence for the accuracy of current molecular models: evidence from spectroscopy and microscopy. Third, we organize the evidence from spectroscopy and microscopy into a formal argument after which we present and reply to objections.

2 Molecular Structures and Scientific Realism

2.1 Scientific Realism and Test Cases

There are a number of distinct yet overlapping views that reflect some type of robust realism with respect to scientific methods and content. The two main realist sortal categories are *entity* realism and *structural* realism, with structural realism further subdividing into *epistemic* and *ontic* versions of structural realism. Common among realists is the claim that the overwhelming success of some scientific domain—where success is taken to involve both empirical adequacy and novel predictions—is strong evidence for the accuracy of theories within that domain. Moreover, theoretical accuracy is taken to include semantic accuracy and the successful reference of referring terms, even when the objects of reference have not been observed directly. Entity realists hold that we should be realists about the objects described by successful theories, whereas structural realists hold that we should be realists about the structures described by successful theories. Epistemic structural realism can be understood as realism about structural relations between entities in theories, while remaining agnostic about the nature of the relata. Ontic structural realism can be understood as going beyond its epistemic counterpart and claiming, "there are no 'things' and that structure is all there is" (Ladyman 2020).

Philosophers of chemistry have connected developments in chemistry to issues concerning scientific realism. Hasok Chang notes that in practice, chemists tended to have a role for unobserved entities in theory construction and interpretation of laboratory results. According to Chang, the chemical atomism that emerged in the nineteenth century resulted in most chemists adopting a type of realism toward atoms, and that "Confidence in the reality of chemical atoms, whose defining property was weight, understandably went up when various chemists could agree on what their weights were" (2016, 236). However, Chang concludes that the discovery of the electron, the development of quantum mechanics, and the way that the concept of orbitals is utilized in describing the chemical bond undermine classical notions of an atom's shape.

Eric Scerri discusses realism in the context of claims made about the observation of orbitals (2001). Zuo et al. claimed to have observed *d*-orbital holes on a copper atom using convergent-beam electron diffraction and X-ray diffraction technologies (1999). Scerri argues that Zuo et al. are conflating electron density with orbitals. Conceptually, an orbital is a mathematical description of a probability space that, by definition, cannot be observed. Electron density, however, can be detected in ways that generate representative images. In a passage that prefigures some aspects of the argument we develop in Section 2 below, Scerri compares the claims about observing orbitals with the claims about observing atoms:

> Readers might be wondering how this situation relates to the fairly ubiquitous claims regarding the observation of atoms using scanning tunneling microscopy (STM) and atomic force microscopy (AFM). Of course atoms are not being directly observed in these studies, since all that is measured is the flow of current across a tip, or the force that the tip exerts when passing across a surface. Is the question of the putative observation of orbitals analogous? My response to this question will be in two parts. The situation with the recent reports is analogous in the sense that electron density is also being indirectly observed (in fact, more indirectly than atoms, since the technique involves subtracting a reference-state density, which is not the case in STM and AFM studies).
>
> (2001, 1493)

Note that according to Scerri, both the images generated by X-ray diffraction in the case of orbitals and the images generated by STM and AFM in the case of atoms are capturing electron densities—and in the case of atoms, the path from measurement to generated image is more direct.

Scerri's point is important to consider in terms of the sort of realism for which we are arguing; we think that the evidence from spectroscopy and microscopy supports entity realism at the level of atoms and their particular connections and orientations in space to form molecules. Recall that we are not making claims about the precise shape of individual atoms or the subatomic properties of bond connections. But we do argue that our current models of molecular structures are correct with respect to (i) the number and kind of atoms they depict, (ii) the bond connections between atoms (i.e., what is bonded to what), and (iii) the orientation and approximate shape of the molecule in space (i.e., its stereochemistry). Precise measurement of electron density provides one way of testing whether molecular models are accurate.

2.2 The Ball in the Box

An additional way to gather evidential support for scientific realism is through confirming, converging data from different measurement technologies.[3] Below, we will look at examples of how recent data from microscopy and infrared (IR) radiation spectroscopy support the realism about molecular structures. But first, we present an analogy to illustrate the connection between measurement data and realist commitments.

Imagine an unopenable box approximately the size of a football helmet. Maybe it was built by an evil genius, or perhaps it was left on the planet by aliens with a cruel sense of humor. Regardless of the box's origin, it cannot be opened, and its construction is such that its contents cannot be imaged with any current technology. However, various tests *can* be performed on it. In addition to measuring the box's dimensions and mass, scientists have performed all types of operations on the box to try and ascertain what is inside. Based on how it responds to various movements, consensus is that it contains a single object. Theories complete with mathematical formalism were developed about what could possibly be in the box. Experiments were devised, building on different assumptions to test theoretical claims. At one point, a team of engineers built what they thought was a close replica of the box, with the added feature that the replica could be opened so that they can put different objects inside it. Researchers performed tests on both the box and the replica in order to determine which objects caused the replica to respond in ways similar to the box's responses. Eventually, scientific consensus converged around the claim that the box's unseen object is spherical, approximately 5 centimeters in diameter, with a mass somewhere between 120.2 and 120.4 grams. Not known for literary creativity, they dub it the "ball in the box." The epistemic situation of researchers in this thought experiment is analogous to the epistemic situation of chemists and physicists developing the first models of molecules in the nineteenth century. Should these imaginary scientists be realists about the ball in the box?

Perhaps at this stage of ball in the box research, realism might seem unwarranted. Or at least, one might want a better look at the experimental results and the competing theoretical alternatives before weighing in with a view. However, now imagine a postdoc working in a lab that specializes in ball in the box studies devises a new theory about the box's exterior construction—a theory which, if accurate, would both explain why previous attempts to image the box's contents failed, and specify the precise frequency and conditions under which sound waves *could* penetrate the box. With visions of a Nobel Prize, she designs an experiment to test her theory, and for the first time someone is able to send sound waves through the box! Comparing interference patterns from sound waves that pass through the box at locations where the ball is deemed not to be, with interference patterns from sound waves that pass through the box at locations where the ball is deemed to be, supports the earlier belief that the internal object is spherical. However, with more precise data, she is able to update the ball's diameter to a more precise measurement of 3.797 centimeters. This epistemic situation is roughly similar to the epistemic situation in contemporary science where models of molecular structures are confirmed through various types of spectroscopy. Does our imaginary postdoc have enough evidence to be a realist about the ball in the box?

One final addition to our box analogy. Now imagine a researcher who, building on the foundational discovery of the postdoc mentioned above, is able to fuse light with sound in a way that enables, for the very first time, light to pass through the box! This scientist is able to generate a computer image based on the data collected. The first grainy "images" of the box's now famous object depict what appears, in fact, to be a ball in the box. This epistemic situation is roughly similar to the epistemic situation we are currently in with respect to some molecules where images have been generated via new technologies in microscopy. Has realism about the ball in the box been vindicated at last?

The purpose of this thought experiment is to present an easy-to-grasp example showing how intuitions concerning realism vary across evidential contexts. Initially, one might think that when presented with a box that is impossible to open, one would not ever have a good epistemic basis for forming beliefs about what is inside. However, the grip that intuition has on one is likely loosened, somewhat, when following the ingenious ways that our imaginary scientists were able to gather good data about the box's contents even without cracking it open. As we will see, what real scientists are able to do to generate models of molecular structures is much more precise and substantial, epistemologically, than what our imaginary researchers did with the ball in the box.

3 The Evidence for Molecular Structures

3.1 Molecular Symmetry and Spectroscopy

Throughout the nineteenth century, chemists made substantial progress in identifying and classifying elements, with the periodic table of elements taking its current shape by the early twentieth century (Scerri 2012). The classification of elements, and the increased understanding of chemical bonds gave rise to three-dimensional models for the structure of every known molecule. With three-dimensional models in hand, chemists were able to organize molecules according to their molecular symmetry (Rosenthal and Murphy 1936; Wigner 1959).

Molecular symmetry is a classification system of molecular structure that can explain chemical behavior in different kinds of light. After giving a brief introduction to molecular symmetry, we will demonstrate how a molecule's symmetry explains how it will respond to IR light. This response is captured by IR spectroscopy, a common chemical identification tool.

Something is said to be symmetric if it appears unchanged under a particular operation or transformation. A symmetry operation is an operation you can perform on an object such that the state of the object before and after the operation appears qualitatively identical. An easy example of a type of symmetry operation is rotating an object 360 degrees; two images of an object, one taken before the rotation and the other after the rotation, look qualitatively identical, regardless of whether the object is the Statue of Liberty or a molecular structure. Thus, a symmetry operation performed on a molecule is a symmetry operation if the resulting state of the molecule is indistinguishable from the initial state.

A symmetry element is a geometrical property of a molecule by virtue of which it is possible to perform a symmetry operation on that molecule (Vincent 2001). In other words, different symmetry elements allow for different symmetry operations. There are five types of symmetry elements: *identity, proper rotation, reflection, inversion,* and *improper rotation*. A molecule can be classified not only in terms of the number of types of symmetry elements it possesses, but also in terms of the number of tokens it possesses within each type (e.g., the number of different proper rotation axes it may have). Group theory and matrix algebra are used to organize molecules mathematically into point groups by virtue of their symmetry elements; molecules in one point group all have the same symmetry elements.

Point group classification allows molecules to be viewed in terms of their geometrical properties. Consider two chemical compounds that have molecular structures that visually look quite different, and that are composed of different elements (see Figures 14.1 and 14.2). Although the molecules look different, they belong in the same point group meaning both molecules possess the same symmetry elements.

At this stage, one might say, "It's all well and good to classify molecules into symmetry categories, but how can we have confidence that we've modeled the molecular structure correctly?" Molecular symmetry predicts that molecules in the same point group will respond to IR light in the same ways. The IR spectroscopy shows that individual molecules behave the way we would expect, based on how we have modeled their structure. Moreover, IR spectroscopy shows that molecules in the same point group behave the same way, which indicates that we have correctly identified their symmetry elements. If we have correctly identified their symmetry elements, we must have correctly identified their

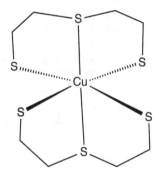

Figure 14.1 bis(1,4,7-trithiacyclononane)copper(II).

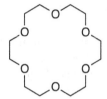

Figure 14.2 18-crown-6 (1, 4, 7, 10, 13, 16-hexaoxacyclo-octadecane).

three-dimensional structure, because symmetry elements are just geo-metrical properties of a particular shape.

One of the classic molecular examples that demonstrate the connection between molecular structure and analytical data are the metal carbonyl complexes.[4] These complexes have an octahedral geometry. The complex has a metal in the center with six groups called "ligands" surrounding the metal. In what follows, we will consider chemical compounds that have molybdenum as the metal atom center with four carbonyls and two triphenyl phosphite groups (see Figures 14.3 and 14.4).

Vibrational spectroscopy is concerned with what happens when certain frequencies of light are absorbed by molecules, and understanding a molecule's spectroscopic data involves understanding both its geometry and polarity. Of course, a molecule's geometry refers to how its constituent parts are connected and oriented in three-dimensional space. A molecule's polarity is connected to the way electromagnetic charge is distributed throughout the molecule. A "dipole moment" is said to occur when a molecule experiences an uneven distribution of charged particles. Together, a molecule's geometry and polarity will predict how it will respond when absorbing IR energy. In certain conditions, a molecule's dipole moment will change from one uneven distribution of charge to another, causing the molecule to vibrate. The change in uneven

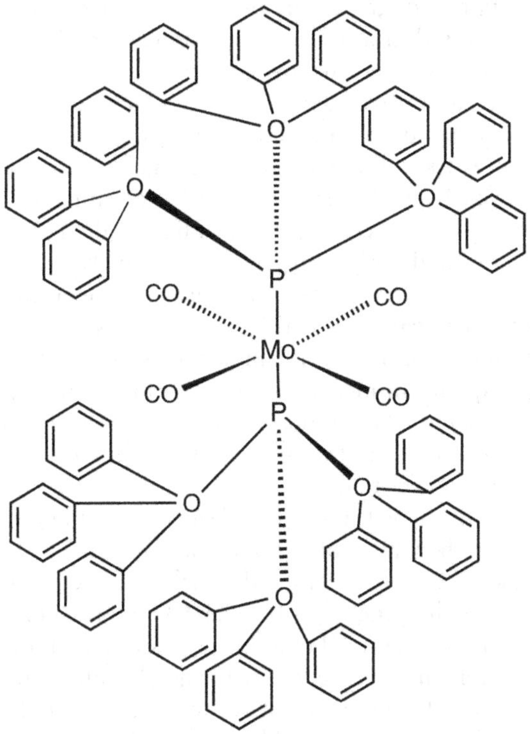

Figure 14.3 Trans form of tetracarbonylbis(triphenylphosphite)molybdenum.

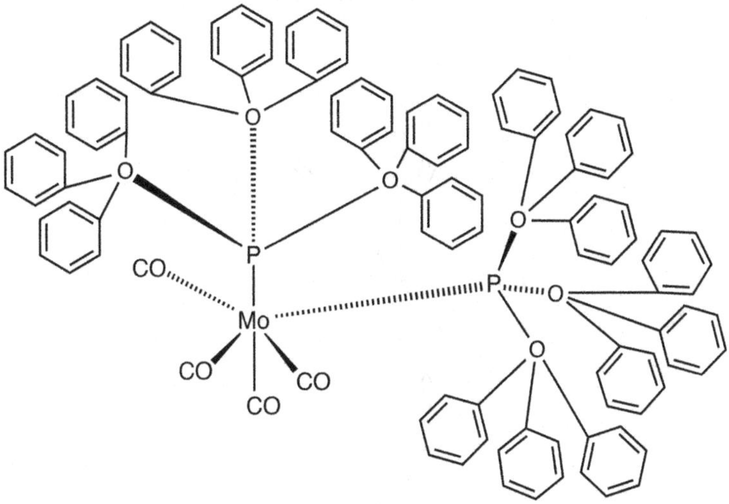

Figure 14.4 Cis form of tetracarbonylbis(triphenylphosphite)molybdenum.

charge distribution brought about by exposing a molecule to IR can cause some atoms within a molecule to stretch further apart, contract closer together, move in synchrony, or move asynchronously. A vibrational mode at a certain frequency is termed "active" in the IR energy region when those bonds are aligned along x, y, or z axes. The IR spectroscopy measures these fluctuations in charge distribution, with each "active band" appearing as a peak on an IR spectrum. The IR spectroscopy correlates the observed frequencies of light (called normal modes of vibration) with molecular structure. Because each chemical has a unique molecular structure, IR spectroscopy results in something like a "fingerprint" of the chemical. Classical mechanics is used to picture what is happening to a model of the specific molecule, whereas the experimental data aligns with quantum-mechanical calculations (Rosenthal and Murphy 1936). In these molybdenum carbonyl complexes, it is the carbonyl group, the carbon atom and the oxygen atom, that absorbs energy at a certain frequency and stretches.

Figures 14.5–14.8 depict a simplified version of the *cis* form of tetracarbonylbis (triphenylphosphite)molybdenum molecule, identically drawn four times. The arrows along the carbonyl (CO) groups are what differentiates each drawing (a)–(d). The CO represents the carbonyl group, the carbon atom bonded to an oxygen atom, and the L represents a ligand, specifically in our case, the triphenylphosphite group. The center metal atom, molybdenum, is not shown, but it is understood to be at the center of the crossed lines. The overall shape is octahedral. The vertical z axis (i.e., the proper rotation axis) has carbonyl groups on both ends. The other two lines are assigned the x and y axes and are perpendicular to the principal rotation axis. Each picture has a different set of arrows representing the way the molecule is stretching when

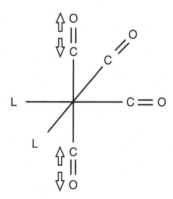

Figure 14.5 IR active synchronous stretches along the vertical z axis pictured
on the metal carbonyl complex structure for *cis* form.

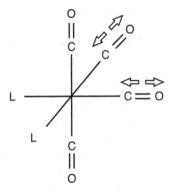

Figure 14.6 IR active synchronous stretches along the perpendicular x and y axes pictured on the metal carbonyl complex structure for *cis* form.

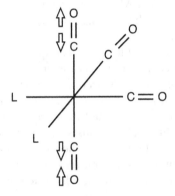

Figure 14.7 IR active asynchronous stretches along the vertical z axis pictured on the metal carbonyl complex structure for *cis* form.

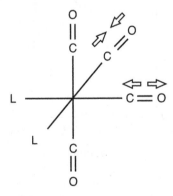

Figure 14.8 IR active asynchronous stretches along the perpendicular x and y axes pictured on the metal carbonyl complex structure for *cis* form.

IR excites the carbonyl groups. Figure 14.5 depicts the carbonyl groups stretching synchronously on the rotation axis. Figure 14.6 depicts the carbonyl groups stretching synchronously along the x and y axes. Figure 14.7 depicts the carbonyl groups stretching asynchronously on the rotation axis. Finally, Figure 14.8 depicts the carbonyl groups stretching asynchronously along the x and y axes. If these images have correctly identified four unique active stretches, IR spectroscopy should indicate four peaks in the IR spectrum. This is because IR spectroscopy is measuring the energy that is absorbed when there is a change in the dipole moment. And that is what we see in Figure 14.9.

We have just demonstrated one particular instance in which there is a correspondence between what a molecular model predicts IR spectroscopy should indicate and what IR spectroscopy does, in fact, indicate. However, this type of correspondence between what a molecular model predicts and what IR spectroscopy indicates *can be demonstrated for every known molecule*. Moreover, similar correspondences can be demonstrated for other characterization methods, including Raman spectroscopy and X-ray diffraction analysis.

To review, chemists are able to collect spectroscopic data on every known molecule that correlates with how chemists have identified each molecule's symmetry elements. Moreover, the classification of molecules according to their symmetry elements is a direct function of how molecules are understood to be oriented in space by virtue of their atomic structure. To put it bluntly, if chemists are mistaken about molecular structure, they would not be able to gather the precise spectroscopic data that correlate with a molecule's symmetry elements. But they are able to gather such precise, correlating data, which demonstrates that they are not mistaken about molecular structure.

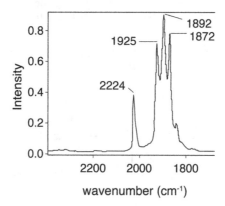

Figure 14.9 Infrared spectrum of *cis* forms of tetracarbonyl molybdenum complex (based on the spectrum from Nareetsile 2005).

3.2 Microscopy

Current microscopy can include three main types: standard-magnitude light microscopes, high-magnitude electron microscopes, and scanning probe microscopes. The first electron microscopes were developed in the 1930s, which allowed scientists to increase the magnitude of microscopes by a 1,000-fold compared with light microscopes. This ability to see objects down to the order of less than a nanometer is due to using a different type of electromagnetic radiation than visible light. In the case of electron microscopes, a beam of electrons is used to create an image because electrons have a shorter wavelength than visible light and thus, the resolution is greater (Bozzola and Russell 1999, 4–5). Instead of the glass lenses that are used in the light microscopes, these microscopes use solenoid magnetic lenses (163). There are two common types of electron microscopes: transmission electron microscope (TEM) and scanning electron microscope (SEM). The TEM creates an image of a cross-section of a sample; electrons pass through the sample hitting a phosphorescent screen, much like how an X-ray works (9). The TEM resolution is down to 0.1 nm (150). In contrast, SEM scans the sample surface with electrons, and the scattered or diffracted electrons create the image. As a result, the SEM image shows surface details down to a resolution of 20–50 nm (204). The scanning probe microscopes were first developed in the 1980s. There are two general types of scanning probe microscopes: scanning tunneling microscope (STM) and atomic force microscope (AFM). Scanning probe microscopy produces images of the material surface by measuring the interactions between the probe tip and the sample surface (Chatterjee, Shrikanth, and Tapas 2010, 624). Molecular structures, including atoms, have been imaged using these types of microscopy.

A Berkeley lab made the science news in 2013 when researchers using AFM were able to image a single molecule in a chemical reaction (Sanders 2013). The molecule, oligo-(phenylene-1,2-ethynlene) ($C26H14$), has different possible arrangements of the atoms. These different arrangements all have the same number and type of atoms (26 carbon atoms and 14 hydrogen atoms), and they can change arrangement when heated. The reactant and products of the reaction were imaged using STM and AFM (Figure 14.10).[5] There are two images and a molecular model on the left side of the arrow representing the reactant molecule. The top image is produced from STM. The overall shape of the molecule can somewhat be seen, but clearer areas of the molecules are shown in the second image that is produced from an AFM. The bottom molecular model is how chemists represent the reactant. There are similarities between the molecular model and the two microscope images. The overall shape of the molecule is similar in both the model and the images. In the second AFM image, the overall shape of the molecule is similar in both the

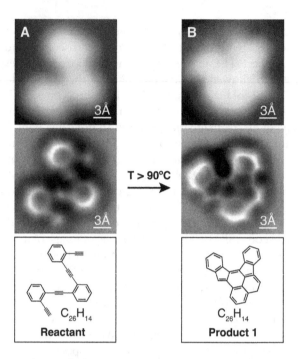

Figure 14.10 Before and after reaction images using scanning tunneling microscope (STM) (top), atomic force microscope (AFM) (center), and classic molecular structure models (bottom). Used with permission from UC Berkeley and Sanders 2013; based on work by Felix R. Fischer and Michael F. Crommie discussed in de Oteyza et al. (2013).

model and the images, including imaged regions of high electron density that correspond to bonds in molecular models. Further, the right side of the reaction (the product molecule) is imaged with the same two types of microscopes used to image the reactant molecule; the bottom picture is the molecular model that chemists use to model the product molecule. The microscope images of the reactant and product molecules look different as well, showing a different arrangement of the atoms.

The *cis* and *trans* forms of the precursor to the reactant (1,2-bis (2-ethynylphenyl)ethyne) were also imaged, confirming the predicted symmetry point groups of each form (C2h has *trans* conformation, and C2v has *cis* conformation: see Figure 14.11). The two forms of the molecule are imaged using AFM (labeled A). The top molecule is the *trans* form of the molecule shown as the molecular model in B, and the *cis* form is the bottom molecule in the image that matches the representation modeled in C. The AFM can differentiate between stereoisomers, which are molecules that only differ in their orientation in space.

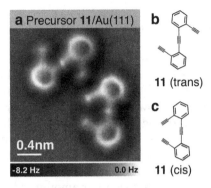

Figure 14.11 Cis and *Trans* forms of precursor molecules captured in (a) AFM image that match the molecular structure models (b and c) (Gorman 2016, 32).

Both Figures 14.10 and 14.11 serve as examples where the modeled structure of a molecule is confirmed through microscopy. High-magnitude microscopes are able to generate images of these molecules that bear a very striking resemblance to molecular models, even with respect to capturing the electron density of double bonds. If chemists had modeled molecular structure incorrectly, then it is very likely that images generated by high-magnitude microscopes would not correspond to the molecular models. However, images and models do correspond, indicating that chemists have, indeed, modeled the molecular structures correctly.

4 The Formal Argument, Objections, and Replies

4.1 The Correspondence Argument

The following argument summarizes and makes explicit our assessment of the evidential force of data from spectroscopy and microscopy for realism about molecular structures:

4.1.1 The Correspondence Argument

1 The best explanation for the correspondence of data from IR spectroscopy with predictions based on models of molecular structure is that the molecular models are accurate.
2 The best explanation for the correspondence between models of molecular structures and emerging images from high-magnitude microscopes is that the molecular models are accurate.
3 Given that the accuracy of molecular models is the best explanation for the high quality and quantity of data from IR spectroscopy and

high-magnitude microscopy, realism *about molecular structures is justified*.

4   Therefore, realism about molecular structures is justified.

Premises (1) and (2) are claims about what constitutes the best explanation for particular observations, and they are the sorts of claims that are common in abductive inferences. Providing a robust defense of these premises would involve showing how the hypothesis that molecular models are accurate fares against competing hypotheses for explaining the relevant observations. We will address this in more detail below when we consider both the use of and the objection to abductive inference. However, before proceeding, more needs to be said about our use of the term "accurate," which does a fair amount of heavy lifting for us.

The concept of accuracy we have in view is tied closely to the type and level of realism for which we are arguing. Molecular models depict a specific number of atoms of various kinds bonded together in a way that entails that molecules have a particular shape in three-dimensional space. A molecular model is accurate to the degree that it accurately describes how a molecule of the substance it is intended to model is, in fact, "put together"—that is, put together with respect to the number and kind of atoms that constitute a molecule of that substance, and how they are bonded together to form a three-dimensional structure. For example, consider the typical model of a carbon dioxide molecule that depicts one carbon atom double-bonded to two oxygen atoms to form a linear, three-dimensional structure. The model's accuracy is a function of whether a typical, actual, molecule of carbon dioxide is, in fact, constituted by one carbon atom double-bonded to two oxygen atoms in the linear fashion depicted by the model.[6]

Whether or not one is inclined to think that premise (3) is true will likely depend on the degree to which one is amenable to abductive inference. That is because (3) is a particular instance of a more general epistemological principle of justification arising from inference to the best explanation:

Abductive Justification Principle: *H*'s being the best explanation for *O* provides epistemic justification for *H*.

The amount of justification conferred on *H* in virtue of *H*'s being the best explanation for some *O* will depend on a number of factors, including (a) how confident are we that the observation in question is accurate, and (b) the width of the "explanatory gap" between *H* and its nearest competitor *H**. We think that (3) fares very well in light of factors (a) and (b).

With respect to (a)—the degree of confidence in the accuracy of observation—the observations in question are twofold: the observation of correspondence between the observed data from IR spectroscopy and predictions based on models of molecular structure, and the observation of correspondence between the observed images of molecular structures from high-magnitude microscopes and the models of those molecular structures. With respect to both observations of correspondence, we can have a very high degree of confidence that the correspondence obtains. In the IR spectroscopy example we looked at, there is a straight line between the molecular model predicting four active vibration bands and four bands being recorded. In the high-magnitude microscopy example, the correspondence between model and image is plain to see. The accuracy of molecular models also fares well with respect to (b)—the explanatory distance between H and its nearest competitor. In fact, it is hard to come up with remotely plausible alternatives for explaining the correspondences described in (1) and (2). It being challenging to come up with remotely plausible alternative explanations by itself does not guarantee that (1) and (2) are true. But in the absence of plausible alternatives, the explanatory gap in this case between H and its nearest competitor is both deep and wide.[7]

4.2 Objections and Replies

4.2.1 Abductive Inference Is Weak Justification

Though strictly speaking, the Correspondence Argument is not an abductive inference (also called "inference to the best explanation"); premises (1) and (2) are claims that employ abductive reasoning in determining what constitutes the best explanation for some phenomenon. And someone might object that because the Correspondence Argument employs abductive reasoning, the type of justification conferred on the conclusion is provisional, defeasible, and low grade. As such, the objector concludes, the Correspondence Argument does not provide strong support for realism about molecular structures.

We agree with the objector that the justification provided for realism about molecular structures is provisional and, in principle, defeasible. However, those features of the argument are not unique to the Correspondence Argument and are just part of the nature of epistemic justification. However, provisionality and defeasibility are distinct from whether or not the quality of justification is low or high grade. One might think that abduction provides low-grade justification, because the strength of inferences to the best explanation involves appeals to philosophical criteria for determining whether one explanation is better than the other, and that these criteria are disputed.

Our response here is twofold. First, typical candidates for properties relevant to identifying a best explanation include fit with empirical data, fit with background knowledge, and simplicity, and all three of those properties fit with the abductive claims in the Correspondence Argument. Second, the best type of inferences to the best explanation are the ones where the link between the postulated best explanation and the empirical phenomena is strong. Consider the well-known example of Urbain Leverrier's discovery of Neptune. Leverrier postulated the existence of a then unseen planet, as this would best explain observed perturbations in the orbital path of Uranus. Leverrier "sent his results to Johann Galle of the Berlin Observatory. In a matter of hours in September of 1846, the same day he received Leverrier's letter, Galle found the new planet, Neptune."[8] In this case, "same day results" reflected a tight connection between the proposed explanation and the empirical confirmation, as Galle confirmed that the existence of Neptune was, indeed, the best explanation for perturbations in Uranus's orbit.

Notice, however, that the connection between the proposed explanation and the empirical confirmation in (1) and (2) of the Correspondence Argument is even stronger and more comprehensive than as evidenced in the Neptune case chronicled above. For example, there are many, many known molecules and thus many, many known correspondences between molecular models, their corresponding symmetry elements, and what the model predicts concerning how the molecule will respond to IR spectroscopy. If any instance of abductive reasoning can yield high-grade justification, this example is one such instance.

4.2.2 Realism About Molecular Structures Requires More Data and Ontological Commitment to Subatomic Particles

Here, the objector says that until one can resolve controversies about orbitals and quantum ontology, one should not be a realist about the molecular structures involving atoms. This objection begins by noting that our best understanding of the ultimate constituents of matter are the subatomic particles specified by the Standard Model of particle physics. The objector continues by noting that our best understanding of how these particles interact at the quantum level are the principles of quantum mechanics and the quantum field theory. As a result, the objector continues, a robust ontology of subatomic particles, orbitals, and quantum fields needs to be incorporated in any attempt to be realists about molecular structures.

We agree with the spirit of this objection in that realism about molecular structures, as we have defined it, would be strengthened if it could be embedded in a plausible and well-defined realism concerning orbitals and quantum fields. However, noting that realism about molecular structures would be strengthened even more if embedded in a plausible realism "one level down" in no way undermines the considerable strength of realism about molecular structures understood at the level of atoms. In the absence of any argument to the contrary, there is no reason to think that realism about molecular structures requires realism about, say, orbitals or quantum fields.

4.2.3 *The Constructive Empiricist Objection*

Here, the objector takes issue with the concept of "accuracy" we have employed in the premises of the Correspondence Argument. The objector points out that one could substitute "accurate" with "empirically adequate" in premises (1) and (2) of the argument, *salva veritate*. So doing would result in the following:

1* The best explanation for the correspondence of data from IR spectroscopy with predictions based on models of molecular structure is that the molecular models are empirically adequate.

2* The best explanation for the correspondence between models of molecular structures and emerging images from high-magnitude microscopes is that the molecular models are empirically adequate.

The objector continues that in comparison to their predecessors, (1*) and (2*) are more plausible, in part because they claim less, ontologically, with respect to unobservable atoms. However, (1*) and (2*) do not provide support for (3), and as such, our conclusion in support of realism about molecular structures is unjustified.

Space limitations prevent a more robust interaction with this objection than we can provide. However, there are two promising lines of response available to rebut the constructive empiricist objection. First, whether or not empirical adequacy as opposed to accuracy really best explains the correspondences described in the premises of the argument will be a function of the concept of explanation in view. Second, how one understands empirical adequacy with respect to the relevant correspondences and the theories in which they are embedded will depend on how one understands the "observable/unobservable" distinction. Several have argued, contra Bas van fraassen, that the standard constructive empiricist distinction between observable and unobservable breaks down when considering technologically assisted observations of the sort considered in our paper.[9]

5 Conclusion

In this chapter, we have defended a version of realism about molecular structures. We began by articulating how we understand realism more broadly—as a type of selective scientific realism about the entities modeled by the molecular structure theory. After surveying the development of the structure theory, we argued that both the evidence from molecular symmetry and IR spectroscopy, and the evidence from high-magnitude microscopy, yields a very precise correspondence between the measurements predicted by molecular models and the measurements obtained through these methods. These correspondences served as the basis for the formal argument we presented in the final section of our chapter: the Correspondence Argument, the conclusion of which is that realism about molecular structures is justified.

Acknowledgments

Work on this project was supported by a grant from the John Templeton Foundation. Many thanks to participants in the SURe 2020 conference who offered helpful suggestions and comments. Thanks are especially due to Julia Bursten and Elay Shech who offered substantial and informative feedback.

Notes

1 For surveys of the historical and contemporary ways of understanding the chemical bond, see Hendry (2008, 2012); Weinberg (2008); and Sutcliffe and Wooley (2012).
2 For a discussion of realism/anti-realism as it pertains to orbitals, see Scerri (2001). For a survey of issues connected to the metaphysics of quantum mechanics, see Lewis (2016).
3 For a discussion of support for realism based on converging lines of evidence, see Section 2.2 of Chakravartty (2017); also, see Salmon (1984, 213–227) for a similar discussion concerning the overdetermination of Avogadro's constant and Norton (2000) for how the overdetermination of the mass to charge ratio supports existence claims about electrons.
4 An accessible overview of this example can be watched at https://www.jove. com/v/10442/application-of-group-theory-to-ir-spectroscopy.
5 For details surrounding the preparation of the reactant including the proposed chemical mechanism as well as the modeling predictions of the products, see Gorman (2016).
6 See the carbon dioxide molecular structure on the PubChem website: https:// pubchem.ncbi.nlm.nih.gov/compound/Carbon-dioxide#section= 2D-Structure.
7 According to P. Kyle Stanford's Problem of Unconceived Alternatives, there is reason to think that the present inability to conceive of plausible alternatives to some theory will be overcome at some future time (2006). We respond to Stanford's argument in Nichols and Penner (2021, 141–145).
8 See Cushing (1998, 153).
9 See Alspector-Kelly (2004); see also Vollmer (2000).

References

Alspector-Kelly, Marc. 2004. "Seeing the Unobservable: Van Fraasen and the Limits of Experience." *Synthese* 140: 331–353. DOI: https://doi.org/10.1023/ B:SYNT.0000031323.19904.45.
Bozzola, John J., and Lonnie D. Russell. 1999. *Electron Microscopy: Principles and Techniques for Biologists*. 2nd ed. Sudbury, MA: Jones and Bartlett.
Chakravartty, Anjan. 2017. "Scientific Realism." *The Stanford Encyclopedia of Philosophy*. Edward N. Zalta (ed.), Summer 2017 Edition. https://plato. stanford.edu/archives/sum2017/entries/scientific-realism/.

Chang, Hasok. 2016. "Scientific Realism and Chemistry." *Essays in the Philosophy of Chemistry.* Eric Scerri and Grant Fisher (eds.), Oxford: Oxford University Press, pp. 234–252.

Chatterjee, Snehajyoti, Shrikanth S. Gadad, and Tapas K. Kundu. 2010. "Atomic Force Microscopy – A Tool to Unveil the Mystery of Biological Systems." *Resonance* 15 (7): 622–642. DOI: https://doi.org/10.1007/s12045-010-0047-z.

Cushing, James T. 1998. *Philosophical Concepts in Physics: The Historical Relation between Philosophy and Scientific Theories.* Cambridge: Cambridge University Press. DOI: https://doi.org/10.1017/CBO9781139171106.

de Oteyza, Dimas G., Patrick Gorman, Yen-Chia Chen, Sebastian Wickenburg, Alexander Riss, Duncan J. Mowbray, Grisha Etkin, Zahra Pedramrazi, Hsin-Zon Tsai, Angel Rubio, Michael F. Crommie, and Felix R. Fischer. 2013. "Direct Imaging of Covalent Bond Structure in Single-Molecule Chemical Reactions." *Science* 340 (6139): 1434–1437. DOI: https://doi.org/10.1126/science.1238187.

Gorman, Patrick. 2016. "Surface and Solution Mediated Studies of Small Molecule Enediyne Reactivity." PhD diss., University of California Berkeley.

Hendry, Robin F. 2008. "Two Conceptions of the Chemical Bond." *Philosophy of Science* 75 (5): 909–920. DOI: https://doi.org/10.1086/594534.

Hendry, Robin. 2012. "The Chemical Bond." *Philosophy of Chemistry.* Andrea I. Woody, Robin F. Hendry, and Paul Needham (eds.), Oxford, UK: North Holland, pp. 293–308. DOI: https://doi.org/10.1016/B978-0-444-51675-6.50022-0.

Ladyman, James. 2020. "Structural Realism." *The Stanford Encyclopedia of Philosophy.* Edward N. Zalta (ed.) (Winter 2020). https://plato.stanford.edu/archives/win2020/entries/structural-realism/.

Lewis, Peter J. 2016. *Quantum Ontology: A Guide to the Metaphysics of Quantum Mechanics.* Oxford: Oxford University Press. DOI: https://doi.org/10.1093/acprof:oso/9780190469825.001.0001.

Nareetsile, Florence M. 2005. "Solventless Isomerisation Reactions of Six Coordinate Complexes of Ruthenium and Molybdenum." PhD diss., University of the Witwatersrand.

Nichols, Amanda J. and Myron A. Penner. 2021. "Selective Scientific Realism and Truth-Transfer in Theories of Molecular Structure." *Contemporary Scientific Realism.* Timothy D. Lyons and Peter Vickers (eds.), Oxford: Oxford University Press, pp. 130–158. DOI: https://doi.org/10.1093/oso/9780190946814.003.0007.

Norton, John D. 2000. "How We Know About Electrons." *After Popper, Kuhn, and Feyerabend: Recent Issues in Theories of Scientific Method.* Robert Nola and Howard Sankey (eds.), Dordercht, NL: Kluwer Academic Publishing, pp. 67–97. https://doi.org/10.1007/978-94-011-3935-9_2.

Rosenthal, Jenny E. and G. M. Murphy. 1936. "Group Theory and the Vibrations of Polyatomic Molecules." *Reviews of Modern Physics* 8: 317–346. DOI: https://doi.org/10.1103/RevModPhys.8.317.

Salmon, Wesley C. 1984. *Scientific Explanation and the Causal Structure of the World.* Princeton, NJ: Princeton University Press.

Sanders, Robert. 2013. "Scientists Capture First Images of Molecules before and after Reaction." *Berkeley News,* May 30, 2013. https://news.berkeley.edu/2013/05/30/scientists-capture-first-images-of-molecules-before-and-after-reaction/.

Scerri, Eric R. 2001. "The Recently Claimed Observation of Atomic Orbitals and Some Related Philosophical Issues." *Philosophy of Science* 68 (3): S76–S88. DOI: https://doi.org/10.1086/392899.

Scerri, Eric R. 2012. "The Periodic Table." *Philosophy of Chemistry*. Andrea I. Woody, Robin F. Hendry, and Paul Needham (eds.), Oxford: North Holland, pp. 329–338. DOI: https://doi.org/10.1016/B978-0-444-51675-6.50024-4.

Stanford, P. Kyle. 2006. *Exceeding Our Grasp: Science, History, and the Problem of Unconceived Alternatives*. Oxford: Oxford University Press. DOI: https://doi.org/10.1093/0195174089.001.0001.

Sutcliffe, Brian T., and R. Guy Wooley. 2012. "Atoms and Molecules in Classical Chemistry and Quantum Mechanics." *Philosophy of Chemistry*. Andrea I. Woody, Robin F. Hendry, and Paul Needham (eds.), Oxford: North Holland, pp. 387–426. DOI: https://doi.org/10.1016/B978-0-444-51675-6.50028-1.

Vincent, Alan. 2001. *Molecular Symmetry and Group Theory: A Programmed Introduction to Chemical Applications*. Chichester, UK: John Wiley & Sons.

Vollmer, Sara. 2000. "Two Kinds of Observation: Why van Fraassen Was Right to Make a Distinction, but Made the Wrong One." *Philosophy of Science* 67: 355–365. DOI: https://doi.org/10.1086/392785.

Weinberg, Michael. 2008. "Challenges to the Structural Conception of Chemical Bonding." *Philosophy of Science* 75: 932–946. DOI: https://doi.org/10.1086/594536.

Wigner, E. P. 1959. *Group Theory and Its Application to the Quantum Mechanics of Atomic Spectra* (J. J. Griffin, translator). London: Academic Press.

Zuo, J. M., M. Kim, M. O'Keeffe, and J.C.H. Spence. 1999. "Direct Observation of *d*-orbital Holes and Cu-Cu Bonding in Cu2O." *Nature* 401: 49–52. DOI: https://doi.org/10.1038/43403.

15 Anti-Fundamentalist Lessons for Scientific Representation from Scientific Metaphysics

Julia R. S. Bursten

1 Introduction

One question worth asking about scientific representation is, "What must the target system, and the world, be like in order for a representation to successfully represent?" This question connects investigation in scientific representation to investigation in scientific metaphysics. It is that connection that will be the subject of this chapter. Regardless of one's metaphysical orientation, it is undeniable that one's beliefs about what counts as successful scientific theorizing and modeling will influence one's beliefs about what science is able to reveal about the ontological structure of the world. Equally undeniable is the banal observation that different answers to the question of what counts as successful representation will license different inferences about what scientific representations can reveal what the world is like.

Consider scientific realism. Many early presentations of scientific realism aimed at defending the truth of statements of the scientific theory and the reality of unobservable objects postulated by those theories. The attention paid to the role of unobservables in justifying theoretical claims led many early realists to draw connections between realism and reductive approaches to inter-theory relations: it was atoms, not tables, whose reality early scientific realists were concerned with defending.

A frequent, though not universal, implication of this attention was that ontology could, and should, be read off the most "fundamental" scientific theories available. When applied to considerations of representation, such a view suggests that representations of unobservables should operate in a fashion analogous to representations of observables, and further that well-wrought representations of unobservables ought to, or at minimum can, serve as maps or guides to ontology. The result of such a chain of reasoning is an approach to representation and scientific metaphysics that suggests that one's ontology is best represented in the contents of our most fundamental physical theories.

This sketch is, of course, a caricature of an extreme version of fundamentalism about ontology. More common are more moderate views that tolerate, or even embrace, theoretical objects from less fundamental

DOI: 10.4324/9781003202905-17

theories. But the caricature is intended to gesture toward a pernicious intuition that persists even in those more moderate views in the literature, namely, that representing reality via scientific theories and models is best accomplished by what Barker and Kitcher (2014) term a "reductionist strategy."[1]

My aim in what follows is to articulate, and to recommend, an approach to scientific metaphysics that is actively anti-reductive. To put the point in Barker and Kitcher's terms, I mean to recommend that research on higher-level objects is sometimes hindered by considering the constituents of those objects and what the scientific investigation of the constituents can tell us about them. Ultimately, and somewhat counterintuitively, I think that this anti-reductive strategy is compatible with certain reductive strategies (that "might" in Barker and Kitcher's definition does a fair amount of work for their view). However, the reductive, or fundamentalist, approach has had an extensive and pernicious influence on the development of scientific metaphysics in a way that the anti-reductive approach has not, and so my emphasis here will be on giving air time to the anti-reductive approach.

Thus, I do not aim at dismantling the fundamentalist intuition behind the reductive strategy definitively. Instead, I simply aim at offering some evidence against it and at highlighting some advantages of alternative routes; this is a project of suggestions. I draw on two recent accounts of scientific metaphysics from philosophers working across a variety of scientific domains and philosophical traditions. Each of these accounts has what I shall call "anti-fundamentalist" leanings: they reject the premise that the fundamental physical theory is the appropriate or best source material for scientific metaphysics. These leanings also engender the use of anti-reductive strategies in articulating the ontology of a science.

In the first account, C. Kenneth Waters (Waters 2017, 2018) analyzes the concept of a gene in order to develop what he calls the "No General Structure Thesis," which calls into question the association between generality and fundamentality in scientific metaphysics. Waters uses this account to critique the variant of scientific realism known as structural realism. Structural realism has historically drawn its notions of generality from fundamental physics, and Waters aims at showing the limits of that strategy through an analysis of generality as conceived through the lens of biology.

The second account is from Robert Batterman's recent monograph (Batterman 2021). Therein, Batterman offers a further critique of fundamentality from within physics itself. He argues that physical theories from many-body physics, rather than fundamental physics, should be used as the source material for scientific metaphysics. I will emphasize, in particular, the use of "minimal-model explanations" in Batterman's account, which builds on work that he began with Collin Rice (Batterman and Rice 2014).

Both Waters' and Batterman's accounts foreground the role of scale in defining ontological categories, and both reject the reductionist ideal that the stuff at the smallest scale is the most fundamental, the most general, or the most real. Consequently, both reject the ideal that the stuff at the smallest scale should be taken as the basis of a scientifically-informed ontology. These are not the first anti-fundamentalist accounts of scientific metaphysics: Nancy Cartwright's (Cartwright 1999) dappled world and John Dupré's disunity of science (Dupré 1995), for instance, may be seen as influential progenitors of both, and Michela Massimi's perspectival realism is a contemporary fellow traveler (Massimi 2018, 2019, 2022). I focus on Waters and Batterman here because they both emphasize a connection between structure, scale, and (non-) fundamentality that I believe plays an important role in subverting the reductionist intuition underlying much philosophical work on the connection between representation and realism.

A final prefatory remark: in Waters' and Batterman's rejections of fundamentalism, and in the analysis below, one may read implications for both the content of philosophical positions in the literature on scientific metaphysics and the method by which those positions are approached, articulated, and defended. Some philosophers endeavor to distinguish fundamentalism about the contents of an ontology from fundamentalism about the method of obtaining that ontology. It is my view that these two varieties of fundamentalism are more intimately interconnected than many such disambiguations tend to take into account, and I believe that Waters' rejection of general structure, in particular, recognizes this interconnection. As such, the remarks that follow are "lumpy," that is, I do not make a special effort to identify the influence of one type of fundamentalism or another as I suggest reasons for rejecting both.

2 Fundamental Physics and Realism about Structure

In this section, I review some tenets of the contemporary group of metaphysical positions that fall under the umbrella of "structural realism." Structural realism is hardly a pinnacle of a fundamentalism in contemporary scientific metaphysics: it was developed in part as a response to certain dissatisfactions with ontologies that more fully embraced the reductive intuition I articulated above. So, it might seem an unusual place to begin a discussion of fundamentalism today. However, structural realism illustrates how the reductive strategy influences even moderately non-fundamentalist ontologies. In addition, since Waters frames his own thesis in contradistinction to certain claims made by one formulation of structural realism, an introductory discussion of structural realism makes it easier to explicate Waters' views below. This review is not intended as a primer on structural realism nor a critical analysis of that family of views.

John Worrall advanced his initial formulation of ontic structural realism in "Structural Realism: The Best of Both Worlds?" (Worrall 1989). The view he put forward then, and which he has defended since, came about as a conciliatory alternative to both scientific realism, the view that the objects posited by scientific theories exist, and scientific anti-realism, a collection of views that range from agnosticism about to denial of the existence of unobservable objects of the scientific theory. The debate is often framed in terms of key arguments presented by each side, namely the no-miracles argument for realism and the pessimistic meta-induction to anti-realism.

The no-miracles argument urges that realism about the unobservable objects of the scientific theory is the best explanation for the success of scientific enterprises, as it is the only explanation that does not make the success of science a miracle. Contrariwise, the pessimistic meta-induction argues (one might say, from hubris) that the vast majority of scientific claims made up to this point are no longer believed, so it is unlikely that current theories have got things right. Therefore, the present state of science gives us no reason to expect that the claims of our theories, such as about the objects that they posit, are true.

Both arguments are often given by way of examples from the history and current state of various sciences. A majority of these examples are physical. This is one of the ways in which naturalistic metaphysics has been historically tied to the physical sciences. For instance, Worrall cites the accuracy of quantum-theoretical calculations of the observed Lamb shift between the 2s and 2p energy levels of hydrogen as evidence toward the no-miracle argument, and the shift from Newtonian to Einsteinian conceptions of gravity as evidence toward the pessimistic meta-induction. Worrall's own structural realism is also demonstrated by way of physical examples from the history of nineteenth-century optics, as it moves from Fresnel's to Maxwell's conception of the phenomena associated with the propagation of light.

Worrall argues that despite changes in the conception and explanation of optic phenomena, a set of mathematical relations persisted. More generally, he argues that these relations are what is preserved over theory change. This is the foundational principle of structural realism, that mathematical or structural relations are preserved over theory change. This approach is meant to reconcile the no-miracles argument with the pessimistic meta-induction. Satisfying the pessimistic meta-induction, we have no reason to expect that the particular phenomena posited by the present scientific theories are real; whereas satisfying the no-miracles argument, we have a non-miraculous explanation for the success of science, namely that science has empirically discovered certain *structural relationships* that really are the stuff of the natural world—and so should be the stuff of ontology.

In Worrall's original formulation and in the canonical development of ontic structural realism[2] in the work of Ladyman, Ross, and Spurrett, structural realism continues to develop by suggesting that fundamental physical theories are the ones that capture the most general structural relationships, and are therefore the best source material for scientific metaphysics. This line of reasoning is evident in Ladyman and Ross's (2007). Therein, Ladyman, Ross, and David Spurrett endorse a condition on scientific metaphysics that they call the Primacy of Physics Constraint, which states, "Special science hypotheses that conflict with fundamental physics, or such consensus as there is in fundamental physics, should be rejected for that reason alone. Fundamental physical hypotheses are not symmetrically hostage to the conclusions of the special sciences" (Ladyman and Ross 2007, p. 44). This is a contemporary philosophical justification of the grounding of scientific metaphysics in fundamental physics—and quite an explicit one, at that.[3]

The Primacy of Physics condition does not exclude "higher-level" physical phenomena from the ontology of physics, as I discuss below. Indeed, Ladyman and Ross's attention to what Shech and McGivern have labeled as the "scale-relativity" of ontology (Shech and McGivern 2019) is one of its great strengths. However, the Primacy of Physics Constraint is an excellent example of how Barker and Kitcher's reductive strategy has been implemented in practice, and as such, it illustrates the influence of the fundamentalist intuition behind the reductive strategy on the formulation of even relatively moderate views in contemporary scientific metaphysics.

In his work analyzing generality from the lens of biology, Waters aims at undoing the association between generality and fundamental physics, and thereby at challenging the Primacy of Physics Constraint. I summarize this work next.

3 Waters' No General Structure Thesis

Waters uses the term "traditional scientific metaphysics" (Waters 2017, p. 85) to pick out views in which the results of scientific investigation are interpreted to inform metaphysical inquiry. Traditional scientific metaphysics uses fundamental physics, by which he means "the most basic theoretical results of physics" (Waters 2017, p. 84) as its source material, and this preference is said to be connected to the supposed generality of fundamental physics. He contrasts this emphasis on "basic theoretical results" with the relatively smaller role that theoretical results from biology have played in traditional scientific metaphysics. In its most straightforward formulation, Waters' No General Structure Thesis (NGST) states: "the world lacks a general, overall structure that spans scales" (Waters 2017, p. 83).

Of the four words titling the view, the one that receives the lion's share of Waters' attention is *general*. Waters is not interested in denying that the world has structure—indeed, he identifies many structures that scientific metaphysics suggests are in the world—but he is interested in denying that there is an overarching, general, or top-down structure of the world to be read off the results of science. Even more strongly, he asserts that taking the results of certain scientific investigations seriously implies that there is *not* a general structure in the world.

Waters develops his view by discussing concepts of the gene. He offers a contrastive analysis comparing classical gene concepts with molecular gene concepts. For pre-1950s research in classical genetics, he argues, a central aim of investigation was to track stable causal relations. Even though many researchers believed the gene to have internal physical structure, tracking these relations did not require understanding or representation of the supposed internal physical structure of a gene. Instead, it required the "difference principle": the idea that the differences in genetic makeup cause differences in phenotype.

Waters' difference principle is an example of what Angela Potochnik has termed a "causal pattern" (Potochnik 2017), and it is a representation of a relationship between phenomena that both Potochnik and Waters (like Woodward before them (2003)) have striven to distinguish from *structural* relations. Waters takes pains to highlight that the difference principle does not care about the internal physical makeup of a gene, and that "the structure of the world that geneticists were manipulating and investigating was not directly reflected in the structure of their concepts and theories" (Waters 2017, p. 91). He argues that the success of the classical gene concept despite its disregard of physical structure poses a problem for structural realism.

Ultimately, Waters concludes (following the received view in the contemporary philosophy of biology) that there is no such thing as *the* concept of a gene. However, the various gene concepts are more causal than they are structural. Consequently, genes are not the fundamental structures of biology; further, biology is not the kind of thing that has fundamental structures. This analysis is used as empirical evidence against the idea that the world is arranged into fundamental, natural classes of structures that can be accessed via traditional scientific metaphysics. Rejecting the fundamentality of structure in biology further enables Waters to reject the generality of structure for traditional scientific metaphysics, that is, the idea that whatever the fundamental structural relationship of the physical world turns out to be, it spans across length scales and thereby guides the unfolding of the world from the very small to the very large. (Although he focuses on length scales, presumably there are analogous arguments to be made for time and energy scales.)

Waters offers a cartographical metaphor to characterize what he sees as the difference between structure and general structure: some cities have overall or general structure, whereas others do not. Cities with a general structure—like Manhattan, Beijing, and Waters' own Calgary, Alberta—have street systems that make it easy to navigate from one part of town to another, using information about street layout and naming conventions. Cities lacking such structure—like Delhi, Cairo, and Pittsburgh, Pennsylvania—tend to be more difficult to navigate. Waters does not name or expand upon the type of structure that some cities have and others lack, other than to call it general. I will expand on this metaphor slightly, in order to develop a point about how structure interacts with scale in urban planning and beyond.

Although there are a few strategies for street layout in urban planning, one of the most ancient and most common is the grid. The streets of Manhattan are famously laid out in a grid consisting of numbered avenues running north–south with the number increasing to the west, crosscut by east–west-running streets with the number increasing to the north. This allows for inductive projection from one's present street location to the location of nearby streets, and even to the location and distribution of streets in other parts of the city.

Pittsburgh, on the other hand, is a mess wrought by geography and time. Grid layout and street naming conventions vary by neighborhood. Main thoroughfares run sort-of parallel to the banks of the two rivers that converge around the city's downtown district, which means they often are angled at around 40° to each other. A road called Beechwood Boulevard snakes through multiple neighborhoods, varying from north–south to east–west and back again, while also overtaking blocks of other roads. Further, the hilly topography of the city makes certain roads appear to intersect on maps, whereas one may be 100 yards higher than another. This topography also creates a system of cross-neighborhood shortcuts that might be likened to the warp pipes in Super Mario videogames. Unlike in Manhattan, in most areas of Pittsburgh, little can be induced about nearby neighborhoods or how to get to them from one's present street address.

Although many Pittsburghers take pride in the convolutions of their city's street network,[4] some might object that there are portions of the city, such as the Strip District, that do conform to a grid plan. Likewise, the residents of Inwood in upper Manhattan might object that their portions of the city are not governed by a numbered grid. These exceptions illustrate (1) that street grids exist at a variety of scales and (2) that grids are not identical with Waters' notion of overall or general city structure. Manhattan happens to have a general structure, which happens to be a grid. Pittsburgh lacks a general structure, but it employs grids in certain sectors. Other cities, like Paris and Washington, D.C., have large-scale wheel-and-spoke street networks that divide the city into wedge-shaped districts. Still others, like Beijing and Santiago and Houston, employ ring roads that enclose a mixture of wheel-and-spoke and grid networks, alongside geographically-constrained networks that look nothing like either grids or wheel-and-spokes.

This navigational meditation complicates Waters' metaphor by pressing on the question of which among these street networks count as having—and which count as lacking—a general structure. I am quite sympathetic to Waters' intuition that cities whose street networks confer ease of navigation through layout and naming conventions are networks that can be said to have a general or overall structure. However, street networks can be the size of a neighborhood or the size of a nation. The U.S. Interstate Highway system, though not a grid, aims at following numbering conventions akin to those in Manhattan: east–west routes take even numbers, ascending from south to north, whereas north–south routes take odd numbers, ascending from west to east. I suspect that Waters would concede that it has a general structure; likewise, I expect that he would admit the same of Pittsburgh's Strip District or D.C.'s system of numbered and lettered streets, crisscrossed by diagonal avenues and sectioned into quadrants. Overall or general structure can (generally) be found, if one selects an appropriate scale or resolution to search for it.

This latter point is friendly to Waters, who ultimately concludes that "the world" lacks a general structure *that spans scales*. It is possible to define a system by reference to a general structure that defines it only if the boundaries of the system are selected such that the general structure applies to it. For Waters, it is an empirical result that the system defined as "the world" lacks a general structure. Geographically, this seems true, although Waters' "world" is more one of biology than geography. An upshot of NGST is the ability to reconceive generality in terms of scale-dependent conceptions of reality.

His alternative articulates that attention should be paid to the length scales at which systems are investigated in order to assess the applicability of structural approaches to those systems. I agree. In fact, I think that Waters does not go far enough in claiming only that general and scale-invariant notions of structure fail in biology. They fail in physics, as well. Batterman's work on the importance of mesoscale descriptions of phenomena in many-body physics illustrates this point. I summarize some of this work in the next section.

4 Minimal-Model Explanations and Mesoscale Metaphysics

Batterman and Rice (2014) contrast minimal-model explanations with a class of theories of explanation that they call "common-features accounts." Common-features accounts, in their view, are theories of explanation in which a model explains "in virtue of meeting some 'accuracy' or 'correctness' conditions," (Batterman and Rice 2014,

p. 356) by which they mean that representational features of the model correlate with represented features of the system. They count mechanistic and causal or difference-making accounts of explanation among common-features accounts, as well as mapping accounts of the role of mathematics in scientific explanation.

Their minimal-model explanations offer a different view of what makes a model explanatory. They argue that rather than pointing to a list of common features between the model and the target system as *justification* of explanatory efficacy, one must, in addition, be able to explain why heterogeneous details of the class of systems being modeled are irrelevant to an explanation of the class of systems. Their physical example comes from fluid dynamics and appeals to the renormalization group—a mathematical strategy for abstracting away from the details of a given represented system—to make their point.

In Batterman and Rice's (not uncontroversial) view, the renormalization group does not represent a common feature among a class of systems being modeled, nor does it signify a definable structural relation between the features of a system. Instead, it is a technique for eliminating the degrees of freedom in computational models. It produces a set of fixed points that are characteristic of the system not by drawing structural relations among the elements, but by eliminating many structural details that are *irrelevant* to understanding the system's behavior. They argue that processes like this delimit the universality class to which a system or set of systems belong. This strategy is not limited to physics: via a discussion of Fisher's sex ratios, they show that it also illustrates how minimal models work in the modeling of biological populations.

Minimal-model explanations disentangle the notion of invariance from that of structure, and they assign explanatory priority to the former. I believe that minimal-model explanations can cut at structural realism in two ways. First, Batterman and Rice show that many common-features accounts appeal to *idealized* structural relationships to explain phenomena. The role of idealization in such accounts calls into question the metaphysical underpinnings of those structural relationships. This is not a new point (Cartwright 1983, for instance, makes it). But what can be taken away here is that in order to maintain a metaphysics grounded in structural relations (physical or otherwise), structural realists will need to provide an account that distinguishes genuine, ontological structural relations from the merely useful ones that, nonetheless, support much of scientific practice.

Waters' NGST points to the despair of accomplishing such a task by highlighting that the generality sought in scientific metaphysics is unlikely to be found in worldly structural relations. Minimal-model explanations suggest further reasons to question the defensibility of structural realism, both by showing that structural relations need not be accompanied by metaphysical ones (as in the mathematical structure of the renormalization group) and by appealing to the features besides structural relations as the explanatory ones (as in explanations appealing to the difference principle). Minimal-model explanations show that the notion of structure—even in physics—is not so general as the structural realists need it to be.

There is an objection here for the structural realist: the physics from which the minimal models account draws is not *fundamental* physics, so it is subject to the same metaphysical asymmetry as the special sciences. Batterman's recent work in scientific metaphysics (Batterman 2021) provides a rejoinder. His examination of many-body and condensed-matter physics considers a wide variety of physical systems and analyzes the role of those systems' mesoscale features (as opposed to macro- or microscale features) in explaining how those systems work and why certain mathematical models are appropriate representations of those systems. He concludes that minimal models capture "natural properties" of the systems they model by representing correlations between the mesoscale features of the target systems. Further, he writes, "[i]t is this fact that justifies taking the mesoscale parameters as the most natural or most joint-carving with respect to the bulk behavior of many-body systems" (Batterman 2021, Ch. 7.3).

Batterman's analysis shows that using the fundamental theories of physics as source material is the wrong starting point for getting metaphysics out of physics. Instead, examining mathematical models of many-body physical systems reveals the explanatory power and ontological naturalness of the mesoscale parameters that represent those systems. Like Waters' NGST, Batterman's middle-out approach produces a rejection of the ideal of a top-down, generalist account of the relations and relata that comprise a traditional scientifically-grounded ontology. In minimal-model explanations, ignoring the fundamental physics of the systems is what allows for unification of many systems under a single universality class. This is an ontological result grounded in strict anti-reductive, anti-fundamentalist approaches, and it shows that in at least some cases, fundamentalism is incompatible with the ontology suggested by the representational contents of our best explanations of physical phenomena.

Waters, in particular, sees results like these as damning for structural realism, due to the fact that that view requires a top-down, generalist conception of "*the* structure" of the world (Waters 2017, p. 101, emphasis in the original) grounded in the "fundamental" science of physics. I agree strongly with the anti-reductive result, and I wish to emphasize the strides that can be made by recognizing and rejecting the influence of the reductive strategy on scientific ontologies. However, I believe that some substance (or a structure representing it, anyway) is still left in structural realism after the exorcism of the fundamentalist spirit. In the next section, I consider what rejoinders might be available for structural realists and identify a difference in the methodological attitude between Waters' and Batterman's work, on the one hand, and that of the structural realists, on the other.

5 Fundamentality and Conceptions of Scientific Metaphysics

Both Waters' NGST and Batterman's middle-out approach suggest that there is no scale-invariant notion of general structure to be gleaned from traditional scientific metaphysics. Following Waters, I have framed these results as problematizing structural realism, especially Ladyman and Ross's canonical formulation of ontic structural realism (Ladyman and Ross 2007). However, a bit more charity is due to the view, which will both offer an olive branch and provide a backdrop for some remarks on the role of scale in scientific metaphysics and scientific representation.

First, there is a reading of Ladyman, Ross, and Spurrett's view that may be able to accommodate Waters' NGST. Here is the view in their own words:

> All scientific disciplines except mathematics and, arguably, some parts of physics, study temporally and/or spatially bounded regions of spacetime. By this we mean that the data relevant to identifying real patterns are, for most disciplines, found only in some parts of the universe. Biology draws data only from regions in which natural selection has operated. [. . .] What accounts for the specific selection of spacetime regions, at specific scales, to which disciplines are dedicated is obviously, to a very great extent, a function of practical human concerns. Scientific institutions organized by dolphins wouldn't devote more than 40 per cent of their total resources to studying human-specific diseases. There is no puzzle as to why far more scientific attention is lavished on the dry parts of the Earth's surface than on the waterlogged parts. [. . . T]here are far more real patterns to be discovered by isolating specific spacetime regions, and scales of resolution on those regions, and treating them as relatively encapsulated from other regions and scales, on Earth than on the Moon. This was true before life on Earth began, and is the main part of the explanation for why life did begin on the Earth but not on the Moon; but the extent to which the ratio of real patterns to possible physical measurements on Earth has increased relative to that on the Moon has been made staggeringly great by the progress of first biological, and then social and cultural, evolution.
>
> (Ladyman and Ross 2007, pp. 45–46)

I want to attend to two features of this view. First, there is no claim to scale-invariance as a working notion of generality; indeed, quite the opposite, as much of what they say in the first paragraph indexes the subjects of their ontology (that is, the "real patterns"[5]) to particular length scales. To the extent that this view leaves room for a real pattern that is identifiable as a general structure, it would be one real pattern among many, and it would not necessarily be one that occupies center stage in most scientific theorizing. This is supported further by Ladyman and Ross's notion of scale-relative ontologies, which appears later in their monograph and which Ladyman, in particular, has further developed in recent years.

A thin notion of general structure is even recoverable from the cartographic metaphor: sure, it is easier to navigate Manhattan than Pittsburgh by using inferences about grids, road naming conventions, and city layout. But both Manhattan and Pittsburgh *have maps*. These maps are instances of real patterns just as much as the pattern of organs in a typical human or the pattern of electronic symmetries in ionized sodium, or the pattern of planets distributed throughout the solar system. On a thin reading of general structure, Pittsburgh's map is just as general as Manhattan's, because both describe street networks, even though Manhattan's street network could also be redescribed by a grid with labeled axes, and Pittsburgh's could not.

Viewing the Ladyman et al. claim as being about only a thin notion of general structure does, I believe, rescue the view from the letter of NGST, but not the spirit. This brings me to my second point. The description of how real patterns come to be limited in scope is noticeably top-down: it begins by invoking the image of all of spacetime, in order to point out that the most real patterns apply to the proper parts thereof. The examples—from biology, then economics, and then psychology and anthropology—are offered in order to capture the intuition that patterns exist at more general, then more specific, scales of resolution, despite that the inclusion of anthropology seems to defy the Matryoshka-like "nesting" of scales developed in the first three examples.

This top-down perspective suggests a methodological approach to scientific metaphysics that is certainly incompatible with the methodological underpinnings of Waters' NGST and Batterman's middle-out approach. Ladyman et al.'s description of the complexity of the Earth relative to the Moon underwrites this methodology: they consider candidates for real patterns by beginning from the scale of planets, and they work their way in; similarly, biological evolution begets social, which begets cultural. The asymmetric dependence of smaller-scoped patterns on larger-scoped ones, generated by the Primacy of Physics Constraint, further underscores this top-down methodological attitude: quantum and relativistic patterns are more reliably repeated across broader regions of spacetime, relative to condensed-matter patterns, let alone anthropological ones.

Even though there may be an escape route for Ladyman et al. from the letter of NGST, I worry that the top-down methodological approach baked into their view is derived from the fundamentalist intuition underlying the reductive strategy. And I worry that that intuition continues to be counterproductive to theorizing about the connection between scientific realism and representation. Waters and Batterman both emphasize the fecundity of their alternative methodological attitudes. For Waters, a focus on biological complexity generates different types of ontological questions than a physics-centered approach; while for Batterman, attention to mesoscale descriptions of many-body systems redirects metaphysical inquiry toward a new understanding of what counts as a "natural" grouping of physical phenomena that does not rely on either bottom-up or top-down description. In the next section, I show how a top-down attitude toward scientific metaphysics may skew one's approach to scientific representation.

6 Anti-Fundamentalist Metaphysics as a Guide to Scientific Representation

An implication of Waters' and Batterman's views that is not explicitly addressed by either but is, I suspect, tacit in both, is that philosophers would do well to reject a part–whole approach to scientific metaphysics and should instead seek alternative logical and conceptual foundations for a scientifically-informed ontology. This implication bears directly on the connection between scientific metaphysics and scientific representation. Rejecting a part–whole conception of metaphysics means that there is no ontological impetus to develop representations of scientific systems based on part–whole models. This leaves room for approaches to modeling that emphasize other varieties of relation between the relata in a model. Massimi's perspectival realism is one such approach; my "conceptual strategies" account of how multiscale models come to represent nanoscale phenomena is another.

Massimi's project surveys a wide array of natural and social scientific practices in order to examine what types of metaphysical commitments are required in order to believe that reliable scientific knowledge is generated through contemporary scientific practices. Her central question is about what a realist commitment implies about the kinds of knowledge that is possible to obtain—and advisable to pursue—from scientific investigation. She contrasts this project with what she sees as the traditional realist project of "mapping the existence of the 'scientific zoo'" (Massimi 2022, Ch. 1). Massimi's approach emphasizes the role of the knower in answering her central question, and the resulting realism is built around the complex relations between knowers, known, and ways of knowing. Unsurprisingly, given these interests, Massimi's account emphasizes the role of representation in defining a scientifically-informed ontology. Her view is, I believe, a viable path forward for realism that also suggests new ways to connect realism and representation while avoiding part–whole approaches to scientific metaphysics.

Through a case study on a computer simulation of a nanoscale crack propagating in silicon, I have argued (Bursten 2018) that there are multiple types of techniques for stitching together the component models in a multiscale model. The case study contains three component models—a macroscale continuum model, a mesoscale classical rigid-body model, and a microscale quantum model—and two "handshaking" algorithms that bridge the macro to the meso, and the meso to the micro. She shows that the two handshaking algorithms are neither strictly logically nor empirically dictated by the component models they aim at bridging: the macro-to-meso handshake involved the manipulation of non-representational features of the macro model in order to make a handshake between the two components possible, whereas the meso-to-micro handshake involved the construction of a fictional entity on which to perform computation. These are distinct conceptual strategies, and neither strategy can be redescribed in terms of part–whole relations between the different component models. Warrant for trusting a model to represent a system well derives from an analysis of how these conceptual strategies function.

To put the point in terms of some of this chapter's earlier discussions, each of the component models in my case are instances of real patterns at different spatial scales of resolution. What my analysis shows is that there is conceptual work to be done in order to build the modeling techniques that allow the model to "zoom out" or "zoom in." The resulting multiscale model does produce a top-down, overall description of the simulated system—namely, a Hamiltonian describing the distribution of energy in the system at a given instant—but modelers only come to formulate, and to trust, this top-down description by developing conceptual strategies that enable the construction of handshake algorithms. I also argue that it is not the case that the most zoomed-in model is the one that best or most truly represents the system. This suggests an approach to scientific modeling that further defies the reductionist intuition about representation. Like the accounts of Waters, Batterman, and Massimi, my analysis generates an anti-reductive account by offering an alternative to part–whole conceptions of systems of scientific interest. Further, Waters, Batterman, and I each explicitly emphasize the role of scale in limiting the scope of claims and inferences about the systems of scientific interest. This is evidenced in Waters' opposition to general structure, Batterman's anchoring of natural categories in the meso-scale, and my attention to the conceptual strategies that stitch dynamics together across scales.

The point of bringing these accounts into friendly dialogue with one another has been to highlight some limitations in approaching scientific metaphysics from the top down—or the bottom up. Both directions of approach fall prey to the part–whole conception of metaphysics. I suspect that conception is responsible for lingering reductionist intuitions about how scientific representation works. I hope these alternative metaphysical starting points can generate alternative starting points for representation, as well.

Although structural realism played the role of a foil here, I suspect that there are versions of a structural-realist approach to metaphysics that can avoid the pitfalls of top-down methodological approaches— Steven French's developments of his own version of ontic structural realism (e.g., French 2014), as well as Ladyman's expansions of his views (Ladyman 2017), may avoid some of the critiques leveled here. Further, I suspect that what is necessary is, at least in part, increased and careful attention to the role of scale in delimiting, defining, and explaining the nature of various structures, which is to be found especially in Ladyman's more recent work. A gesture toward such a view can be seen in the thin notion of general structure articulated via cartography above. A more robust exploration of how scale-dependent notions of structure might reorient the methodology of structural realism is a project for another day.

Acknowledgments

Thanks to an anonymous reviewer and to Elay Shech for comments on an earlier draft of this chapter, and to Jennifer Jhun, Collin Rice, Chris Grimsley, Stephen Perry, Bob Batterman, and Ken Waters for the discussions that led to these remarks.

Notes

1 They distinguish the more moderate reductionist strategy from a stricter reductionism, advocating for the former while decrying the latter (Barker and Kitcher 2014, pp. 54–55).
2 "Structural realism" denotes a family of philosophical positions. Ontic structural realism is one cluster of positions that contrasts with, for example, epistemic structural realism.
3 Sandra Mitchell has referred to this variety of physicalism as "physicsism" (Mitchell 2009, p. 33).
4 Indeed, there is even a fan page for "The Nonsensical Roads of Pittsburgh," https://pghroads.tumblr.com/.
5 A term due to Daniel Dennett (Dennett 1991).

References

Barker, Gillian, and Philip Kitcher. 2014. *Philosophy of Science: A New Introduction*. Oxford University Press.
Batterman, Robert. 2021. *A Middle Way: A Non-Fundamental Approach to Many-Body Physics*. Oxford University Press. https://doi.org/10.1093/oso/9780197568613.001.0001.
Batterman, Robert, and Collin C. Rice. 2014. "Minimal Model Explanations." *Philosophy of Science* 81 (3): 349–376. https://doi.org/10.1086/676677.
Bursten, Julia R. 2018. "Conceptual Strategies and Inter-theory Relations: The Case of Nanoscale Cracks." *Studies in History and Philosophy of Science Part B* 62: 158–165. https://doi.org/10.1016/j.shpsb.2017.09.001.

Cartwright, Nancy. 1983. *How the Laws of Physics Lie.* Cambridge University Press. https://doi.org/10.1093/0198247044.001.0001.

———. 1999. *The Dappled World: A Study of the Boundaries of Science.* Cambridge University Press. https://doi.org/10.1017/CBO9781139167093.

Dennett, Daniel C. 1991. "Real Patterns." *The Journal of Philosophy*, 27–51. https://doi.org/10.2307/2027085.

Dupré, John. 1995. *The Disorder of Things: Metaphysical Foundations of the Disunity of Science.* Harvard University Press.

French, Steven. 2014. *The Structure of the World: Metaphysics and Representation.* Oxford University Press. https://doi.org/10.1093/acprof:oso/9780199684847.001.0001.

Ladyman, James. 2017. "An Apology for Naturalized Metaphysics." In *Metaphysics and the Philosophy of Science: New Essays*, edited by Matthew H. Slater and Zanja Yudell, 141. Oxford University Press. https://doi.org/10.1093/acprof:oso/9780199363209.003.0008.

Ladyman, James, and Don Ross. 2007. *Every Thing Must Go: Metaphysics Naturalized.* Oxford University Press. https://doi.org/10.1093/acprof:oso/9780199276196.001.0001.

Massimi, Michela. 2018. "Four Kinds of Perspectival Truth." *Philosophy and Phenomenological Research* 96 (2): 342–359. https://doi.org/10.1111/phpr.12300.

———. 2019. "Realism, Perspectivism, and Disagreement in Science." *Synthese*, 1–27. https://doi.org/10.1007/s11229-019-02500-6.

———. 2022. *Perspectival Realism.* Oxford University Press.

Mitchell, Sandra D. 2009. *Unsimple Truths: Science, Complexity, and Policy.* University of Chicago Press. https://doi.org/10.7208/chicago/9780226532653.001.0001.

Potochnik, Angela. 2017. *Idealization and the Aims of Science.* University of Chicago Press. https://doi.org/10.7208/chicago/9780226507194.001.0001.

Shech, Elay, and Patrick McGivern. 2019. "Fundamentality, Scale, and the Fractional Quantum Hall Effect." *Erkenntnis*, 1–20. https://doi.org/10.1007/s10670-019-00161-y.

Waters, C. Kenneth. 2017. "No General Structure." In *Metaphysics and the Philosophy of Science: New Essays*, edited by Matthew H. Slater and Zanja Yudell, 81–108. Oxford University Press. https://doi.org/10.1093/acprof:oso/9780199363209.003.0005.

———. 2018. "Ask Not 'What Is an Individual.'" In *Individuation, Process, and Scientific Practices*, edited by Otávio Bueno, Ruey-Lin Chen, and Melinda Fagan, 91–113. Oxford University Press. https://doi.org/10.1093/oso/9780190636814.003.0005.

Woodward, James. 2003. *Making Things Happen: A Theory of Causal Explanation.* Oxford University Press. https://doi.org/10.1093/0195155270.001.0001.

Worrall, John. 1989. "Structural Realism: The Best of Both Worlds?" *Dialectica* 43 (1–2): 99–124. https://doi.org/10.1111/j.1746-8361.1989.tb00933.x.

Part III

Understanding, Representation, and Inference

16 Factivity, Pluralism, and the Inferential Account of Scientific Understanding

Jaakko Kuorikoski

1 Introduction

The current philosophical disputes about the concept of understanding concern whether it can be reduced to either that of explanation or that of knowledge, and they mostly use intuitions pumped via thought experiments or episodes in the history of science as evidence. A mostly unarticulated presupposition in these debates is that understanding is a special internal epistemic state, and that the philosophical task is to characterize this state and the grounding of its epistemic value, to find the "underlying metaphysics of understanding". Epistemologists have asked whether this state can be reduced to the state of knowledge, whereas philosophers of science have mostly queried whether something of epistemic interest remains to be said about science, once we have understood the logic of scientific explanation.

These battles about the true concept of understanding are fought mostly on the pages of philosophy journals, whereas important disputes *using* the concept of understanding are constantly taking place between scientific paradigms, traditions, and fields. Within the social sciences, grand methodological debates have been cast in terms of the proper mode of understanding action, meaning, and society. Echoes of these discussions still reverberate in surprising places, such as in the plea for microfoundations as a necessary ingredient for providing understandable economic stories with our best current computational macroeconomic models. The methodological identities of new sub-fields of science are often built on specific conceptions of scientific understanding. Prominent examples have been computational social science, championed early on with the motto "if you did not grow it, you do not understand it" (Epstein 1999), and systems biology with its promise of delivering a new kind of understanding of biological complexity. At present, the ever-increasing role of computation in general and machine learning in particular in almost all fields of science is forcing us to radically rethink the place of human understandability in science. Historical ruptures in foundational metaphysics, such as the shift from Aristotelian species-essences to Darwinian population thinking or from the classical mechanical conception of the

DOI: 10.4324/9781003202905-19

physical reality to the quantum field theory have also constituted fundamental changes in the very criteria of understandability. For a philosophical theory of understanding to be worth its salt, it should be able to shed light on such claims, provide clarity for the scientists and their audiences about what is at stake in questions of understandability, and, if possible, help mediate such disputes across the sciences.

Two grand philosophical traditions of conceptualizing understanding in general intersect these debates. The representationalist or intellectualist tradition understands understanding as the private possession of correct mental representations of the objects of understanding. In contrast, the pragmatist tradition, drawing on Kant, Wittgenstein, and Sellars, understands understanding as a matter of doing. In Ryle's parlance (1946), representationalist understanding amounts to knowledge that and the pragmatist understanding to knowledge how. This chapter restates and defends a pragmatist and thoroughly deflationist view of understanding developed in a series of papers (e.g., Kuorikoski 2011; Kuorikoski and Ylikoski 2015; Ylikoski 2009; Ylikoski and Kuorikoski 2010), which equates (explanatory) understanding with the ability to draw correct counterfactual what-if inferences about the object of understanding.

This view is broadly in line with the "received view" in philosophy of science, according to which understanding can be explicated in terms of knowledge of explanations, but it is more committed to a specific theory of explanation than, for example, Khalifa's explanation-knowledge-science model of understanding (2017). There is nothing especially "deep" about understanding. Our view is also in line with what could even be called an emerging consensus in philosophy of explanation, according to which the distinctive feature of this explanatory knowledge is its modal dimension and to understand a phenomenon is to be able to correctly situate it within a space of possibilities.

Our motivation has not been to test intuitions against thought experiments and historical cases to formulate necessary and sufficient conditions for someone to be in a special epistemic state, but to develop a philosophical theory of understanding that is capable of explicating and arbitrating disputes and controversies concerning understanding both within and across the sciences. I argue that in order to do this, a philosophical theory of understanding needs to account for (1) the social-epistemic function of the attributions of understanding, (2) the factivity of understanding, and (3) moderate explanatory pluralism. The account of understanding as a set of modal inferential abilities fulfills these desiderata.

A background commitment for this argument is that the representationalist and the pragmatist differ not only in terms of their proposed analyses, but also in terms of the very goal of the exercise. If understanding is taken to be a special epistemic *state* for which an analysis must be *found*, the resolution of scientific disputes supposedly amounts to finding out which side *really* has more true understanding. The relevant evidence for this project mostly consists of the intuitions of the philosopher and cherry picked instances of the use of the concept by scientists. In contrast, I take the aim of the pragmatist project to be to explicate the *function* of the attributions of understanding. The pragmatist does not seek to uncover a hitherto unrecognized special state grounding epistemic progress – the inner essence of understanding – but to characterize what is being done when someone is said to understand something. This methodological stance is very much indebted to the later Wittgenstein and draws heavily from the work of Robert Brandom. The resolution of disputes relating to understanding are not resolved by finding out who really is in the proper state, but by explicating *what is at stake* in such disputes. And what is at stake is not what happens inside the heads of scientists, but what can be *done* with rival models, theories, or paradigms.

In epistemology, such a methodological shift has previously been put forward by Edward Craig, who in his *Knowledge and the State of Nature* (1990) proposed that the function of knowledge attributions is to publicly certify good sources of information. Michael Hannon (2019, Ch 9) has applied this "function first" approach to the concept of understanding and proposed that the attributions of understanding pick out good *explainers*. However, although in broad agreement, the picture to be painted below provides two important further insights on understanding. First, the *inferentialist* analysis drawing on Brandom highlights the way in which attributing understanding is not just about labeling reliable epistemic agents, but also about *undertaking* epistemic commitments. Analogous to knowledge, understanding is a device for simultaneously referring to *and using* the epistemic perspective of others. Second, the *pragmatist* perspective emphasizing doing instead of representing provides a natural connection for the attributions of explanatory understanding with other forms of understanding, something that the function of picking good explainers lacks. This perspective also bestows, I think quite rightly and contrary to Hannon's view, a higher epistemic value on understanding than on knowledge. In the following, I begin from this broadly pragmatist understanding of the grammar of understanding and then proceed to the inferentialist analysis on the social and regulative function of understanding. After this, I will address the questions of factivity and pluralism.

2 The Grammar of Understanding

As a further illustration of the difference between the representationalist and the pragmatist, consider the philosophical project of analyzing "knowledge". The traditional epistemologists' analysis is that to be in the state of "knowing that P", the agent has a justified true belief that *P*, further amended with one's anti-Gettier condition of choice. In contrast, Brandom's inferentialist analysis of knowledge concerns the function of, what is *achieved with*, knowledge attributions: to assert that an agent *S* knows that *P* is to simultaneously attribute the doxastic commitment "that P" to S, as well as to undertake the commitment that *P* on oneself. Knowledge is not a special inner state, a kind of super-belief, but a score-keeping *device* in the social regulation of belief (Brandom 1994, 1995). I propose that the concept of understanding is a similar scorekeeping device used in the coordination and regulation of the *use* of information.

The claim that understanding is not an internal state is admittedly unintuitive; surely, understanding refers to the hidden psychological processes that are causally responsible for the manifest inferential abilities and not to the abilities themselves (e.g., Strevens 2017)? In discussing the grasping of the meaning of a mathematical operation in the Investigations, Wittgenstein (1953[2001]) warns us from drawing quick philosophical conclusions from this beguiling intuition:

> PI §154: Try not to think of understanding as a 'mental process' at all.—For that is the expression which confuses you. But ask yourself: in what sort of case, in what kind of circumstances, do we say, 'Now I know how to go on,' when, that is, the formula has occurred to me?—In the sense in which there are processes (including mental processes) which are characteristic of understanding, understanding is not a mental process.

This point generalizes to other forms of understanding, such as understanding the meaning of words and symbols: although we are tempted to think of understanding as an internal *state*, "the grammar" of understanding is that of an *ability*. Whether someone understands a word, a procedure, or a theory is grounded on manifest performance in appropriate linguistic, practical, and epistemic tasks, judged against the public criteria of correctness. Understanding is attributed according to *doing*, not possessing. These criteria are *normative* in that understanding is evaluated according to whether the agent demonstrates the ability to do things *correctly* or in a *right* way. Internal psychological states are neither correct nor incorrect, they just are (Baker and Hacker 2005).

Linguistic understanding is evaluated according to the abilities in using expressions of a language *correctly* within the rules of the appropriate language games. This basic pragmatist's understanding of understanding is further developed in Brandom's inferentialism (1994), according to which full conceptual understanding is constituted by the ability to correctly embed the concept in question within a web of inferences – to use the concept proficiently in the game of giving and asking for reasons. I suggest that also explanatory understanding or understanding-why is a species of understanding in general in that it functions according to this common grammar of understanding.

As explanatory understanding or understanding-why is the intended product of explanation, the specific inferential ability characteristic of it should be the one afforded by successful explanations. And according to a widely accepted account of explanation, the key ability is to answer what-if-things-had-been-different questions. This is the ability to infer what would happen (have happened) to a phenomenon if a well-defined change were (had been) introduced into a specific part or aspect of the phenomenon/system (Woodward 2003). *Explanatory understanding therefore amounts to the set of abilities for making correct w-inferences about the object of understanding* (Rice 2016; Ylikoski and Kuorikoski 2010) – or to put it in another way, the ability to successfully navigate the space of possibilities within which the actual phenomenon is located (Le Bihan 2017). The key ingredient distinguishing explanatory information from merely descriptive information, and the difference between a true *dependency* and mere regularity, is this modal import.

Further, explanatory inferences should track *ontic* dependencies holding objectively between different aspects of the phenomenon in question. This is to be contrasted to epistemic dependencies reflecting evidential and inferential relations between our representations and conceptualizations of phenomena. The ontic requirement reflects a fundamentally realist commitment, according to which understanding is achieved by discovering how the world works, not just by organizing our store of descriptive information in a more efficient way. In the case of causal explanation, these w-inferences concern the effects of hypothetical causal interventions and a key conceptual role of the concept of intervention is precisely to ensure that the corresponding w-inferences track ontic and not epistemic dependencies. These interventionist ideas can also be applied, albeit with some modifications, to other ontic explanatory dependencies such as constitution (Kuorikoski 2012; Reutlinger 2016; Rice 2015; Ylikoski 2013). The epistemology and metaphysics of this modal dimension of understanding are wrought with difficult philosophical questions, which will be put aside here. I will simply take this commitment to a specific theory of explanation as given for now, as the current focus is on understanding the grammar of understanding, but the issue is taken up again in the last section.

As a set of abilities, understanding is gradual: one can have the ability to reach a smaller or a larger set of what-if conclusions about aspects of *P*, these conclusions may be more or less approximately accurate, and these inferences can be more or less reliable. The range and reliability of these inferences measure the depth of understanding. As with explanatory power, this depth of understanding comes in a variety of dimensions, such as precision and non-sensitivity, some of which may exhibit systematic trade-offs in various contexts (Ylikoski and Kuorikoski 2010).

This graduality of understanding has been put forward as one reason for distinguishing between understanding and knowledge (e.g., Elgin 2017). The total set of the correct w-inferences can be taken as a rough measure of the level of understanding about a phenomenon (what epistemologists sometimes classify as "objectual understanding"). One of the key take-home messages of contrastive theories of explanation has been that one can meaningfully explain only specific aspects of phenomena (or events), not phenomena as a whole, and that explicating a contrast class for the explanandum is a powerful way of disambiguating what is intended or can be explained with the explanatory resources at hand. As has generally been acknowledged in the literature, understanding is, therefore, a natural epistemic currency for evaluating broader cognitive achievements than single explanations.

So, is understanding something over and above the knowledge of explanations or causes? Grimm (2017) and Khalifa (2017) argue that at least everything epistemically relevant about understanding can be analyzed without residue in terms of knowledge of explanations. This is probably right – perhaps depending on how stringent conditions are placed for possessing knowledge of an explanation. Modulo certain epistemological subtleties relating to justification, defeasibility, and epistemic luck, a reliable ability to make correct w-inferences about P is based on knowledge of the corresponding explanatory dependency, and I do not want to get bogged down in epistemological disputes concerning what exactly having knowledge entails. Further, what exactly is understood about P can be explicated by listing what aspects of P can be explained with the knowledge at hand.[1] Even in cases in which the understanding subject may not be able to propositionally articulate all the premises on which her w-inferences rest, such understanding could, in principle, always be explicated in terms of implicit and tacit knowledge of explanatory dependencies. The main point made here, however, is that we should not really be focused on an analysis of understanding as a special epistemic state to begin with, but on the social epistemological function of the attributions of understanding.

3 The Regulative Role of Understanding

Although the Wittgensteinian insights about the grammar of understanding are hopefully persuasive, they might not be enough to overcome the recalcitrant commonsense intuition that understanding is an internal psychological state from which the verbal and behavioral manifestations of understanding causally flow. However, we do not need to rely solely on past philosophical authority. In fact, there is no need to deny that there is a common and perfectly legitimate understanding of understanding as an internal mental state. There is even some experimental empirical evidence for lay conceptions of explanation and understanding being solidly psychologistic in this sense (Wilkenfeld and Lombrozo 2020). But if the ultimate objective of the present philosophical theorizing is not to analyze a pre-theoretic folk concept of understanding, but to provide conceptual tools for elucidating scientific controversies related to a robustly epistemic use of "understanding", then such an internalist concept is simply irrelevant – or so I will attempt to argue in the following.

The main reason for favoring the inferential ability understanding of scientific understanding, which emphasizes the social and public nature of understanding, is that the advancement of science is not really constituted by changes in individual psychological or doxastic states. In contrast to the near solipsist epistemology of sense experience and belief that has traditionally preoccupied philosophers interested in knowledge, science is a fundamentally social enterprise and scientific knowledge a public, communal achievement (Rice this volume). Vast amounts of public resources are not annually spent on changing the beliefs of individual scientists or providing scientists with satisfying feelings of understanding, but on expanding a shared body knowledge and the practical capabilities flowing from it. Understanding is the aim of science and the currency in which this advancement in public knowledge and technical capability is evaluated (Dellsén 2016, see also Rice 2019). Focusing on the public and collective aspects of understanding is all the more important, when the objective of the theory of understanding is to help explicate scientific controversies. Although participants in such disputes undoubtedly entertain strong hypotheses about what goes on inside the heads of the other participants, these internal processes are not what is ultimately at stake. What is at stake is what follows from the rival models, theories, or paradigms – what can be *done* with the proposed conceptual and empirical frameworks.

Lots of important things, undoubtedly, happen in the minds of scientists, but investigating these psychological processes is the job of empirical cognitive scientists. Nevertheless, the inferential account of understanding is not built upon a categorical distinction between the context of discovery and the context of justification. As a large body of important naturalist philosophy of science has demonstrated, reconstructions of research heuristics and logics of discovery are at least as important for the understanding of science as more traditional theories of confirmation. However, even heuristics become relevant for the body of scientific knowledge only when they enter the discursive and public game of giving and asking for reasons. Historians and sociologists of science entertain psychological hypotheses about the private thought processes behind important discoveries, but these thought processes become science only when they are translated into public reasons in the pages of journals and books.

Evaluating understanding purely in terms of demonstrable inferential abilities forces the scientist to put her inferential money where her epistemic mouth is. A fully reductive and deflationist concept of understanding as a set of abilities implies that in the analysis of understanding-related controversies across the sciences, what is at stake can be exhaustively accounted for in terms of the range, scope, accuracy, and reliability of w-inferences made on the basis of the explanatory information available. Such an analysis draws the focus from subjective intuitions onto legitimate disagreements about whether the w-inferences afforded by rival theories are in fact correct, reliable, and relevant to anything of importance. In many cases, there probably are residual intuitions or feelings of understanding that fall outside of these demonstrable abilities in the use of the information, but these intuitions at least require an independent argument if they are to be granted epistemic relevance.

These last points bring us to the following, broadly Brandomian analysis of the regulative *function* of attributions of explanatory understanding: an attribution of understanding bestows *inferential authority* to the agent in question. When A states that S understands P, not only is A prepared to undertake the doxastic commitments of S relating to P (as in an attribution of knowledge that P to S), but also the inferential implications of those commitments drawn by S. By stating that S understands P, A proclaims that she trusts, and thus suggests that others should also trust, that S can reliably *use* her information about P. Like knowledge, understanding is therefore *a social status*. Of course, derivatively, an agent can publicly endorse her own inferential trustworthiness by attributing understanding to herself. As the actual inferential abilities and their reliability are not immediately epistemically accessible to the agent, she may even hold erroneous beliefs about her own understanding. As is already widely acknowledged, a sense of understanding is only an imperfect indicator of true understanding.

As Wilkenfield, Plunkett, and Lombrozo have demonstrated (2016), something like this regulative role of understanding is present in lay attributions of understanding: attributions of understanding why P pick out individuals who possess explanatory depth concerning P. This enables *deference* to these individuals in problems related to P. Ordinary lay attributions of understanding differ from attributions of knowledge precisely in the higher standard of "depth" demanded of the former, which, in turn, is related to a broader set of questions that the understander is expected to be able to answer. These empirical results chime nicely with the present suggestion that the attribution of understanding confers not just doxastic authority that P holds, but also inferential authority about what follows from P. As the scope of these commitments that the attributor of understanding is willing to undertake is broader than in the case of knowledge, the standards that must be fulfilled for this attribution to be bestowed are correspondingly higher.

4 An Inferentialist Argument for the Factivity of Understanding

There are two main challenges for a factive analysis of understanding, that is, an account according to which the information responsible for the understanding has to be true for understanding to be genuine. First there are the ubiquitous idealizations and other falsities in scientific models by which explanations are produced. It is not just that idealizations of negligible and irrelevant features are often introduced to models for reasons of pragmatic convenience and tractability, but that idealizations and other distortions of the most explanatorily relevant features are often important for the explanatory qualities of models. Further, full understanding of complex phenomena usually demands the use of multiple models, and the necessary idealizations and other distortions may result in these models making mutually inconsistent assumptions about the key explanatory factors. This discussion on idealizations, explanation, and truth has a long history in the literature on scientific models, but the later discussion on understanding has especially emphasized the necessity of multiple and mutually inconsistent models in our understanding of complex phenomena (Elgin 2017; Potochnik 2017; Rice 2019).

The second challenge is the fact that most theories and models in the history of science have been proven false. Yet it would be hard to deny that many of them have contributed to the growth of scientific understanding despite their falsity (de Regt 2015, 2017). Copernicus' heliocentric model did advance understanding of the solar system, even though planets do not have circular orbits. Newton's theory of gravity was a monumental advance in scientific understanding, even though Einstein showed that the postulated attractive force does not really exist.

Although identifying understanding with a set of abilities offers an easy way to simply evade the factivity question – abilities or skills are not really truth-apt to begin with (de Regt 2015) – I will, nevertheless, defend a broad notion of factivity of understanding. This is because at least some form of objectivity, if not outright factivity, is a desideratum for a theory of scientific understanding with the ambition of addressing scientific debates. It is a truism that disagreement is possible only against a broader backdrop of agreement. Genuine disagreements about the right way of understanding phenomena are possible only if the disagreeing parties are committed to there being some facts about the phenomena to begin with and that these facts have some bearing on understanding. Many critics of economics have accused economic models of being fictitious and it is clear that these critics would not be impressed if economists simply acknowledged that, indeed, their models and the understanding they produce were *nothing but* fictions with no relation to economic phenomena.

As the explanatory models themselves are not constitutive of understanding, falsities in models do not automatically render understanding non-factive. Models, as such, are not explanations; they are epistemic tools used in the formulation of explanations (Lawler 2021). As Collin Rice has argued, even though falsities and the use of multiple and mutually inconsistent models are often necessary for successful w-inferences, it is these inferences and not the idealizations that constitute understanding and therefore it is only the correctness of these inferences that matter for the factivity of understanding (Rice 2019; see also Kuorikoski and Ylikoski 2015). It is hard to see any other way that these correct w-inferences could reliably be made other than that they are at least implicitly based on true premises about the explanatory dependencies in question. Therefore, the factivity of understanding seems to directly imply a certain level of representational accuracy about the explanatory dependencies in question, although any one model from which these premises are drawn may distort these even quite dramatically (Kuorikoski and Ylikoski 2015; cf. Rice 2019).

The argument from the history of science can be at least partly addressed by resorting to the observation made by structural realists, that what tends to remain stable and cumulatively grow across radical metaphysical ruptures in the history of science is the store of relational inferences about the phenomena in question (e.g., Worrall 1989, see also Potochnik in this volume). A similar point, although from a different philosophical perspective, has recently been made by Michela Massimi (2021), who notes that historical perspectival disagreements about the deep nature of the phenomena under investigation often eventually resolve into an agreement on the right inferential patterns. One need not commit to the structuralist answer to the pessimistic meta-induction or structuralist ontology to appreciate how this historical narrative provides hope for a factive account of understanding: what tends to remain relatively invariant and even grow cumulatively across historical discontinuities in underlying ontology are the inferential capabilities related to the robustly demonstrable empirical aspects of the phenomena under investigation. There is no room here to provide an even rudimentary overview of the historical episodes discussed, for example by de Regt, but I hypothesize that our judgments about what *specific aspects* of the physical phenomena in question were *genuinely* understood on the basis of now defunct theories and ontologies correspond closely to the set of w-inferences about these phenomena we still consider correct. Even though Newtonian mechanics is strictly speaking false and rests on an ontology that we no longer accept, it still affords many counterfactual w-inferences about physical systems that we believe to be true enough. Note that this is *not* an argument for scientific realism per se. Accepting the factivity of understanding does not necessarily commit one to the optimistic realist thesis that most of our explanatory theories are, in fact, true and our scientific understanding thus mostly genuine. Factivity states that understanding is genuine only if the corresponding explanatory information is true.

To sum up, as many authors have suggested, our intuitions seem to accord with a factive concept of understanding and the ubiquity of distortions and falsehoods in past and present scientific models is perfectly consistent with this factivity requirement. But *why* do we have this factive concept? What explains the apparent fact that only understanding based on true – or true-enough – w-inferences feels genuine? Are we simply all indoctrinated to a veristic *ideology* of scientific realism, according to which our standards of correctness should demand truth as an intrinsic epistemic good?

Pragmatists of various breeds have traditionally offered instrumental defenses of the value of truth in science in general: true-enough theories facilitate our practical endeavors through prediction and manipulation of nature. However, this general instrumentalist defense seems to have very little to do with understanding in particular. The broadly Brandomian inferentialist perspective on the function of understanding provides a closer, and I believe more illuminating, link between understanding and truth. The factivity of understanding is a natural consequence of its function in the social allocation of inferential reliability and authority *about* our shared reality. As a comparison case, let us again first consider knowledge. The social status of knowledge is a device for simultaneously attributing and undertaking doxastic commitments: when A claims that S knows that P, A attributes that S believes that P, that S is entitled to her belief, and acknowledges that she (A) herself also believes that P. These attributions are factive, because they socially articulate the possibility of having different epistemic perspectives on a single shared reality. If A becomes aware of reasons denying her entitlement to believe that P, she must also remove her endorsement of S knowing that P, since although they have different epistemic perspectives, both A's and S's commitments were *about* the one and the same P. The "norm of truth" is not something separate and over and above the concept of belief, but an internal consequence of the socially and conceptually articulated epistemic perspectives on the common objective reality (Brandom 1994, 1995).

Analogously, explanatory understanding is factive because it is a status conferring inferential authority *about* a phenomenon approached from different perspectives and itself independent of those perspectives. When A attributes understanding of P to S, she makes an open-ended endorsement of the conclusions of the potential w-inferences about P made by S. If A becomes aware of reasons that conclusively remove her entitlement for some modal conclusion about P drawn by S, she has to also remove her attribution of understanding to S (or at least the part of S's understanding pertaining to the said conclusion). After all, from her epistemic perspective, S has made an incorrect w-inference and must therefore either be in the possession of false explanatory information, or not capable of using that information correctly.

5 Understanding and Explanatory Pluralism

It is undeniable that the standards of what is a good explanation and what therefore constitutes genuine scientific understanding differ across disciplines and have undergone dramatic changes during the history of science. Because of this variety in the kinds of explanations across different fields of science, many philosophers have advocated that there are also deep differences in the very grounds that make them all explanations to begin with. As already argued by Stephen Toulmin (1961), history of science exhibits a series of shifts not only in explanatory theories, but also in what kinds of theories and phenomena are considered as understandable. Newtonian mechanics not only provided dramatic new explanatory resources over Aristotelian physics, but it also shifted what was considered to be in need of explanation from motion to changes in motion. Natural selection is not only an important new explanatory model-template, but also the Darwinian population thinking has (mostly) dispelled the Aristotelian conception that class-membership is somehow explanatory in itself. De Regt has argued that in order to stay true to these fundamental shifts, we need to be pluralists about the very criteria of understanding as well. For de Regt, causal reasoning and unification, among other things, are just different, alternative *means* for producing understanding. Similarly, Khalifa et al. (2020) argue that many forms of explanation seem to explain by virtue of something else than exhibiting ontic dependencies.

This presents us with an apparent dilemma. On the one hand, for a philosophical account of understanding to be useful in the explication of controversies within the sciences, accommodating at least some amount of pluralism about explanation and understanding is obviously an important desideratum. A theory that renders whole disciplines unexplanatory and the understanding they have provided as illusory is not only descriptively inadequate, but also prima facie illegitimate to much of the intended audience of the analysis. When faced with the original covering law account of explanation, most social scientists have quite rightly dismissed the account as irrelevant to their scientific pursuits rather than obediently reform their epistemic practices to fit the nomothetic ideal of science.

On the other hand, thoroughgoing pluralism risks robbing the philosophical account of its analytical teeth. For example, de Regt's only general criteria for what can legitimately count as an explanation, empirical adequacy and internal consistency (2017, 92), are so weak that it is hard to see how anyone claiming to be engaged in science could disagree with them. If the account only describes and systematizes the actual usage of the word "understanding" and the intuitions accompanying it in various sciences, then it can, at best, describe and document understanding-related controversies within the sciences. Sometimes this can be valuable, but do we really need more explication of the fact that the conception of understanding as emphatic understanding and thick description advocated by the ethnographer differs from the conception of understanding as the systematic derivation and unification from a small set of axioms of rational choice espoused by the microeconomic theorist? Perhaps such fundamental disagreements really are fundamentally insolvable and these epistemic communities really do have incommensurable forms of life. But even if this were so, simple pluralism without any commitment to an account of what really makes an explanation provide understanding provides no criteria by which to judge when such disagreements are "real" and when they are only apparent.

The moderate monism of the ontic dependency conception of explanation outlined above at least seems to strike a decent balance between scope, descriptive accuracy, and analytical power – what Khalifa et al. (2020) call the dependency-monism dilemma. The account does face important challenges in accounting for the explanatory value of many important kinds of explanations (ibid.). Various distinctively mathematical explanations apparently appeal to necessary and thus unchangeable mathematical facts as explanatory factors. Optimality, equilibrium, and renormalization group explanations seem to explain precisely by showing that the explanandum does not depend on causes or other initial conditions. Complexity and self-organization in the life and social sciences may render the consequences of hypothetical local changes fundamentally untractable. Further, there are still important unanswered questions relating to the explanatory credentials of various interpretive approaches in the social sciences. Nevertheless, counterfactual dependency accounts have been able to accommodate much of this diversity in kinds of explanations by carefully (re)defining the explananda and the explanantia, by proposing different kinds of ontic explanatory dependencies, or even by reducing this diversity by concluding that some epistemic achievements are not really explanatory (although this might entail some serious bullet-biting, as in Kuorikoski [2021]).

The obvious and perfectly legitimate question now concerns the basis of this philosophical commitment. The first possible answer points to the slew of persuasive arguments for the counterfactual ontic dependency conception made in the philosophical literature on scientific explanation. According to standard philosophical reflective equilibrium methodology, the weight of the intuitive judgments supporting these arguments ought to be balanced against the descriptive accuracy of the theory in terms of the actual usage of the concept in the sciences. De Regt, however, is adamant that in the study of scientific understanding, the intuitions of philosophers should *never* override the expert intuitions of the scientists themselves (de Regt 2017). This stance seems all the more reasonable if the ambition of the theory is to explicate and mediate debates within the sciences; on what grounds can the philosopher claim that she is in the possession of a privileged epistemic view-from-nowhere from which to judge the practices of the working scientists?

First, it is not immediately obvious as to why scientists ought to be regarded as the final authorities on such meta-level issues as the nature of understanding. The evaluation of understanding is a metacognitive task (Ylikoski 2009), and scientists may be systematically biased in such tasks without this fact undermining their first-order epistemic authority about their primary subject matter. Second, and more importantly, intuitions of *both* philosophers and scientists are mostly the product of accidents of personal intellectual histories and disciplinary indoctrination, not of an explicit and reason-based discussion about the epistemic goals of a particular scientific enterprise. Fortunately, to have analytical and explicative value, the ontic dependency account of explanation and the view of understanding as the ability to make w-inferences about the object of understanding do not, in principle, need *any* backing from the intuitions of philosophers or scientists. This is because the analytical value ultimately derives from three points: a categorical difference between inferences based on dependencies and those based on associations; a categorical difference between ontic and epistemic dependencies; and the difference between merely possessing information and the ability to use that information. The disagreeing scientist is free to insist that something else than w-inferences, such as deriving a description of a phenomenon from initial conditions and natural laws, is what understanding is *really* about, but these categorical distinctions and their implications still stand. Also, if the validity of these distinctions is acknowledged, categorizing an epistemic achievement as an explanation or as something else on the basis of a philosophical theory is not totally question-begging (pace Khalifa et al. 2020). Once we acknowledge these distinctions and agree on the scope, range, and reliability of what can be *done* with the information at hand, then it does not really matter whether we call this understanding or schmunderstanding. Unification and predictive power are obviously important epistemic achievements, but why muddle the conceptual waters by also calling them understanding? If some other preferred essence of explanation and understanding holds some further epistemic values over and above the inferential conception advocated here, the burden of proof is upon the reader to demonstrate this.

Note

1 Dellsén (2020) argues that there are three types of objectual understanding in which understanding is disconnected from the knowledge of explanations: understanding of explanatorily brute phenomena, understanding of

the explanans rather than the explanandum, and increase of understanding by knowledge that something is not an explanation. Although space does not permit a comprehensive treatment of these cases here, I do not think that these constitute counterexamples to the link between understanding and explanation presented here.

References

Baker, Gordon P. and Peter Hacker 2005. *Wittgenstein: Understanding and Meaning, Part I: Essays*. 2nd. revised ed. Blackwell.

Brandom, Robert. 1994. *Making It Explicit: Reasoning, Representing, and Discursive Commitment*. Harvard University Press.

Brandom, Robert. 1995. "Knowledge and the Social Articulation of the Space of Reasons." *Philosophy and Phenomenological Research* 55(4): 895–908. https://doi.org/10.2307/2108339.

Craig, Edward. 1990. *Knowledge and the State of Nature: An Essay in Conceptual Synthesis*. Oxford University Press.

Dellsén, Finnur. 2016. "Scientific Progress: Knowledge versus Understanding." *Studies in History and Philosophy of Science Part A* 56: 72–83. https://doi.org/10.1016/j.shpsa.2016.01.003.

———. 2020. "Beyond Explanation: Understanding as Dependency Modeling." *The British Journal for the Philosophy of Science* 71(4): 1261–1286. https://doi.org/10.1093/bjps/axy058.

De Regt, Henk W. 2015. "Scientific Understanding: Truth or Dare?" *Synthese* 192(12): 3781–3797. https://doi.org/10.1007/s11229-014-0538-7.

———. 2017. *Understanding Scientific Understanding*. Oxford University Press.

Elgin, Catherine. 2017. *True Enough*. MIT Press. https://doi.org/10.7551/mitpress/9780262036535.001.0001.

Epstein, Joshua 1999. "Agent-based Computational Models and Generative Social Science." *Complexity* 4(5): 41–60. https://doi.org/10.1002/(SICI)1099-0526(199905/06)4:5<41::AID-CPLX9>3.0.CO;2-F.

Grimm, Stephen R. 2017. "Understanding and Transparency." In Grimm et al. *Explaining Understanding: New Perspectives from Epistemology and Philosophy of Science*, 228–245. Routledge.

Grimm, Stephen, Christoph Baumberger and Sabine Ammon, eds. 2017. *Explaining Understanding: New Perspectives from Epistemology and Philosophy of Science*. Routledge. https://doi.org/10.4324/9781315686110.

Hannon, Michael. 2019. *What's the Point of Knowledge?: A Function-first Epistemology*. Oxford University Press. https://doi.org/10.1093/oso/9780190914721.001.0001.

Khalifa, Kareem. 2017. *Understanding, Explanation, and Scientific Knowledge*. Cambridge University Press. https://doi.org/10.1017/9781108164276.

Khalifa, Kareem, Gabriel Doble and Jared Millson. 2020. "Counterfactuals and Explanatory Pluralism." *The British Journal for the Philosophy of Science* 71(4): 1439–1460. https://doi.org/10.1093/bjps/axy048.

Kuorikoski, Jaakko. 2011. "Simulation and the Sense of Understanding." In *Models, Simulations, and Representations*, eds. Paul Humphreys and Cyrille Imbert, 250–273. Routledge.

———. 2012. "Mechanisms, Modularity and Constitutive Explanation." *Erkenntnis* 77(3): 361–380. https://doi.org/10.1007/s10670-012-9389-0.

———. 2021. "There Are No Mathematical Explanations." *Philosophy of Science* 88(2): 189–212. https://doi.org/10.1086/711479.

Kuorikoski, Jaakko and Petri Ylikoski. 2015. "External Representations and Scientific Understanding." *Synthese* 192(12): 3817–3837. https://doi.org/10.1007/s11229-014-0591-2.

Lawler, Insa. 2021. "Scientific Understanding and Felicitous Legitimate Falsehoods." *Synthese* 198(7): 6859–6887. https://doi.org/10.1007/s11229-019-02495-0.

Le Bihan, Soazig. 2017. "Enlightening Falsehoods: A Modal View of Scientific Understanding." In Grimm et al. *Explaining Understanding: New Perspectives from Epistemology and Philosophy of Science*, 127–152. Routledge.

Massimi, Michela. 2021. "Realism, Perspectivism, and Disagreement in Science." *Synthese* 198(25): 6115–6141. https://doi.org/10.1007/s11229-019-02500-6.

Potochnik, Angela. 2017. *Idealization and the Aims of Science*. University of Chicago Press. https://doi.org/10.7208/chicago/9780226507194.001.0001.

Reutlinger, Alexander. 2016. "Is there a Monist Theory of Causal and Noncausal Explanations? The Counterfactual Theory of Scientific Explanation." *Philosophy of Science* 83(5): 733–745. https://doi.org/10.1086/687859.

Rice, Collin. 2015. "Moving Beyond Causes: Optimality Models and Scientific Explanation." *Noûs* 49(3): 589–615. https://doi.org/10.1111/nous.12042.

———. 2016. "Factive Scientific Understanding without Accurate Representation." *Biology & Philosophy* 31(1): 81–102. https://doi.org/10.1007/s10539-015-9510-2.

———. 2019. "Understanding Realism." *Synthese* 198(5): 4097–4121. https://doi.org/10.1007/s11229-019-02331-5.

Ryle, Gilbert. 1946. "Knowing How and Knowing That: The Presidential Address." *Proceedings of the Aristotelian Society* 46: 1–16. https://doi.org/10.1093/aristotelian/46.1.1.

Strevens, Michael. 2017. "How Idealizations Provide Understanding." In Grimm et al., *Explaining Understanding: New Perspectives from Epistemology and Philosophy of Science*, 53–65. Routledge.

Toulmin, Stephen. 1961. *Foresight and Understanding*. Harper & Row.

Wilkenfeld, Daniel A., Dillon Plunkett and Tania Lombrozo. 2016. "Depth and Deference: When and Why We Attribute Understanding." *Philosophical Studies* 173(2): 373–393. https://doi.org/10.1007/s11098-015-0497-y.

Wilkenfeld, Daniel A. and Tania Lombrozo. 2020. "Explanation Classification Depends on Understanding: Extending the Epistemic Side-effect Effect." *Synthese* 197(6): 2565–2592. https://doi.org/10.1007/s11229-018-1835-3.

Wittgenstein, Ludvig. 1953[2001]. *Philosophische Untersuchungen / Philosophical Investigations: The German Text, with a Revised English Translation*. Blackwell.

Woodward, James. 2003. *Making Things Happen: A Theory of Causal Explanation*. Oxford University Press. https://doi.org/10.1093/0195155270.001.0001.

Worrall, James. 1989. "Structural Realism: The Best of Both Worlds?" *Dialectica* 43(1–2): 99–124. https://doi.org/10.1111/j.1746-8361.1989.tb00933.x.

Ylikoski, Petri. 2009. "The Illusion of Depth of Understanding in Science." In *Scientific Understanding: Philosophical Perspectives*, eds. Henk De Regt, Sabina Leonelli and Kai Eigner, 100–119. Pittsburgh University Press.

Ylikoski, Petri. 2013. "Causal and Constitutive Explanation Compared." *Erkenntnis* 78(2): 277–297. https://doi.org/10.1007/s10670-013-9513-9.

Ylikoski, Petri and Jaakko Kuorikoski. 2010. "Dissecting Explanatory Power." *Philosophical Studies* 148(2): 201–219. https://doi.org/10.1007/s11098-008-9324-z.

17 Scientific Representation and Understanding

A Communal and Dynamical View

Collin Rice

1 Introduction

Numerous philosophers have recently focused on how various kinds of scientific representations are used to produce understanding of natural phenomena. In this chapter, I argue that both scientific representation and scientific understanding are *communal* and *dynamic*. That is, both are the products of larger groups of individuals interacting over long periods of time. As a result, in order to determine what a scientific model (or theory) represents and what it enables scientists to understand we must pay attention to the assumptions, interpretations, and goals of the broader scientific community and track how these contextual features change over time. For the purposes of this chapter, I will focus on scientific models as the paradigmatic example of scientific representation, and I will assume that most (if not all) models represent in some way. However, I contend that the communal and dynamical views I present here apply equal well to other kinds of scientific representations, for example, theories or simulations.

Although these claims might seem uncontroversial, as we will see, several philosophical accounts of what scientific models represent and how they produce understanding have focused on the interests, goals, and cognitive properties *of individuals* using particular models *in isolation*. For example, philosophers tend to analyze what features a given model represents simply by looking at what is claimed in the assumptions of the model and (perhaps) an individual modeler's interpretation of which parts of the model are intended to accurately represent which features of the target system. It is then determined, based primarily on what the model accurately represents, what understanding can be acquired via the model.

In contrast, in Section 2, I argue that we can only determine what a scientific model represents by embedding it within a dynamical social context that includes the goals of the broader scientific community, the historical development of the model, and the various ways that the model has been used in the past. In particular, I draw on resources from Ruth Millikan's (1984, 1989) teleological account of representation to capture these social and historical features of scientific representation. Then, in

DOI: 10.4324/9781003202905-20

Section 3, I argue that the production of scientific understanding also depends, in essential ways, on the communal/social and dynamical/historical aspects of scientific practice. Section 4 concludes by exploring some of the implications of these claims for how we ought to think about scientific progress.

2 Scientific Representation: A Communal and Dynamical View

The communal and dynamical view of representation I defend here contrasts with several views that have appealed primarily (or exclusively) to the mental states, goals, or purposes of individual model users/builders. For example, Craig Callender and Jonathan Cohen (2006) distinguish between fundamental and derivative representations as follows: "among the many sorts of representational entities . . . , the representational status of most of them is derivative from the representational status of a privileged core of representations" (2006, 70). They then argue that "scientific representation is just one more specific case of derivative representation" (2006, 75). The fundamental representations, on this "General Gricean" view, are those studied by the philosophy of mind: the mental states of individuals (2006, 71–74). In short, "the varied representational vehicles used in scientific settings . . . represent their targets . . . by virtue of the mental states of their makers/users" (2006, 75). Consequently, they argue that constructing a scientific representation "requires only an act of stipulation to connect representational vehicle with representational target" (2006, 79). In a similar way, Roman Frigg and James Nguyen suggest that scientific representations are generated when an agent, *A*, chooses an object as the base of the representation and then adopts a particular interpretation of what the model represents (2017, 169).

Rather than appealing to individuals' stipulative fiats, Ronald Giere (2004) defends a similarity-based view of scientific representation that focuses on the *activity of representing* by analyzing how "*S* uses *X* to represent *W* for purposes *P*" (Giere 2004, 743). Giere then tells us that "*S* can be an individual scientist, a scientific group, or a larger scientific community" (2004, 743). Thus, in contrast with Callendar and Cohen, Giere's view allows that the purposes of the larger scientific community might determine what a scientific model represents. However, Giere's description of *S* suggests that the purposes of the larger scientific community are just one *optional* way to establish the content of a scientific representation. Thus, in Giere's view, the purposes of an individual scientist will often be sufficient for creating a scientific representation.

Following Giere, Michael Weisberg's (2007, 2013) view focuses on the model user's *construal* to determine which similarities matter for evaluating the representational success of the model. Specifically, "The construal tells us which parts of the model correspond to parts of the real phenomenon and which parts can be ignored" (Weisberg 2007, 220). Weisberg notes that communities often have standard conventions for how to interpret a model, but he also suggests that individual modelers "make decisions about which aspects of their models are to be taken seriously. Their intended scope specifies which aspects of potential target phenomena are intended to be represented by the model" (Weisberg 2013, 40). Moreover, in other work, Weisberg tells us that the construal of the model "depends on the intentions of the model user" (Weisberg 2007, 221). Similarly, Peter Godfrey-Smith tells us that "two scientists might use the same model for the same target system, but with different resemblance relations in mind. I call these 'construals' of the model" (2006, 733). Indeed, such appeals to the "goals of the modeler" or the "aims of the model builder" are widespread in the scientific modeling literature.

All the above views suggest that a (if not, the) primary determiner of the representational content of a model is the goal or purpose of the individual model builder/user. In contrast, I argue that mere stipulation, or construal, on the part of an individual scientist is insufficient for determining the content of a scientific representation. This is because only if the consumers of the model within the scientific community accept the proposed interpretation, assignment, or construal, will the model be able to contribute to accomplishing the aims of the scientific community. More specifically, I contend that the users that really matter for scientific representations are the broader scientific community that *consumes* the scientific representations rather than individual model builders that produce them. Consequently, the purposes that matter the most are those of the broader scientific community rather than the purposes of the original model builder/user. In order to be a *scientific* representation, the model must be accepted by the larger scientific community and used for the purposes of that community. What the above accounts miss are the crucially important ways in which the background assumptions, goals, and interpretations of the scientific community both constrain and largely determine what a scientific representation represents. In other words, I propose that we need to replace Giere's schema with the following:

> C interprets X as a (scientific) representation of T for purposes P in social context S constrained by X's history of use H within C.

Here, C is a scientific community, group of scientists, or research program. X is a representational vehicle (e.g., a model, theory, physical object, computer simulation). T is a real or possible target system(s). P are the purposes or goals of the scientific community. S is the social context that includes the community's background knowledge, goals, and values that determine which features are relevant and which can be ignored. And H is the set of historical uses of the representation within the community that furnishes conventions for interpreting the representation, the inferences/results it produces, and the justifications for various idealizing assumptions used within the representation. This schema makes explicit that it is scientific communities that interpret models and it is the purposes and histories of those communities that matter for determining the goals/purposes of a scientific representation. Rather than a special case, I argue that these communal and historical features play an essential role in determining the representational content of every scientific representation.

The communal and dynamical view I defend here incorporates several ideas present in Brandon Boesch's critique of Callendar and Cohen's view (Boesch 2017). In particular, Boesch notes that what Callendar and Cohen's view misses are the communal aspects of how scientific representations obtain their content (Boesch 2017, 970). What is more, Boesch correctly notes that "Scientists do not merely start using a model however they like, without recourse to the history of the use of the model" (Boesch 2017, 978). Boesch goes on to argue that the importance of these communal and historical features of scientific representation is that they *license* scientists' use of the model for scientific purposes such as answering questions about, explaining, or understanding a phenomenon (Boesch 2017, 978).

Despite these points of agreement, this chapter goes beyond Boesch's view by analyzing in more detail some of the specific communal and dynamical aspects of scientific representations that enable them to play particular roles within scientific practice. Using Boesch's terms, the view I present here aims at filling in the details of precisely which communal and dynamical features determine the content of scientific representations and license them to perform specific aims of science. In particular, although Boesch briefly mentions that scientific representations can be used to produce understanding, he does not discuss the particular ways that representation plays (or does not play) a role in allowing a model to produce understanding. In contrast, the relationship between representation and understanding (and its implications) is the focus of the later sections of this chapter.

In order to develop these general ideas a bit further, I propose an account of scientific representation based on Ruth Millikan's teleological theory of representation (1984, 1989). Although I do not have the space to work through all the details of Millikan's views here, a few general features of her account will prove particularly useful in capturing the social/communal and historical/dynamical aspects of scientific representation.

First, Millikan's approach requires us to focus our attention on the *consumers* of representations rather than on their producers (Millikan 1989, 283). As Millikan explains:

> Let us view the system, then, as divided into two parts or two aspects, one of which produces representations for the other to

consume. *What we need to look at is the consumer part, at what it is to use a thing as a representation.*

(Millikan 1989, 285, emphasis added)

I suggest that because scientific communities are the consumers of scientific models and the results derived from them, it is the interpretations and goals of the broader community that we need to focus on to determine the representational content of a scientific model rather than the intentions of the individual producers of those models (whose interpretations or goals might run contrary to those of the broader community).

A second feature of Millikan's view is that it requires us to consider the history of a representation (and the context in which it has been developed) in order to determine its content (Millikan 1984, 18; 1989, 284). Specifically, in order to determine the content of a representation, we need to determine its function, which is determined by its historical use. According to Millikan's etiological view, the function of something is what earlier things of this type have done that has contributed to their survival and reproduction, which, in turn, explains their current use. In the case of representations (and removing the biological benefits involved in natural selection), this etiological view suggests that a representation means whatever earlier versions of the representation have been taken to mean (by the consumers) that has enabled them to positively contribute to the goals/aims/functions of the system in which those representations are used. This aspect of Millikan's view helps us recognize that the historical development of a scientific representation and its past uses are essential for determining its current representational content.

A third feature of Millikan's view is that the purposes/functions of a given representation are determined by the purposes of the larger producer-consumer system of which it is a part. In particular, the functions (or goals) that matter are the ways in which consumers use representations to generate benefits for the overall system (that includes both the producers and the consumers of those representations). In short, the etiological function of a representation is determined by the benefits that its use (or the use of other similar/related representations) has provided for the overall system in the past.

In sum, Millikan's view involves three crucial features that need to be incorporated into accounts of how scientific models represent:

1 The account focuses our attention on the interpretations, assumptions, and uses of *consumers* (rather than producers).
2 The account is *etiological/historical/dynamical* in that it focuses on the past uses and development of a representation to account for its representational content.

3 The account focuses our attention on *the goals/purposes/benefits of a larger system* that consumes/uses the representations to accomplish certain goals; that is, the function of the representation is to produce certain benefits for the overall system.

Let us apply these features of Millikan's view specifically to scientific models. With respect to the first feature, I argue that the interpretations, background assumptions, and justifications adopted by the scientific community that uses/consumes a model are essential for determining what the model represents. Concerning the second feature, I propose that the content of a scientific model can only be determined by looking at the historical context and development of the model and the ways it has been interpreted/used in the past. It is the histories of a scientific model that tell us why a particular interpretation of its representational content has been adopted by the community. For novel models (or theories), I suggest that these contributions are made by the past uses of other similar (or related) representations, or the history of the use of models and theories within the discipline or research program more generally. In short, looking at the historical uses of the model (and other related representations) enables us to determine what the model has been taken to represent that has enabled it (or other related representations) to contribute to the aims of scientific inquiry. Moreover, I contend that these historical aspects make important contributions to the representational content of a model even if current scientists have different representational aims. Users' current aims will certainly contribute something as well, but what the model represents for the scientific community cannot be isolated from its history of uses and interpretations. Finally, concerning the third feature, I suggest that the goals/purposes/functions that matter the most for determining what a scientific model represents are the benefits provided for the broader scientific community rather than individual model builders/users. In sum, what the model represents is largely determined by what the community of scientists has taken that model to represent (both accurately and inaccurately) that has led to its past production of benefits for the scientific community; for example, predictions, explanations, or understanding. The production of these benefits for the community then results in the representation (and its interpretation) being copied/reused in later scientific work. Indeed, given that the broader community determines what the aims of science are and the standards for evaluating when they have been achieved, what makes a representation a *scientific* representation is that it is designed to be consumed by the scientific community so as to achieve the goals of that community.

Putting these pieces together, I contend that consideration of each of the following five components is necessary to determine the representational content of a scientific model (this list is not intended to be exhaustive):

1 The beliefs, interpretations, and goals of the model builder/user (the producers)
2 The beliefs, interpretations, and goals of the larger scientific community (the consumers)
3 Background assumptions furnished by science's best current (and past) scientific theories (these influence how a model is built, used, and interpreted by the community)
4 The community's understanding furnished by other models (these beliefs partially constitute the social context in which the model is interpreted/used by the community)
5 Justifications provided for past uses of idealizations within the model (this clarifies the reasons why the model has been used in the past that justify its continued use in certain contexts)

What this list shows is that although the beliefs and goals of an individual model builder/user are part of the story, they are nowhere near sufficient for determining the representational content of a scientific model (or theory). Only by placing the model and its user(s) within a larger social and historical context will we be able to determine what it represents. Although some of the representational content will be determined by the representational aims of current users, many parts of that content will be fixed (or heavily constrained) by past interpretations and uses of the model (or other similar models).

Since other philosophers have described the ways in which the model builder's intentions can contribute to a model's representational content, I will begin by looking at the beliefs, interpretations, and goals of the larger scientific community. First, as Weisberg notes, "communities of modelers have standard conventions for reading model descriptions" (2013, 40). Indeed, scientific communities routinely adopt various rules and conventions about how a given (type of) model ought to be interpreted, justified, or used. What is more, these conventions often provide what Christopher Pincock refers to as "anchors" that link mathematical models to their target systems in specific ways and can structure entire research programs (2012, 492). Consequently, the representational content of a scientific model is typically partially determined, and thereby highly constrained, by the interpretations, assignments, and conventions adopted by the communities that use the model for a variety of scientific purposes. Of course, the particular goals of individual model builders/users will certainly have *some* role to play here. For example, what the modeler is interested in might determine which communities of scientists they take their target audience to be and which types of models they choose to develop. Nonetheless, because the scientific community adopts particular conventions for interpreting and using certain models, an individual scientist is limited in the ways they can stipulate, construe, or map their models onto their targets and still have the model accomplish its purposes within the larger community.

The next feature of the scientific community that partially determines the content of a scientific model are the modeling assumptions it incorporates from the community's best theories (both past and present). As Giere and others have noted, scientific theories often provide "general principles" that "act as general templates for the construction of more specific abstract objects that I would call 'models'" (2004, 745). In addition, scientific theories will often furnish various assumptions about what is relevant or irrelevant to the phenomenon. For example, the kinetic theory of gases tells us that what is relevant for understanding changes in gas behavior are the macroscale averages of the system (e.g., its mean kinetic energy) and that the particular motions and locations of the individual particles are irrelevant. As a result, the scientific community in which various gas models are formulated implicitly assumes that the overall statistical properties of the system (e.g., its pressure and temperature) are what the model aims to represent about real gases and that the motions and interactions of individual particles are not part of what the model aims to (accurately) represent. In short, the construal that specifies which parts of the model ought to be interpreted as relevant and which ought to be ignored is heavily influenced by the theories that the scientific community has adopted.

Within a particular community (or research program), another feature that determines the representational content of a scientific model is the understanding furnished by results derived from other models. We might call this the community's "background knowledge" into which the model, its interpretation, and its results are embedded (Weisberg 2013, 135). For example, a modeler in cognitive science might choose to model (and measure) certain areas of the brain due to previous studies demonstrating the role that those regions play in producing the behaviors of interest. This can also happen across scientific disciplines, for example, when economic modelers incorporate information from evolutionary biology concerning the features of human cognition (Rice and Smart 2011). Thus, it is not just the history and the development of a particular model that helps determine its representational content, but also the history, development, and understanding produced by other models (and theories) that furnish various assumptions or parameter values used for constructing and interpreting the model. Consequently, what a model is taken to represent depends heavily on the community's current state of scientific understanding.

Finally, one of the ways that the community helps determine the representational content of a scientific model is by providing justifications for the use of various idealizing assumptions employed within the model.

Indeed, many idealizing assumptions that are repeatedly reused within a research program are justified by the modeling results of previous scientists (Pincock 2012). Moreover, the ways these idealizations are interpreted and justified are often carried along with the model across different applications (or contexts) that depend on the ways in which the model has been developed, used, and understood by the broader scientific community (Knuuttila and Loettgers 2014; Rice and Smart 2011).

The features briefly surveyed here show why the history, purposes, interpretations, and uses of the larger scientific community are absolutely essential for determining what a scientific model (or theory) represents. What is more, the above view provides a clear way to demarcate scientific representations from non-scientific representations. These representations are scientific representations, because they are developed within the social/historical context of a scientific community for the purposes/goals of that community. The above features also show that what a model represents can change as different theories, assumptions, interpretations, and goals are adopted by the scientific community. Indeed, as Weisberg notes,

> Modelers might initially deem some features of models and targets important, but, as science progresses, these might be judged to be irrelevant. Similarly, new properties of targets might come to be recognized as especially important. These changes in practice and interest will occasion...a reevaluation of the model-world relationship.
>
> (Weisberg 2013, 149)

This entails that scientific representation is inherently dynamic (rather than static). Finally, given that the features discussed above can be realized in multiple different ways, there is plenty of room in this view for pluralism about the kinds of things that represent (e.g., mathematical equations, physical objects, or computations), what they represent (e.g., real or possible systems), and how their content is specifically determined within different research programs. Despite this variety, in each case, what a scientific model (or theory) represents will essentially depend on the history, assumptions, interests, and goals of the broader scientific community. Moreover, as those features of the community change, so will the representational content of the model (or theory). Therefore, scientific representation is both communal and dynamic.

3   Scientific Understanding: A Communal and Dynamical View

Similar to the above account of scientific representation, in this section I argue that philosophical accounts of scientific *understanding* need to move beyond their focus on the cognitive states of individuals and the representational capacities of individual models and, instead, must consider the conditions required for scientific communities (i.e., groups) to understand natural phenomena via multiple conflicting models used at different points in the history of science.

To see why this shift is important, it is worth noting that many accounts of scientific understanding have appealed to the mental states or cognitive limitations of individuals. A primary reason for this is because much of the epistemology literature concerning understanding has sought to determine how individual agents understand and how that cognitive state differs from knowing (Grimm 2006; Khalifa 2017; Pritchard 2009; Zagzebski 2001). As a result, following the literature on knowledge, philosophers writing about understanding routinely adopt a definition that begins with "*S* understands *p* if and only if . . . ," where *S* is an individual agent and what comes after the "if and only if" specifies the conditions that must be met for that agent to understand. Further, some epistemologists have explicitly denied that the etiological aspects of understanding are important. For example, Jonathan Kvanvig argues that "Understanding does not advert to the etiological aspects which can be crucial for knowledge. What is distinctive about understanding, once we have satisfied the truth requirement, is internal to cognition" (2003, 198–199).

Philosophers of science, too, have routinely appealed to the cognitive states of individuals in order to analyze scientific understanding (e.g., see Khalifa 2017; Potochnik 2017; Strevens 2013).[1] As an example, following Sandra Mitchell (2012), Bill Wimsatt (2007), and others, Angela Potochnik argues that the reason scientists need to use idealizations to understand natural phenomena is because the causal complexity of the world greatly outstrips what our limited minds can comprehend (Potochnik 2017, 1–2). This approach to idealization, and its role in generating understanding, implies that what science can understand is largely determined by what individual scientists are able to grasp. Indeed, Potochnik tells us that "Scientific understanding is generated via the production of scientific explanations. Successful explanation explicitly depends on the features of human psychology and cognition as much as it depends on features of the world" (Potochnik 2017, 20). More specifically, Potochnik appeals to the cognitive psychology literature to argue that agents (scientifically) understand by grasping general patterns that are based on causal relationships (Potochnik 2017, 113–114). Indeed, several philosophers have argued that the features of individuals' psychology provide insights into the nature of scientific understanding.

A final way in which philosophers have tied scientific understanding to the cognitive states of individuals is by arguing that in order to scientifically understand a phenomenon an individual must grasp an explanation of that phenomenon. For example, both Michael Strevens and J. D. Trout argue that an individual has scientific understanding of a phenomenon just in case they grasp a correct scientific explanation of that phenomenon (Strevens 2013, 510; Trout 2007, 584–585). Even those accounts that claim there can be understanding without explanation have largely focused on the understanding of individuals (Lipton 2009; Rohwer and Rice 2013).

In addition, most accounts of how scientific models enable scientists to understand have appealed to the relatively static representational capacities of individual models. In particular, most philosophers suggest that whether or not a scientific model is able to produce understanding depends primarily on what the model accurately represents, captures, or exemplifies (e.g., see Elgin 2017; Potochnik 2017; Strevens 2013; Weisberg 2013). This implies that what scientific understanding can be produced via a given model is largely determined by the representational capacities (or accuracy) of that particular model.

Although there are certainly important epistemological questions about how individuals understand via individual models, I contend that this focus has forced philosophers to miss two of the most distinctive aspects of *scientific* understanding. First, scientific understanding is produced by, and codified within, scientific communities that are composed of diverse individuals with different experiences, values, background assumptions, and goals. What is more, the understanding produced by these scientific communities almost always depends on various interactions among diverse members of the community that enable the scientific community's understanding *to go well beyond what could be grasped by any individual scientist.* Consequently, appealing only to the cognitive states and limitations of individuals misses the distinctively communal methods that science uses to produce understanding. Second, a scientific community's understanding of a natural phenomenon is typically produced via the use of multiple conflicting representations over relatively long spans of time. As a result, scientific understanding is essentially a diachronic phenomenon that cannot be analyzed merely by looking at the representational capacities of a single model (or theory) in isolation.

Following several other accounts in the literature (Grimm 2006; Le Bihan 2017; Rice 2016; Saatsi 2019; Woodward 2003), I suggest that the kind of information that produces scientific understanding of a phenomenon is modal information about how the phenomenon would (or would not) change in various counterfactual situations. In other words, scientific models produce understanding by enabling scientific communities to answer a range of what-if-things-had-been-different questions. This idea is derived from various accounts of scientific explanation that appeal to the outcomes of interventions (Woodward 2003). However, I see no reason to restrict the modal information that constitutes a scientific community's understanding to information about the results of interventions. Instead, I suggest that scientific understanding can also be deepened by grasping what would be the case in very distant counterfactual situations that tell us little about the actual features of real-world systems, or how intervening on those features would change the system (Rice 2021). I do not have the space to lay out all the details of this modal account or a full defense of all its features. However, I aim at showing that adopting a modal approach to scientific understanding enables us to capture crucial aspects of both the communal/social and dynamical/historical ways that science produces understanding.

Let us begin with the communal aspects of scientific understanding. It is crucial to note that scientific understanding is only produced via a model when scientists employ their experiences, values, background assumptions, and goals to interpret, interact, and *use* a model for particular purposes. Merely determining the representational capacities of the model is insufficient for determining the ways that these uses/interactions generate understanding. Moreover, since different scientists will bring different background assumptions, interests, and values to their interactions with the model (and the larger scientific community), different scientists will be able to extract different sets of modal information from a given scientific model. As Henk de Regt notes, when scientists engage with scientific representations, their "preferences are related to their skills, acquired by training and experience, and to other contextual factors such as their background knowledge, metaphysical commitments, and the virtues of already entrenched theories" (2009, 592). The key point I want to emphasize here is that these contextual values, skills, training, experience, background knowledge, metaphysical commitments, etc. are all heavily influenced and constrained by the broader scientific community. In other words, these important features of the ways that different scientists interact with their models are part of a much larger social (and historical) context (de Regt and Dieks 2005). Further, it is *the interactions between* different members of the larger scientific community that integrates these various pieces of modal information into a body of information that constitutes the scientific community's overall understanding of a natural phenomenon. Crucially, this set of modal information will (almost always) go well beyond the set of modal information that is grasped—or is graspable—by any individual scientist. In short, only by looking at the social context that structures the ways scientists use their models and how they interact with one another can we determine what the scientific community is able to understand via a given (set of) model(s).

What is more, in contrast with the views that analyze scientific understanding via accurate representation of the actual world, a modal account of scientific understanding easily accommodates the epistemic contributions made by groups of scientists using multiple conflicting models in order to understand the same phenomenon (see Chakravartty 2010, Morrison 2011, or Weisberg 2013 for some examples). Specifically, different models can be used to explore different counterfactual situations that are of interest to different modelers within the community. For example, one group of scientists might use a model focused on genetic factors to determine what would happen in various situations where those factors are changed and environmental factors are absent. Another group of scientists (with different interests and goals) might use a scientific model that makes fundamentally different assumptions and idealizations to explore counterfactual situations in which various environmental factors are changed and genetics is largely ignored (or held fixed). In short, different groups of scientists will use different (and often conflicting) idealized models to answer the what-if-things-had-been-different questions that are of interest to them. The counterfactual information used to answer these questions can then be combined via various interactions among the members of the scientific community—even if no individual scientist is interested in, or able to grasp, all of the counterfactual information that constitutes the community's overall understanding of the phenomenon.

Considering the role of modal information in scientific understanding also enables us to capture the etiological/historical aspects of the processes by which science produces understanding. First, we should remember that the interests, assumptions, and interpretations of the scientific community can change over time. This means that scientists will be interested in using their models to explore different counterfactual situations as different what-if-things-had-been-different questions become interesting to them. For example, looking at the history and development of a scientific model will enable us to see how scientists' background assumptions about what is relevant/irrelevant to the phenomenon of interest have influenced which pieces of modal information they investigate with their models. As science progresses, the same model might be used to extract very different pieces of modal information as the uses, interpretations, and goals of the scientific community (or research program) change. Therefore, what understanding a given scientific model provides cannot be determined synchronically by looking just at what a model represents at a particular time.

The second etiological/historical aspect of scientific understanding involves the contributions of past models and theories. These representations have been disconfirmed/replaced and are typically in conflict with the representations currently adopted by the scientific community. Nonetheless, I suggest that much of the understanding that a scientific community has regarding a phenomenon is derived from these past scientific representations. For example, despite being replaced by alternative theories during the modern synthesis, much of our current understanding of biological traits has been derived from Darwin's original theory of selection. Similarly, Bohr's model of the hydrogen atom has greatly contributed to the scientific community's understanding of atoms despite being replaced by later quantum mechanical models (Bokulich

2011). One way to account for the understanding produced by these past theories and models is to note that learning about counterfactual situations—including rather distant possible worlds—can improve (or deepen) scientists' understanding of a phenomenon. Consequently, past representations can contribute to scientific understanding by providing (true) modal information about what would happen in various counterfactual situations.

In summary, scientific understanding of a phenomenon is not accomplished by a single modeler using a single model in isolation. Only by considering how diverse communities of scientists interact over time to develop bodies of understanding via the use of multiple (potentially conflicting) representations can we capture the essential features of *scientific* understanding.

4 Representation, Understanding, and Scientific Progress

Although both scientific representation and scientific understanding are communal and dynamic, I contend that the understanding that a scientific model is able to produce for a community is largely independent of whether it accurately represents the relevant or interesting features of its target systems (Rice 2021). This suggestion runs contrary to several views that have suggested that scientific understanding is produced by accurately representing the difference makers for the phenomenon (e.g., Strevens 2008, 2013). There are several reasons that accounts of scientific understanding ought to move beyond such appeals to accurate representation (or instantiation) relations:

1 Much of science's understanding is produced via idealized models that directly distort features that are known to make a difference and are of interest to the scientists using the model (Rice 2017, 2018).
2 Scientific models that tell us about merely possible systems can greatly improve scientific understanding (Le Bihan 2017; Rice 2021; Saatsi 2019).
3 Much of science's current understanding of phenomena has been derived from past scientific representations that are known to be inaccurate and are in conflict with our current models/theories of those phenomena (Bokulich 2008; Potochnik 2017; Rice 2021; Saatsi 2019).
4 Science often produces understanding via the use of multiple conflicting models for the same phenomenon (Chakravartty 2010; Elgin 2017; Morrison 2011; Potochnik 2017; Rice 2021).

These observations about scientific practice provide strong reasons for thinking that what a community is able to understand does not depend on the representational accuracy of the models it uses to produce that understanding. Further, they suggest that producing more accurate representations will not necessarily produce more understanding. Rather, given the modal nature of scientific understanding, producing large sets of incompatible models (many of which may distort the difference-making features of interest) will typically produce a deeper understanding of a phenomenon than developing representations that accurately represent different aspects of the phenomenon in consistent ways. What is more, in contrast with realists' suggestion that science makes progress by constructing ever-more accurate representations of real systems, I contend that science makes progress by evolving toward the development of (sets of) representations that are progressively *more useful for producing understanding*. Consequently, science makes progress by developing (conflicting) sets of representations that expand and deepen the community's understanding of natural phenomena.

I have argued that incorporating these communal and dynamical features of scientific practice is necessary for capturing the essential (though not unique) features of scientific representation and understanding. Only by incorporating these features will philosophers be able to generate accounts of these concepts that more accurately describe the ways that evolving scientific communities construct multiple conflicting representations in order to understand natural phenomena.

Note

1 In fact, philosophers of science often assert that scientific understanding is just a species of knowledge (Salmon 1989; Woodward 2003).

References

Boesch, B. 2017. "There Is a Special Problem of Scientific Representation." *Philosophy of Science 84*: 970–981. https://doi.org/10.1086/693989.

Bokulich, A. 2008. *Reexamining the Quantum-Classical Relation: Beyond Reductionism and Pluralism*. Cambridge: Cambridge University Press. https://doi.org/10.1017/CBO9780511751813.

Bokulich, A. 2011. "How Scientific Models Can Explain." *Synthese 180*: 33–45. https://doi.org/10.1007/s11229-009-9565-1.

Callender, C. and Cohen, J. 2006. "There Is No Special Problem About Scientific Representation." *Theoria 55*: 67–85.

Chakravartty, A. 2010. "Perspectivism, Inconsistent Models, and Contrastive Explanation." *Studies in History and Philosophy of Science 41*(4): 405–412. https://doi.org/10.1016/j.shpsa.2010.10.007.

de Regt, H. 2009. "The Epistemic Value of Understanding." *Philosophy of Science 76*(5): 585–597. https://doi.org/10.1086/605795.

de Regt, H. and Dieks, D. 2005. "A Contextual Approach to Scientific Understanding." *Synthese 144*: 137–170. https://doi.org/10.1007/s11229-005-5000-4.

Elgin, C. Z. 2017. *True Enough*. Cambridge, MA: MIT Press. https://doi.org/10.7551/mitpress/9780262036535.001.0001.

Frigg, R. and Nguyen, J. 2017. "Scientific Representation Is Representation-as." In *Philosophy of Science in Practice: Nancy Cartwright and the Nature of Scientific Reasoning*, eds. H.-K. Chao and J. Reiss, 149–179. Berlin: Springer. https://doi.org/10.1007/978-3-319-45532-7_9.

Giere, R. 2004. "How Models Are Used to Represent Reality." *Philosophy of Science* 71: 742–752. https://doi.org/10.1086/425063.

Godfrey-Smith, P. 2006. "The Strategy of Model-based Science." *Biology and Philosophy* 21: 725–740. https://doi.org/10.1007/s10539-006-9054-6.

Grimm, S. 2006. "Is Understanding a Species of Knowledge?" *The British Journal for the Philosophy of Science* 57: 515–535. https://doi.org/10.1093/bjps/axl015.

Khalifa, K. 2017. *Understanding, Explanation and Scientific Knowledge*. Cambridge: Cambridge University Press. https://doi.org/10.1017/9781108164276.

Knuuttila, T. and Loettgers, A. 2014. "Magnets, Spins, and Neurons: The Dissemination of Model Templates Across Disciplines." *Monist* 97: 280–300. https://doi.org/10.5840/monist201497319.

Kvanvig, J. L. 2003. *The Value of Knowledge and the Pursuit of Understanding*. New York: Cambridge University Press. https://doi.org/10.1017/CBO9780511498909.

Le Bihan, S. 2017. "Enlightening Falsehoods: A Modal View of Scientific Understanding." In *Explaining Understanding*, eds. S. Grimm and C. Baumberger, 111–135. New York: Routledge.

Lipton, P. 2009. "Understanding Without Explanation." In *Scientific Understanding:Philosophical Perspectives*, eds. H. W. de Regt, S. Leonelli, and K. Eigner, 43–63. Pittsburgh: University of Pittsburgh Press. https://doi.org/10.2307/j.ctt9qh59s.6.

Millikan, R. 1984. *Language, Thought, and Other Biological Categories*. Cambridge, MA: MIT Press.

Millikan, R. 1989. "Biosemantics." *The Journal of Philosophy* 86(6): 281–297. https://doi.org/10.2307/2027123.

Mitchell, S. 2012. *Unsimple Truths: Science, Complexity and Policy*. Chicago: University of Chicago Press.

Morrison, M. 2011. "One Phenomenon, Many Models: Inconsistency and Complementarity." *Studies in History and Philosophy of Science Part A* 42(2): 342–351. https://doi.org/10.1016/j.shpsa.2010.11.042.

Pincock, C. 2012. "Mathematical Models of Biological Patterns: Lessons from Hamilton's Selfish Herd." *Biology and Philosophy* 27: 481–496. https://doi.org/10.1007/s10539-012-9320-8.

Potochnik, A. 2017. *Idealization and the Aims of Science*. Chicago: University of Chicago Press. https://doi.org/10.7208/chicago/9780226507194.001.0001.

Pritchard, D. 2009. "Knowledge, Understanding, and Epistemic Value." In *Epistemology (Royal Institute of Philosophy Lectures)*, ed. A. O'Hear, 19–43. Cambridge: Cambridge University Press. https://doi.org/10.1017/S1358246109000046.

Rice, C. 2016. "Factive Scientific Understanding Without Accurate Representation." *Biology and Philosophy* 31(1): 81–102. https://doi.org/10.1007/s10539-015-9510-2.

Rice, C. 2017. "Models Don't Decompose That Way: A Holistic View of Idealized Models." *British Journal for the Philosophy of Science* 70(1): 179–208. https://doi.org/10.1093/bjps/axx045

Rice, C. 2018. "Idealized Models, Holistic Distortions and Universality." *Synthese* 195(6): 2795–2819. https://doi.org/10.1007/s11229-017-1357-4.

Rice, C. 2021. *Leveraging Distortions: Explanation, Idealization and Universality in Science*. Cambridge, MA: MIT Press. https://doi.org/10.7551/mitpress/13784.001.0001.

Rice, C. and Smart, J. 2011. "Interdisciplinary Modeling: A Case Study of Evolutionary Economics." *Biology and Philosophy* 26: 655–675. https://doi.org/10.1007/s10539-011-9274-2.

Rohwer, Y. and Rice, C. 2013. "Hypothetical Pattern Idealization and Explanatory Models." *Philosophy of Science* 80(3): 334–355. https://doi.org/10.1086/671399.

Saatsi, J. 2019. "Realism and Explanatory Perspectives." In *Understanding Perspectivism: Scientific Challenges and Methodological Prospects*, eds. M. Massimi and C. D. McCoy, 65–84. New York: Taylor and Francis. https://doi.org/10.4324/9781315145198-5.

Salmon, W. C. 1989. *Four Decades of Scientific Explanation*. Pittsburgh: University of Pittsburgh Press.

Strevens, M. 2008. *Depth: An Account of Scientific Explanation*. Cambridge, MA: Harvard University Press.

Strevens, M. 2013. "No Understanding Without Explanation." *Studies in History and Philosophy of Science* 44: 510–515. https://doi.org/10.1016/j.shpsa.2012.12.005.

Trout, J. D. 2007. "The Psychology of Explanation." *Philosophy Compass* 2: 564–596. https://doi.org/10.1111/j.1747-9991.2007.00081.x.

Weisberg, M. 2007. "Who Is a Modeler?" *The British Journal for the Philosophy of Science* 58: 207–233. https://doi.org/10.1093/bjps/axm011.

Weisberg, M. 2013. *Simulation and Similarity: Using Models to Understand the World*. Oxford: Oxford University Press. https://doi.org/10.1093/acprof:oso/9780199933662.001.0001.

Wimsatt, W. 2007. *Reengineering Philosophy for Limited Beings*. Cambridge, MA: Harvard University Press. https://doi.org/10.2307/j.ctv1pncnrh.

Woodward, J. 2003. *Making Things Happen: A Theory of Causal Explanation*. Oxford: Oxford University Press. https://doi.org/10.1093/0195155270.001.0001.

Zagzebski, L. 2001. "Recovering Understanding." In *Knowledge, Truth, and Duty: Essays on Epistemic Justification, Responsibility, and Virtue*, ed. M. Steup, 235–252. New York: Oxford University Press. https://doi.org/10.1093/0195128923.003.0015.

18 Representation and Understanding Are Constitutively Communal but Not Constitutively Historical

A Reply to Rice

Jaakko Kuorikoski

Collin Rice (this volume, Chapter 17) argues that representation and understanding are deeply communal and dynamic achievements and that existing philosophical accounts have thus far given insufficient attention to these contextual aspects. I could not agree more. I also agree with his diagnosis of one of the roots of this bias: the presumption that the internal mental states of individual scientists are the primary basis of representation and understanding. As my own chapter also argues for the communality of understanding, I focus here on representation. On representation, I am again in full agreement with Rice (and with Callender and Cohen 2006) in that scientific representation should not require a completely novel and distinct philosophical theory of representation, separate from that of language, pictures, and art. Scientific representation is not a sui generis phenomenon and, in hindsight, it is somewhat strange that philosophers of science have sometimes proceeded in discussing representation almost as if it were, without much reference to more general theories of representational content.

Rice builds his argument on the teleosemantic theory of (primitive) mental representation and biological information, according to which the content of a representation is determined by the function responsible for its prevalence and survival in the population (i.e., what philosophers of biology call the Wright function). Within the teleosemantic picture, he emphasizes the importance of history and the consumer side of representational practice, and understandably downplays the role of selection and the original focus on specifically mental representation. In this commentary, I want to briefly offer an alternative, more social, perspective from which to think about the nature of aboutness in general, the inferentialist theory of the conceptual content of Robert Brandom (1994), and to see whether any different conclusions follow. I do not offer any arguments in favor of inferentialism over teleosemantics, but simply leave it on the table as an alternative philosophical framework yielding not only broadly similar but also importantly distinct insights

DOI: 10.4324/9781003202905-21

about the communal and historical nature of scientific representation. Most importantly, inferentialism provides an alternative philosophical foundation for the thesis that scientific representation is, indeed, *constitutively* communal. It is not just that analyzing community-wide practices is epistemically necessary if we want to determine what representationally used artifacts, such as models, represent. It is that models have their representational content *in virtue* of these communal practices. Representation is grounded on social normative practices, not on the intentional states of individuals.

Perhaps more familiar representationalist approaches explain representational content and any inferential implications thereof in terms of some special relation between the representational vehicle and its target (such as representation, reference, primitive mental intentionality, or similarity), whereas inferentialism explains representation in terms of the inferential practices. For example, a map is used as a representation of some piece of terrain *in virtue of* some of its features being used as premises in inferences with conclusions pertaining to the terrain in question (Brandom 1994, 518–519). There is no representation relation prior to and independent of these inferential practices. Like maps, scientific representations are not *arbitrary* in that these inferences are not solely a matter of stipulation, but they are instead partly grounded on independent features of the representing artifact. In his argument for the communal nature of representation, Boesch (2017) calls this autonomous aspect of representation *licensing*. From the inferentialist perspective, however, licensing conflates two equally important but distinct senses of allowing for new inferences: representations must be able to (causally) afford the user to make new inferences about the target, and these inferences are (normatively) permitted when they are arrived at by appropriately using community sanctioned rules of inference.

There certainly must be a complex historical causal connection between the terrain and the piece of paper with markings on it: physical objects with complex and intricate features enabling inferences about some other distinct set of physical facts do not usually come into existence by accident. Many kinds of structure preserving mappings can also be constructed between the terrain and the markings. These facts can function as parts of the causal explanation for why the piece of paper can afford non-trivial inferences to certain kinds of cognitive agents. Nevertheless, these "natural" facts do not, as such, make the paper to be *about* the terrain. It is the fact that the paper is *used* in a discursive practice governed by rules of inference that makes it to be about anything. And as Wittgenstein concluded, there really are no such things as completely private rules and hence no private meaning or aboutness. Representation is therefore not just embedded in, tied up in, or constrained by the social practices of science, as argued by Rice. Representation is strongly communal in that it is *constituted* by the social and normative practices of inference.

Let us now return to scientific representation. Drawing on the teleo-semantic theory, Rice provides a long list of factors that determine the representational content of a model, including past and present intentions, beliefs, theories, other models, and justifications for idealizations. Although I agree that all of these factors can be important, care is in order here about what is meant by determination. Are these the factors that need to be known, the *evidence* required, for the full delineation of the content of a given representation? Or are these the factors *in virtue of which* the representation has the content that it does (regardless of whether we have correctly delineated it or not)? Whichever is the intended meaning, I think the list is implausibly long. Take an every-day example of a representation: the London underground map. The map is an epistemic representation for its users in that they can use it to infer facts about the underground that are novel to them. There probably is much of interest to learn about the history of this iconic map, which can *explain* particular design choices in the map, but the map itself has been specifically designed to be as self-standing as possible. At least in principle, the ideal is that all the representational potential, everything that can be validly inferred about its target, could be exhaustively explicated with only minimal reference to its intended use, the interpretive practices of the community, and preferably without much reference to its history. For example, although the colors of the different lines all have interesting historical roots, today they do not really represent anything *about* the lines themselves.[1]

In the same way, at least mature epistemic representations in science, such as the foundational basic model templates in economics and physics, are relatively self-standing in this sense. Although philosophers often deride the largely ahistorical (and by implication unreflective) nature of most economics and physics degree programs, it should be noted that this independence from history is also an important epistemic achievement by the fields in question. Students of economics can learn the basic principles of equilibrium analysis without having in-depth knowledge of Marshall, Hicks, or Walras, because the norms of inference governing the use of these models have stabilized, are almost universally shared, and there is relatively little room for interpretive negotiation about whether particular results are derived correctly or incorrectly. Granted, this stabilization can lead to a situation in which the use of the representation is regimented to the extent that this may cloud what the canonical representation is supposed to be a representation of (is the Arrow-Debreau general equilibrium a model of "a competitive market economy," and if it is, what is the extension of this predicate in the real world?). But even in this case, it is unclear why the intentions, caveats, and justifications of past modelers would automatically qualify as reasons relevant for how the model *ought* to be interpreted today.

I, therefore, agree that every factor in Rice's list can sometimes be used as evidence to argue for a particular model having certain kind of representational content or potential, but surely they cannot always be all individually necessary for understanding what a model is about. All of the mentioned factors probably are part of the full historical explanation of why any particular model came to have its current representational properties, but this does not mean that we always need to know everything on the list to know what these properties are. I will next further argue that from the inferentialist perspective, the history of a representation also cannot be the reason *in virtue of which* the representation has the content that it has: although scientific representation is constitutively social, it is not constitutively historical.

A community at any given time may uphold some inferential norms of model use in such a way that the norms are not explicitly discussed in the community and perhaps no one is even explicitly aware of their existence. These norms are the historical product of generations of model development and are implicitly instilled in new researchers in their education. In fact, I believe that all fields of science exhibit such strongly implicit norms and that the existence of these norms is an important reason why studying the history of a particular field is a powerful way of expanding the space of epistemic possibilities within that field, to see why things that seemed obvious and necessary are, in fact, contingent. As sociologists and historians of science have emphasized, the way our models look now is not fully determined by the combination of past empirical findings and some fictitious pure epistemic rationality. Economics is an obvious and much discussed example, but the same point holds for the natural sciences. Given that the ability to locate a phenomenon within a space of alternative possibilities is constitutive of understanding, understanding the history of a field and its representational practices is thus at least helpful, if not always strictly necessary, for understanding the current representational practices of that field.

Nevertheless, I hold that if all of these implicit norms were somehow made explicit, the historical roots of these norms would cease to be relevant to the determination of the present representational content. Current representational practices screen off past practices. This follows from two assumptions, both of which are sometimes philosophically contested. First, the representational content of an artifact is fully grounded (supervenes) on the normative practices of using the artifact in the game of giving and asking reasons about its target (inferentialism). Second, the naturalist conviction is that these normative practices are, in the end, nothing more than naturalistically explicable social behavior, albeit of a very sophisticated sort.[2] This social behavior is physically realized causal interaction and, as such, any memory it exhibits must be grounded on causal facts physically realized at all moments in time. If the history is to have an effect on representation, it does so by already having affected present practice. Further, if we also identify the *understanding* created by these representations with the expansion of the modal inferences afforded by the said representations – as I suggest in my chapter – then understanding cannot be constitutively historical either.

Let us indulge in a thought experiment. Imagine that new archival discoveries conclusively showed that Volterra had originally intended his model to represent something completely different, say the interrelations of two species competing for the same resources (this would not make much sense, but hey, this is a thought experiment) and that this intention was generally acknowledged within the relevant ecological community of the day. In this scenario, the whole business with Adriatic fish populations and predators was a result of a bizarre series of misunderstandings and poor translations of original texts and correspondence. Would this curious historiographical discovery of long forgotten intentions and stipulations have any bearing on the representational properties of current ecological models? My immediate but admittedly theoretically tainted intuition would be a resounding no, but if the original intentions really were part of the *constituting* base of current representational properties, the answer would have to be yes. The same, arguably unintuitive, implications seem to follow from many other representationalist semantic theories, such as causal theories of reference. By anchoring meaning firmly to the causal history of (a population of) representations, teleosemantics thus bestows substantial authority on the current use of models to their past creators and users (and by implication to the study of the history of science). I leave it to the reader to evaluate whether this is credible and justified.

Notes

1 With the possible exception of the green of the District line, which connects many places that at least used to be green or named after former greens.
2 It is contentious as to whether Brandom would fully subscribe to either of these premises. The difficult question of what his "conceptual realism" entails and is committed to cannot be addressed here.

References

Boesch, Brandon. 2017. "There Is a Special Problem of Scientific Representation." *Philosophy of Science*, 84(5): 970–981. https://doi.org/10.1086/693989.

Brandom, Robert. 1994. *Making It Explicit: Reasoning, Representing, and Discursive Commitment*. Harvard University Press.

Callender, Craig and Jonathan Cohen. 2006. "There Is No Special Problem about Scientific Representation." *Theoria. (Revista de Teoría, Historia y Fundamentos de la Ciencia)*, 21(1): 67–85.

Rice, Collin. this volume. "Scientific Representation and Understanding: A Communal and Dynamical View."

19 Which Modal Information and Abilities Are Required for Inferential Understanding? A Reply to Kuorikoski

Collin Rice

Jaakko Kuorikoski's "Facitvity, Pluralism and the Inferential Account of Scientific Understanding" does important work in clarifying several key requirements for philosophical theories of understanding. Specifically, Kuorikoski argues that a pragmatist approach to theorizing about understanding suggests that our account of understanding ought to capture: (1) the social-epistemic function of attributions of understanding, (2) the factivity of understanding, and (3) explanatory pluralism. In order to satisfy these criteria, Kuorikoski defends a modal account of understanding based in the ability to correctly draw inferences about counterfactual situations. I am certainly sympathetic with such a modal account of understanding (e.g., see Rice 2016, 2021). However, I have some reservations about Kuorikoski's pragmatist framing of the criteria for a theory of understanding and the details of the arguments offered in support of the modal approach.

My first concern is that the pragmatist approach advocated by Kuorikoski may not be as independent of the contrasting representationalist project as he suggests. Kuorikoski contrasts these two approaches as follows. The representationalist focuses on identifying the necessary mental states required for understanding, whereas the pragmatist focuses our attention on determining what is at stake in disputes about who understands. Although I certainly agree that philosophical accounts ought to account for such social functions of attributions of understanding and should be responsive to the factive aims and explanatory pluralism we find in scientific practice, it is a bit unclear as to whether we can accomplish those tasks without identifying something like criteria for the necessary epistemic states that individuals/communities must satisfy in order to count as understanding. For example, it seems that one way to determine the social function of disputes about understanding is just that we are debating *who satisfies the necessary epistemic criteria* that are the focus of representationalist theories of understanding. More specifically, if we adopt a modal (or counterfactual) approach to understanding, it seems we might cast disputes concerning understanding as attempts to determine who has the correct information (or beliefs) about what would occur in various counterfactual situations. Relatedly,

DOI: 10.4324/9781003202905-22

the pragmatic ability to correctly draw counterfactual inferences about the object of understanding will likely depend on the agent having the prerequisite beliefs about the state(s) of the system and the results of various changes to its features. That is, the abilities Kuorikoski identifies as essential to understanding might have the achievement of certain epistemic states as a prerequisite. In addition, a representationalist might account for the factivity of understanding by merely stipulating that (at least most of) the mental states that constitute a body of understanding *must be true* (i.e., by adopting some kind of externalism about the epistemic states that constitute understanding). This would also allow the representationalist to account for understanding's normative dimension that Kuorikoski suggests involves grasping something "correctly" or "in the right way." Finally, a representationalist might agree that the explanatory pluralism we find in science shows that one's understanding can be expanded by grasping multiple explanations that are constituted by different true beliefs about various counterfactual situations. In sum, it is not clear that Kuorikoski's pragmatist approach will be able to successfully accomplish its aims without appealing to the kinds of mental states that are the focus of the representational approach. Rather than two separate research programs, I suggest that these two approaches will need to work together to account for the diversity of social and epistemic functions of the concept of understanding.

A second issue concerns using the number of what-if-things-had-been-different questions (or w-questions) that can be accurately answered as "a rough measure of the level of understanding about a phenomenon" (Kuorikoski this volume, Section 2). Although I think one can make the case for something like this claim, it will require us to determine whether certain distant possibilities that are disconnected from our current interests can, nonetheless, deepen our understanding. Specifically, there are many ways that a phenomenon could have been different that seem unimportant or, at least, uninteresting to humans. In particular, there are myriad ways that the system might have been different that will have no effect on the occurrence of the phenomenon of interest. Although irrelevant factors can, and often do, play essential roles in scientific explanations (Batterman 2002; Batterman and Rice 2014; Rice 2021), it also seems clear that there will be several irrelevant factors whose changes are simply not of interest. For example, how gases adjust their temperature in response to changes in pressure will remain the same regardless of whatever haircuts the scientists taking the measurements happen to have, but such an answer to a w-question, though accurate, seems to do little to deepen one's understanding of gas behavior. Consequently, it seems that the context might play a crucial role in limiting just which pieces of modal information deepen understanding (Potochnik 2017; Woodward 2003), or it might instead determine the degree to which those answers deepen our understanding (Rice 2021).

Another issue is that we will likely need to take into account the *variety* or *range* of the counterfactual inferences that an individual has the ability to make when evaluating the depth of their understanding. Suppose Jacob can correctly make 100 inferences about how changing the pressure of a gas from 0 to 100 psi would change its temperature. In contrast, Julia can correctly make only 75 inferences, but those inferences include 25 ways that changing the pressure would change the temperature, 25 ways that changing the temperature would change the pressure, and 25 ways that changing the volume of the container would change the pressure. Although Julia can correctly make fewer counterfactual inferences, it is not obvious that Jacob's understanding is deeper than Julia's. All this is to say that it is far from clear that the total set of correct counterfactual inferences can be taken as a rough measure of the degree to which one understands because several other factors will influence whether and to what degree those inferences deepen one's understanding. Presumably, these further considerations are what Kuorikoski is referring to when he notes that depth of understanding involves a variety of dimensions, but I suggest that these details need to be worked out in a bit more detail for us to accomplish the pragmatists' goal of determining the social function(s) of attributing understanding. After all, comparing the depth of two individuals' (e.g., the expert's and the novice's) understanding seems to be an important part of the social function of understanding.

Another potential place of disagreement concerns the connection between the ability to make correct counterfactual inferences and some kind of representational accuracy about the explanatory dependencies. I certainly agree with Kuorikoski that it is the accuracy of the answers to w-questions, and not the accuracy of the idealized models used to make them, that matters for understanding. However, Kuorikoski seems to suggest a close connection between the two when he says "factivity of understanding seems to directly imply a certain level of representational accuracy about the explanatory dependencies in question" (this volume, 226). This statement seems right if we interpret it as a claim about *what is grasped* by the agent or community that understands. But I contend that it would be incorrect if we interpret it as claiming that the *models/theories* used to draw those inferences are required to accurately represent those explanatory dependencies. The key point is that w-inferences are not simply "read off" of the counterfactual dependence structure of scientific models and theories. Instead, as Tarja Knuuttila and Martina Merz note, "understanding through models comes typically by way of building them, experimenting with them, and trying out their different alternative uses" (2009, 154). As a result of these interactions, scientists are often capable of extracting correct answers to w-questions from (a set of conflicting) models that drastically misrepresent the explanatory dependencies. Therefore, although we may want to attribute accuracy regarding an understander's beliefs about the explanatory dependencies, I do not think we should adopt a strong link between the factivity of that understanding and the representational accuracy of the models and theories used to produce it. Kuorikoski does seem to distance himself from such a link at several points in the paper, but given the goals of the pragmatist approach I think it is important to emphasize that part of the abilities (or skills) attributed to a person when we attribute understanding to them is the ability to *extract* or *discover* correct w-inferences from their *interactions with* scientific models and theories.

It is also worth noting a crucial distinction between Kuorikoski's suggestion that explanatory understanding amounts to the ability to make correct modal inferences and the requirement that those inferences must track ontic *dependence* relationships (Kuorikoski this volume, 221). As I have argued elsewhere, many scientific explanations trade in modal inferences that involve the *lack of dependence* (Batterman and Rice 2014; Rice 2021). In particular, this occurs in cases where scientists aim at explaining the stability of some pattern across very different systems. For example, certain patterns of fluid flow, biological evolution, or neural systems reoccur across systems that are very different with respect to their physical-mechanical parts and interactions. When explaining the stability of these patterns, scientists often use renormalization, homogenization, or equilibrium reasoning to show that the occurrence of the phenomenon does *not* depend on various features that are different across those systems (Rice 2021). Grasping the irrelevance of those features is then what enables scientists to understand why the pattern is stable across very different systems. As a result, we should not restrict a modal account of (explanatory) understanding to only those w-inferences that track ontic dependencies.

My final objection concerns the degree to which pluralism about explanation and pluralism about understanding might stand or fall together. Kuorikoski suggests that the history of science shows that we ought to be pluralists about explanation, but he then aims at offering a unified (or monistic) account of (explanatory) scientific understanding. I certainly agree that the existing literature on explanation presents us with a plethora of different explanations. Despite this diversity, one might argue that there is, nonetheless, something that unifies each of these explanations that enables them to all count as scientific explanations. In fact, Kuorikoski suggests just such a feature: all scientific explanations might be required to track at least some relationships of counterfactual dependence (and perhaps some counterfactual independencies as well). In other words, there may be some unifying features that enable us to provide a monistic account of explanation that would more directly relate to the information/abilities involved in Kuorikoski's monistic account of scientific understanding. Another option here is to maintain genuine pluralism about scientific explanation, but to then argue that this suggests that we ought to be pluralists about scientific understanding as well. This alternative view might then argue that the different kinds of explanations we find in scientific practice generate different kinds of understanding for the agents who grasp them. Given space constraints, I will not argue further for either of these options here. However, I will note that it is unclear why the same history and literature that has led to a consensus concerning explanatory pluralism would not provide similar reasons for adopting pluralism regarding the understanding acquired via those explanations.

References

Batterman, R. W. 2002. *The Devil in the Details: Asymptotic Reasoning in Explanation, Reduction, and Emergence.* Oxford: Oxford University Press. http://dx.doi.org/10.1093/0195146476.001.0001.

Batterman, R. W., and Rice, C. 2014. "Minimal Model Explanations." *Philosophy of Science 81*(3): 349–376. https://doi.org/10.1086/676677.

Knuuttila, T., and Merz, M. 2009. "Understanding by Modeling: An Objectual Approach." In *Scientific Understanding: Philosophical Perspectives*, eds. H. W. de Regt, S. Leonelli, and K. Eigner, 146–168. Pittsburgh: University of Pittsburgh Press.

Kuorikoski, J. this volume. "Factivity, Pluralism and the Inferential Account of Scientific Understanding."

Potochnik, A. 2017. *Idealization and the Aims of Science.* Chicago: University of Chicago Press.

Rice, C. 2016. "Factive Scientific Understanding Without Accurate Representation." *Biology and Philosophy 31*(1): 81–102. https://doi.org/10.1007/s10539-015-9510-2.

Rice, C. 2021. *Leveraging Distortions: Explanation, Idealization and Universality in Science.* Cambridge, MA: MIT Press. https://doi.org/10.7551/mitpress/13784.001.0001.

Woodward, J. 2003. *Making Things Happen: A Theory of Causal Explanation.* Oxford: Oxford University Press. https://doi.org/10.1093/0195155270.001.0001.

20 Maps, Models, and Representation

James Nguyen and Roman Frigg

1 The Map Analogy

Scientists from across different fields construct, investigate, and draw conclusions from scientific models. Such models form the basis of much of our scientific knowledge. An immediate philosophical question then is: how do models perform the function that they do, the function of providing information about the parts and aspects of the world (their target systems) that we are ultimately interested in? We call a representation that licences inferences about its target in this way an *epistemic representation*, and we have argued that models perform their function by being epistemic representations of their targets. Of course, this just pushes the question back a level: what does it mean for a model to be an epistemic representation of a target?

In the philosophy of science, it has become popular to draw on the idea that cartographic *maps* provide an appropriate analogy for understanding how scientific models, and indeed science more generally, work(s). In his 1994 presidential address to the Philosophy of Science Association, Giere argues that if we are to understand scientific representation, he "would suggest beginning with maps, e.g., a standard road map. Maps have many of the representational features we need for understanding how scientists represent the world" (1994, 11). Kitcher devotes Chapter 5 of his *Science, Truth, and Democracy* to the analogy, and writes: "I want to clarify the picture of the sciences I have been developing by looking at the core field, the academically rather unfashionable discipline of cartography" (2001, 55). Winther notes that both a scientific theory and a scientific model can be seen as "a map of the world" (2020, 29 and 46). And a few decades earlier, Toulmin noted that "the problems of method facing the physicist and the cartographer are logically similar in important respects, and so are the techniques of representation they employ to deal with them" (1953, 105).[1]

This suggests that both models and maps are epistemic representations and that models represent their targets as maps represent their territories. We call this the *models-as-maps analogy*. This analogy suggests that philosophers of science interested in representation can turn

DOI: 10.4324/9781003202905-23

to cartography to provide them with a worked-out account of representation, ready to be used in the sciences. But cartographers seem rather unconvinced about the viability of this purported lateral knowledge transfer. In their seminal *The Nature of Maps*, cartographers Robinson and Petchenik note with dismay that "while some cartographers and geographers have cast about for things to which they can liken the map [. . .] scholars in other fields tend to use the map as the fundamental analogy" (1976, 2), and they add, "maps clearly are involved in communication, and it would seem much could be learned from other analyses of other types of communication" (*ibid.*, 3). This suggests that there is no clear account of how maps represent and that the direction of knowledge transfer ought to be from other domains to cartography, rather than vice versa.

If nothing else, this diversity of opinion highlights that there is a question about the work that the models-as-maps analogy does, and about how far the analogy reaches. We suggest that this question is productively addressed by distinguishing two levels at which the analogy can be seen as operating. At the first level, the analogy is seen as providing an account of how models represent. At the second level, the analogy is taken to illuminate features of representation, in particular the nature of accuracy and the purpose relativity and historical situatedness of representations, which have wide-ranging philosophical implications.

Our claim is that the analogy fails at the first level. In line with what Robinson and Petchenik suggest, we turn the models-as-maps analogy on its head. Rather than attempting to use maps to learn about scientific representation, we explore how our preferred account of scientific representation (the "DEKI account") can be used to help us understand how maps work. This is not just an exercise in the philosophy of cartography; it also further develops the DEKI account by demonstrating how its conditions work. By design, these conditions are skeletal in the sense that they need to be filled in, or concretised, in any particular instance of epistemic representation. Thus, maps provide an illustration of one way in which this can be done. By contrast, we believe that the analogy works productively at the second level. By understanding how maps represent, we can deepen our understanding of how representations function more generally and draw interesting conclusions about some of their features.

We begin by discussing a concrete example of using a map and illustrate the mistakes that are made if the map is read naïvely. This, we submit, shows that maps presuppose rather than provide an account of representation (Section 2). This leaves open what account that is. We submit that the DEKI account fits the bill. We introduce the account (Section 3) and show how it works in the case of maps (Section 4). We then revisit claims that have been built on the models-as-maps analogy and assess their validity (Section 5).

2 Reading Maps Naïvely

There is a naïve view that maps are somehow "natural" representations that show their territories as they "really are". Although none of the authors mentioned in the previous section holds such a view, showing what is wrong with it leads the way to a better understanding of maps.

Consider the following imaginary scenario. Like other European countries, there have been strong separatist movements in Sweden. Eventually a referendum is called, and the proposal to split the country is successful. Specifically, the decision is to create the two independent states of North Sweden and South Sweden. All parties agree that the border should be drawn on a purely geographic basis by dividing the country in the middle along the north-south axis. Asbjörn is the government minister tasked with drawing the border. To do so he reaches for his map of Europe (shown in Figure 20.1) and sets out to determine Sweden's north-south midpoint.

Figure 20.1 Map of Europe.

Inspecting the patch of the map marked "Sweden", Asbjörn finds that the point marked "Treriksröset", with Euclidean coordinates (x_1, y_1) on the map's surface, is the furthest to the top and hence represents the northernmost point of Sweden. Similarly, he finds that the point marked "Smygehuk", with the Euclidean coordinate (x_2, y_2) on the map's surface, is the closest to the bottom and hence represents the southernmost point in Sweden. He then determines that the border should be given by the horizontal line through $\frac{1}{2}(y_1 + y_2)$, which is the solid line seen in Figure 20.2. This seems like a natural way of completing his task of dividing the country halfway on the north-south axis. Points to the top of the map represent locations to the north; points to the south represent locations to the south; and distances on the map correspond to distances in the world. So, surely the midpoint between the northern and the southern tip must be in the middle between the points on the map that represent the northmost and southmost locations.

Natural as this may seem, Asbjörn's procedure draws the border in the wrong place. His technique for dividing Sweden would have been correct with a map whose projection preserves the even spacing between east-west parallels. But the map in Figure 20.1 does not have this feature. It has been made with the Mercator projection, which preserves bearings (i.e., angles) but distorts distances.[2] In particular, as we approach the top of the map, the same distances on the top-bottom axis represent ever smaller distances on the north-south axis on the globe.

How bad is Asbjörn's mistake? Are we quibbling about epsilons? To answer this question, let us consider another minister, Berit, who knows about the Mercator projection MP:

$$x = sR\,(\lambda - \lambda_0), \quad y = sR\,\ln\tan\left(\frac{\pi}{4} + \frac{\varphi}{2}\right), \tag{20.1}$$

and its inverse MP^{-1}:

$$\lambda = \frac{x}{sR} + \lambda_0, \quad \varphi = 2\tan^{-1}(e^{\frac{y}{sR}}) - \frac{1}{2}\pi, \tag{20.2}$$

where λ is longitude (in radians); λ_0 is a central meridian (in radians), which in our case is Greenwich and so $\lambda_0 = 0$; φ is latitude (in radians); the coordinate system on the globe is chosen such that an increase of λ involves a shift east, and an increase of φ involves a shift north; R is the radius of the globe; and s is a scale factor.

Berit takes the coordinates of the northernmost and the southernmost points, (x_1, y_1) and (x_2, y_2) respectively, and then uses the inverse projection to determine the radian-valued coordinates (λ_1, φ_1) of the northernmost point and (λ_2, φ_2) of the southernmost point. The concrete values she finds, converted to degrees, are 69.06°N 20.55°E and

55.34°N 13.36°E for Treriksröset and Smygehuk, respectively. This means that the midpoint on the north-south axis between them (on the globe) lies at $\frac{1}{2}(\varphi_1 + \varphi_2) = 62.20°\text{N}$. This is where the border should be according to the agreement reached between the parties.

Feeding this value into the projection for the y coordinate on the map, Berit determines that the midpoint is the dotted line shown in Figure 20.2. As we can see, these lines do not coincide with one another. In fact, the solid line (which goes through the midpoint on the map itself) corresponds to a latitude of roughly 63°N. Calculating the distance between them shows that Asbjörn's naïve interpretation of the map, which failed to take into account that the relevant projection is the Mercator projection, resulted in a border roughly 89 km too far to the north! The error is significant, even for a country that is roughly 1,600 km from north to south.

The problem is that the map of Europe does not warn Asbjörn that his attempt to divide a territory in the middle by dividing a distance on the map in the middle will result in a grave error. The map's projection does not, as it were, jump off the page when you look at it. *Per se*, the map is a piece of paper with certain shapes drawn on it, and you have to know what projection has been used to produce the map in order to use it correctly.

This is not just a toy example cooked up to illustrate our point. Sismondo and Chrisman report that errors of this kind are common:

> At least half of a sample of 137 international maritime boundaries appear to have been plotted as equidistant lines on the chart without accounting for differences in scale [. . .]. Even in the relatively equatorial situation of Australia and Indonesia, the agreement specifies positions that are 4 nautical miles (7.4 km) south of the actual line of equidistance [. . .]. This amount of error is large enough for a sizable oil platform or two.
>
> (2001, S42)

The point generalises: the way in which a map represents its terrain cannot be read off the map directly. You could interpret it naïvely, and uniformly scale every measurement on the map to a measurement on the terrain. Alternatively, if you are aware of the distortion introduced by the projection, you can take this into account in your inferences from the map to the terrain. Although the former may look more "natural" than the latter, you would be well served to avoid such an interpretation if you want the results of your map-to-terrain reasoning to be accurate. In order to employ the latter kind of interpretation, you need to know the details of the projection used to create the map, and the conventions associated with its use.

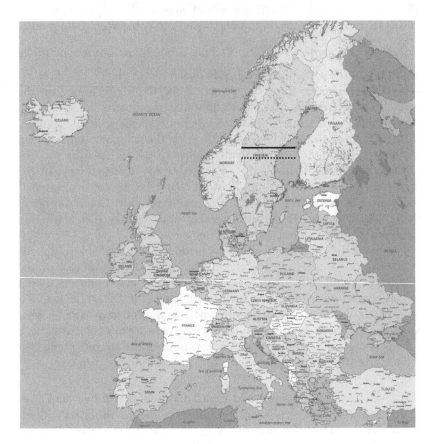

Figure 20.2 Europe according to the Mercator projection. The solid line is the midpoint on the map; the dotted line is the midpoint as calculated using the details of the projection.

This point has been recognised by philosophers writing on maps. Giere, for instance, notes: "Maps require a large background of human convention for their production and use. Without such they are no more than lines on paper" (1994, 11).[3] This is right as far as it goes, but it leaves important questions open. What are the conventions we must be aware of, and in what way does the mode of production of a map matter? Answers to these questions come from an account of representation (and will clearly also involve an investigation into the conventions and practices associated with particular maps, which will, in turn, be the subjects of such an account) rather than from the map itself (whatever that may mean), which undermines the utility of the models-as maps analogy.

Indeed, as noted in Section 1, we are going to turn this point on its head and show that our DEKI account, originally offered as an account of how models represent, provides the required understanding of how maps represent.

3 Analysing Maps with DEKI

The DEKI account is designed to answer the question: in virtue of what is a model (or a map) M an epistemic representation of a target (or territory) T? It understands M as the ordered pair of an object X and an interpretation I, hence $M = \langle X, I \rangle$, and it postulates that M is an epistemic representation of T if and only if:

  i M denotes T (and parts of M may denote parts of T);
  ii M exemplifies features $F_1, ..., F_n$;
  iii M comes with key K that associates $F_1, ..., F_n$ with a collection of features $Q_1, ..., Q_m$; and
  iv M imputes $Q_1, ..., Q_m$ to T.

These four elements – Denotation, Exemplification, Keying-up, and Imputation – give the account its name.[4] We have presented them in detail elsewhere (Frigg and Nguyen 2018, 2020, Chs. 8 and 9). Our purpose here is to briefly summarise them, and then discuss how they play out in the case of maps.

Let us begin with the internal structure of M. At a basic level, a model is an object X: a system of pipes, two imaginary perfect spheres, or an oval block of wood. *Per se* these things are just objects, like the tables and chairs in our offices. What turns a "mere" object into a model is that it is endowed with an *interpretation I*. A system of pipes becomes a model when the flow of water through pipes is interpreted as the flow of money through an economy (the Phillips-Newlyn model); the two spheres become a model of hydrogen if they are interpreted as the electron and the proton (the Bohr model), and the oval block becomes a model when it is interpreted as a ship.[5] Hence, a model is the pair of an object X with an interpretation I: $M = \langle X, I \rangle$. Crucially, this does not presuppose that models have targets. Interpretations presuppose a conceptual scheme in terms of which the interpretation is phrased, but they do not presuppose that anything real is singled out. One can interpret a flow of water as the movement of elfs, the block of wood as a UFO, and the two spheres as Vulcan and one of its moons. This will still turn the objects into models, albeit targetless models.[6]

This carries over to maps. *Per se*, a map is a piece of paper (or, increasingly, an image on a digital screen) exhibiting certain lines and shapes. The piece of paper, X, becomes a map only once it is endowed with an interpretation I according to which surfaces enclosed by solid lines are interpreted as countries; the lines themselves as borders; the blue surfaces as water; and so on. Without such an interpretation, a map is merely a coloured piece of paper. Like models, maps need not be maps of a real territory. Maps of Atlantis, the world according to the Game of Thrones, and Winnie-the-Pooh's Hundred Acre Wood are maps, but not ones that represent a real territory.

If representation is not built into the notion of *M*, what does it take for an *M* to be a representation of a *T*? In the first instance, we want to know what makes *M* be about *T*. Condition (i) addresses this point by appealing to the notion of *denotation*.[7] Models and maps can denote their targets just as proper names denote their bearers, predicates denote objects in their extension, and photographs denote their subjects. The Phillips-Newlyn model denotes the Guatemalan economy, the Bohr model denotes hydrogen, and the ship model denotes RMS *Queen Mary*. In addition to *M* as a whole denoting *T* as a whole, parts of *M* can denote parts of *T*. The flow of water on the right in the Phillips-Newlyn model denotes foreign trade, the small sphere in Bohr's model denotes an electron, and so on. The same is true of maps. The map in Figure 20.1 denotes Europe. In addition, every point on the map denotes a point on the globe, namely the point specified by the inverse projection, given in Equation 20.2. Some of these points are given names. If, say, a point is labelled "Stockholm", this means that the point denotes the place on the globe where the city of Stockholm is located.[8]

Denotation is necessary but insufficient for epistemic representation. It is necessary because it establishes the bare sense in which *M* is about *T*, and parts of *M* are about parts of *T*. It is insufficient because denotation alone is too weak to ground epistemic representation: the fact that *M* denotes *T* does not enable us to use *M* to generate claims about *T* (one cannot draw inferences about London from investigating the features of the syntactic object "London").

Explaining how a representation can function epistemically proceeds in several steps. The first step, condition (ii), involves the concept of *exemplification*. Exemplification is a mode of reference that occurs when an object refers to a feature it instantiates. This is established relative to a context. We can define it as follows: *M* exemplifies a certain feature *F* in a certain context iff *M* instantiates *F* and the context highlights *F*, where a feature is highlighted if it is identified in the context as relevant and epistemically accessible to the users of *M*. An item that exemplifies a feature is an *exemplar*. Standard examples include samples (the beer you try at the brewery exemplifies its flavour) and swatches (the colour swatch in the paint shop exemplifies its colour).[9] Exemplification is selective; the colour swatch does not exemplify rectangularity, even though it instantiates it. Only selected features are exemplified, and which features are selected depends on the context.

Exemplars provide epistemic access to the features they exemplify. This is because they instantiate the features that they exemplify in ways that makes them salient, which, in part, depends on the context in which they are embedded. The paint chip makes a particular shade of blue salient and thereby acquaints those using the chip with that shade of blue because in that context the chip's colour is salient and accessible to an observer. Epistemic representations exemplify certain features. The Phillips-Newlyn model exemplifies a certain level of taxation, the Bohr model exemplifies certain energy levels, and the ship model exemplifies a certain resistance when moving through water. As these examples indicate, instantiation is here understood as instantiation under interpretation I. The Phillips-Newlyn model is a system of water-pipes interpreted in terms of economic properties.[10] Under this interpretation, the model can instantiate, and hence exemplify, economic properties like having a low-tax fiscal regime. We are not committed to restricting instantiated, and exemplified, features to the features possessed by X as a "bare" object, which have the absurd consequence that the Phillips-Newlyn model could only exemplify water-and-pipe features. The point is pertinent to maps. The map in Figure 20.1, under the standard interpretation of a map, does not exemplify lines and colours: it exemplifies there being borders between countries and landmasses having coastlines; and it exemplifies certain points being at a certain distance from each other. In general, then, M will exemplify certain features $F_1, ..., F_n$.

That M denotes T and exemplifies features $F_1, ..., F_n$ is still not sufficient to make M an epistemic representation of T. To get to that point, two further steps are needed. For M to represent T as being such and such, a user of the representation has to *impute* features to T (by this we just mean that the user ascribes features to the target). But which features are these? A natural suggestion here is that these features are simply the ones that are exemplified by M in the representational context. Indeed, this is how some kinds of epistemic representation work: taken as a model of your new front door, the colour swatch in the paint shop exemplifies a certain shade of red and imputes exactly this shade of red to your door. But, at least in a scientific context, this is the exception rather than the rule. It is rarely the case that epistemic representations represent their targets as having *precisely* the features that the former exemplify. Someone who knows how to use a mechanics model from physics will not conclude that a real skier will have the particular trajectory that it follows in the model that assumes air resistance and friction to be absent: they know that the skier is subject to friction and air resistance, and they can take this into account when they use the model to reason about the target.

This leads us to the notion of a *key*. In the abstract, a key can be thought of as a function: it takes as inputs the exemplified features $F_1, ..., F_n$, and it delivers as outputs the features $Q_1, ..., Q_m$ that the user of the representation imputes (or ascribes) to the target system.[11] As stated, this is an abstract notion, and this is by design. We submit that keys are one of the main locations that encode the disciplinary conventions associated with the use of epistemic representations. In the case of modelling, these keys are the sorts of things that students learn when they learn to use their models. In some cases, these keys might involve weakening the isolated exemplified feature of the model to the claim that the target only has a disposition to behave that way (Nguyen 2020); in others, they may result in certain limit-based reasoning: the model feature might be related to the imputed feature via taking a limit (Nguyen and Frigg 2020). One of the important upshots of the DEKI account is the demand that in any particular instance of epistemic representation, we have to understand the key that accompanies it.[12]

Like models, maps exemplify certain features. For instance, the map shown in Figure 20.1 exemplifies there being a land-boundary between the countries labelled "Norway" and "Sweden"; it exemplifies "Norway" being to the left of "Sweden"; it exemplifies the dot labelled "Stockholm" being higher up than the dot labelled "Gothenburg"; and it exemplifies the dot labelled "Stockholm" and the dot labelled "Umeå" being 20 cm from each other. We noted that exemplification is selective, and we can see an example of this here. Blue is exemplified (indicating water), whereas the other colours are not: countries are individuated by solid lines indicating borders, and their colour is a mere convenience that does not contribute to the map's content.[13] Similarly, the texture of the paper on which the map is printed (or the make of the screen on which it is displayed), the typeface of the letters used to label points and areas, and the fact that it has been printed in Germany are all features of the map, but they are not exemplified.

We should not assume that the features exemplified by a map are imputed unaltered to the territory represented. Some are; some are not. A place being to the left of another place, or a place being further up than another place, are meaningless when imputed to the world because there is no unique left-right or up-down on the globe, and no one would take the map to say that the cities of Stockholm and Umeå are 20 cm from each other. Like with models, we need a key to tell us how the transition from the map to the territory works.

The bedrock of the key are the part-part denotations given by the inverse transformation in Equation 20.2, which specifies for each point on the map which point on the globe it stands for.[14] The first element of the key is the rule that when a certain point on the map is singled out by the map's interpretation as having a certain characteristic – being on a border, being on a coastline, being a city centre, etc. – this characteristic is imputed to the globe-point that the map-point stands for. If, say, the interpretation of the map specifies that solid black lines are borders, and a certain point lies on a black line, then the map imputes to the point it denotes the feature of being on a border.

The projection also helps us keying up features like *to the left of* and *higher up than*. *To the left of* corresponds to lower values of x, and the inverse projection for the x-axis, $\lambda = \dfrac{x}{R} + \lambda_0$, tells us that higher values of

x correspond to points further east. This means that the map-property *to the left of* is keyed up with the globe-property *to the west of*. Although this is a frequently used convention, it is in no way necessitated by the situation. The mapmaker could have used a projection with the inverse $\lambda = -\dfrac{x}{R} + \lambda_0$, in which case *to the left of* would be keyed up with *to the east of*. Likewise, the fact the *higher up than* corresponds to higher values of y, together with the fact that the inverse projection for the y-axis, $\varphi = 2\tan^{-1}(e^{\frac{y}{R}}) - \dfrac{1}{2}\pi$ is a monotonic function of y, implies that the map-property *higher than* is keyed up with the globe-property *to the north of*.

Things get more involved when we turn to distances. In the idiom of DEKI, Asbjörn adopted a key according to which map-distances scale linearly with globe-distances, that is, a key according to which a map-distance d_m is keyed up with the globe-distance $d_g = c\,d_m$, for some scale factor c. As we have seen, this key returns wrong results. In fact, there is no (map-wide) scaling factor for map-distances at all! Map-distances in Mercator maps are not keyed up with globe-distances. The correct key says that the distance between the map-points (x_1, y_1) and (x_2, y_2) is given by $G[MP^{-1}(x_1, y_1), MP^{-1}(x_2, y_2)]$, where G is the great-circle-distance on a sphere (i.e., the length of the line-segment between two points on a great circle drawn through these points).[15]

Considerations of the same kind are also needed to determine the surface of an area. We cannot simply measure the surface of part of the map and expect it to scale linearly. We will have to project the boundaries of the relevant territory back onto the globe with the inverse transformation, and then determine the measure of the relevant surface on the sphere.

This shows that the map requires a key for its use, and that this key is more than just a trivial identity that says something like "whatever is true in the map is true on the globe". And the key is only part of what is needed to use the map. As we have seen, all elements of the DEKI account do essential work in explaining how maps work: the interpretation turns a pattern of lines and shapes into a territory-representation; denotation turns this territory-representation into a representation of a particular territory; exemplification singles out relevant features; the key transforms these into other features; and an act of imputation says that the territory has these other features.

4 Philosophical Lessons

In Section 1, we have seen that the models-as-maps analogy can be seen as operating at two levels: at the first level, it can be seen as providing an account of how models represent, whereas at the second level, it is taken to illuminate features of representation, in particular the nature of accuracy, the purpose relativity and historically situatedness of representations, and the possibility of total science. We now assess how well the analogy fares with these.

A time-honoured position in the literature on scientific representation appeals to the notion of similarity: a model M accurately represents its target T in virtue of M and T being similar to one another in the appropriate respects, and to the appropriate degree.[16] Given the popularity of the models-as-maps analogy, it is not surprising to see this being motivated by the idea that maps function in the same way. For example, Giere, who defends a similarity-based account of scientific representation, presents a map of Pavia and argues: "How does this map represent Pavia? The answer is: by being spatially similar to aspects of Pavia" (1999, 45), and then, to illustrate the context sensitivity of what is meant by similarity, further discusses contexts such as using a map of the London Underground to be such where "the important similarities are those between these topological features of the map and of the whole metro system" (*ibid.*, 46).

We submit that an appeal to similarity is either misleading or empty, and that the models-as-maps analogy does not do any productive work at the first level (at least in the sense that the first level analogy does not motivate an account of representation that appeals to similarity as common to both maps and models). It would seem to be an obvious consequence of the notion that maps are similar to their territories that map-distances are proportional to territory-distances. Although this is correct in a standard city map, it leads, as we have seen in Section 2, to a significant error in maps like the ones shown in Figure 20.1. One might now turn around and say that this is not how similarity should be understood: the map being similar to its territory here simply means that the transformation in Equation 20.1 holds.[17] There is a question as to whether this equation, or indeed other projections, can meaningfully be regarded as a kind of similarity. The vagueness of the notion of similarity makes it difficult to say. Let us set this issue aside. The more pressing problem is that this way of approaching the issue makes similarity otiose. One first has to know the projection and all the conventions used, and only when everything has been spelled out one can turn around and say "see, they are similar".[18] Thus understood, similarity does not do any work and it becomes a success term that gets attached to a finished product when things work out as envisaged.[19] It would seem to be more productive to think about maps in terms of DEKI, which is explicit that the account treats denotation and keys as blanks to be filled on every occasion, and then filling them with the appropriate projection.

In contrast with the first level, the analogy is largely correct at the second level. In the remainder of this section, we discuss some pertinent second level points and explain how they bear out in an analysis of maps based on DEKI.

Let us begin with accuracy. There is a temptation to say that a map on which Iceland appears larger than Romania, where in reality Romania has more than twice the surface area of Iceland, is inaccurate. Kitcher, rightly, protests that such verdict would be "foolish" because "[a]ssociated with any map there are conventions that determine which aspects of the visual image are to be taken seriously" (2001, 56). As we have seen in the previous section, maps come with a key (which can be either implicit or explicit, cf. footnotes 11 and 12), and disregarding the key leads to wrong results. Calling a map with these features inaccurate relies on a naïve reading of the map, and we have seen that such a reading is illegitimate. The accuracy of a map has to be judged relative to a key, not relative to visual appearance: a map is accurate when the territory has the features that the key outputs. The same holds true of models, which we should not expect to be "like" their targets in some pre-theoretical and unreflective sense: a model is accurate if attributing features provided by the key to the target results in true statements. This does not (or at least need not) involve "looking alike" or being similar.[20]

Maps are made for a particular purpose, and there is no such thing as a map that is good for everything. If you are hiking the Scottish highlands, you would be well advised to use a map that displays the topography of the terrain depicted and the paths that thread up and down the munros. In contrast, if you are on a scenic drive from Inverness to Fort William, you would be better off with a map that marks roads and speed limits. As Kitcher puts it: "we understand how maps designed for different purposes pick out different entities within a region or depict those entities rather differently" (2001, 56). The same is true of scientific models. As Morrison (2011) and Massimi (2018) have shown, scientists often produce different models of the same target, where the models pick out different aspects and features of the target depending on what aims the scientists pursue. Kitcher generalises this point when he notes that "the aim of the sciences is to address the issues that are significant for people at a particular stage in the evolution of human culture" (*ibid.*, 59). DEKI is compatible with this idea. It sees scientists as having complete freedom both in the choice of model-objects and in the choice of keys, and these choices can be seen as inevitably historically contingent and relative to our aims and purposes.[21]

If correct, this has important consequences for the project of science as a whole. If all representations that science produces are purpose relative and historically situated, then there is no such thing as the perfect map. Borges (1999) reminds us in his notorious story about cartography that a perfect map would have to coincide with its territory point by point, resulting in a map of the Empire as large as the Empire itself. Such a map would be useless, and soon abandoned. Maps are selective in what they represent. There are two ways of thinking about the scientific project corresponding to Borges' map. First, one might hope that we will eventually discover the "final theory" in fundamental physics, and such a theory would, in principle at least, suffice to provide a complete representation of the fundamental structure of the world.[22] Second, and more boldly, one might hope that we will eventually discover the "complete theory", which contains not only a complete representation of the fundamental structure of the world, but everything else too, from molecular bonds to social practices to political systems and everything in-between.

The realisation that every representation, be it geographical or scientific, is purpose relative and historically situated should at least cast doubt on the latter project. Giere submits that "[t]here is no such thing as a universal map" (1994, 11). Kitcher goes into more detail in his discussion of the "ideal atlas" (2001, 60). It could not be a single map, since, as noted, nothing could perform this role except for the terrain itself. But perhaps it could be a collection of "fundamental maps" (corresponding to a "final theory"), from which "all spatial information can be generated, and that they collectively provide a unified presentation of the wide diversity of kinds of knowledge drawn from our actual ventures in cartography (and, presumably, projects we might have undertaken)" (*ibid.*). Kitcher argues that a brief glance at the vast diversity of maps produced in human history should make us immediately sceptical about the possibility of such a compendium, notwithstanding the fact that an ideal atlas would also have to encode the information about the projects and investigations we have not, but might have, embarked on. Thus, he concludes, the models-as-maps analogy should force us to reconsider the idea of a complete theory, and, along with it, the idea that our scientific theories and models are converging on it. We agree.

Even if it is granted that science does not aspire to a complete theory, some may hold onto the possibility of a final theory. Just because we may not be able to represent every fact in the world, this does not preclude the possibility of us, eventually at least, representing the fundamental facts on which the others (making some strongly reductive assumptions) ultimately depend. And perhaps the way in which we will represent these facts, will not turn, in any philosophically significant way, on the contextual aspects of our representations. We will not take sides here, beyond noting that even if the hope in a final theory is still alive, this does not tell against the idea that we need to construct partial and purpose relative representations to help us explore non-fundamental domains. As Weinberg (1993, 18), an enthusiastic adherent to the goal of a final theory, notes:

> Of course a final theory would not end scientific research, not even pure scientific research, nor even pure research in physics. Wonderful phenomena, from turbulence to thought, will still need explanation whatever final theory is discovered. The discovery of a final

theory in physics will not necessarily even help very much in making progress in understanding these phenomena (though it may with some).

So, the models-as-maps analogy pours cold water on the dream of a complete theory, even if a final theory remains a live option. Our (non-fundamental) representations will always be partial, and they will have purpose relative and historically situated aspects. But as we have seen, this does not mean that they have to be inaccurate, and there is a clear sense in which some interpretations (like Berit's) are to be favoured over others (Asbjörn's). Giving up on a complete theory does not mean anything goes.

Acknowledgements

We are extremely grateful to Jared Millson, Mark Risjord, and Elay Shech for comments on a previous draft of this chapter, and for a stimulating dialogue. We also thank Kareem Khalifa, Insa Lawler, and Elay Shech for inviting us to participate in this project.

Notes

1 Similar observations have also been made by Boesch (2019), Bolinska (2013), Contessa (2007), Giere (1999), and Sismondo and Chrisman (2001). There is an interesting question as to whether a similar relation holds between maps and linguistic representations. This is, unfortunately, beyond the scope of this paper. For a recent discussion of this question, see Aguilera (2021).
2 For a survey of projections, see Monmonier (1991, Ch. 2); for an in-depth discussion, see Pearson (1990). For a mathematical definition of the Mercator projection, see Pearson (1990, Ch 5.VII); for a discussion of its history, see Winther (2020, Ch. 4).
3 See also Sismondo and Chrisman (2001, S42), Kitcher (2001, 56–57), and Toulmin (1953, 108).
4 One thing to note here is that we are not assuming that epistemic representation is, in some sense, "mind-independent", or "naturalisable" (we are grateful to Elay Shech for encouraging us to be explicit about this). As such, the conditions we use to explicate it already include some intentional notions. For more on this, see Frigg and Nguyen (2020, 39–40).
5 For a discussion of these models in the context of the DEKI account, see Frigg and Nguyen (2020, 2016), and Nguyen and Frigg (2022), respectively.
6 In Goodman's (1976) terms, models are Z-representations, where the Z is given by the interpretation.
7 Thus, when we say "M is *about* T" we mean it in the minimal sense that M denotes T. And following, for example, Goodman (1976), by "denotation" we mean the bare relation between a symbol and what is symbolised, without invoking the "meaning" or "descriptive content", which may (or may not) be associated with the symbol. See Salis, Frigg, and Nguyen (2020) for a further discussion of this point. Thanks to Elay Shech for encouraging us to be explicit about this.

8 Although note that Stockholm is, of course, a region, rather than a point, on the globe, so points on the map denote locations in Stockholm, and these point-wise denotation relations then associate a region on the map with Stockholm, considered as a region on the globe. Thanks to Mark Risjord and Jared Millson for highlighting this.

9 Note that exemplification is a semantic, rather than an epistemological notion. If, for example, the beer sample was from the bottom of the bottle and thus exemplified consisting of a large amount of yeast slurry, this does not entail that all of the beer in the bottle will exhibit this feature. Exemplification may allow us to successfully infer features of the sampled from features of the sample, but it does not guarantee it. We are grateful to Elay Shech for encouraging us to be explicit about this.

10 One might worry that non-concrete models like the Bohr model cannot, strictly speaking, instantiate features. This worry is addressed in Frigg and Nguyen (2020, Ch. 9).

11 Our use of the term "key" in the DEKI account is inspired by the sorts of representations we are investigating here: maps! However, it should be noted that there may be a difference between the "key" as used in DEKI and the explicit "key", or "legend" (we will use the latter term to disambiguate between the two), that is literally written down on a map. Many simple maps (such as city maps for tourists) may not come with an explicit legend, and thus the keys that are used to interpret them are implicit in the conventions and practices associated with the maps (we are grateful to Mark Risjord and Jared Millson for pointing this out). However, it is also worth noting that many maps do, in fact, explicitly contain the sort of information that keys (in the sense of DEKI) require in their legends; for example, Ordinance Survey maps designed to guide walkers around regions in the United Kingdom are explicit that they are constructed from the Transverse Mercator Projection, and they include information relevant for the key, such as the distinction between Magnetic North, True North (on the globe), and Grid North (on the map).

12 In the scientific context, the key may be implicit in the practice of the discipline in which a model is embedded. But it is worth noting that scientists are free to experiment with different keys, and that their choice of the key may be the result of an investigative back-and-forth between a model and its target. Under this understanding of keys, although some may be more accurate than others, from a semantic point of view, scientists are free to choose them as they see fit. Shech (2015) uses the term "code" to describe a related notion, but as he notes, it has a dual meaning: "[o]n the one hand, a code, understood as a key, legend or guide, is needed in order to make use of a representational vehicle for surrogative reasoning. On the other hand, understood as a cryptogram or cipher, the code of a representation is not always known and so it must be 'deciphered,' so to speak" (p. 3469). Our notion of a key, thus, corresponds to the former meaning. For a related discussion, see footnote 21, our response to Millson and Risjord in this volume, and Frigg and Nguyen (2020, Ch. 8).

13 A coloured version of our map in Figure 20.1 can be seen at https://stock. adobe.com/uk/images/europe-political-map-high-detail-color-vector-atlas-with-capitals-cities-towns-names-seas-rivers-and-lakes-high-resolution-map-of-europe-in-mercator-projection/238888167?prev_url=detail.

14 We emphasise here that Equation 20.2 associates *points* on the map with *points* on the globe. It does not tell us anything about how *features* on the

map represent *features* on the globe. To illustrate this, consider the fact that "roads" on maps (i.e., coloured lines) are typically wider than they "should" be, given the maps' scales. According to our discussion, this is because the points on the edge of the coloured line (i.e., the points that seem to represent the road as wider than it is) are associated with the points on the globe that are not, in fact, paved. But this is the result of the colouring on the map, not the point-to-point denotation relations given by Equation 20.2, and moreover the way that map keys associate colours on a map's surface with a road typically do not require that a map user infer that roads are wider than they, in fact, are (the *width* of the coloured line is neither exemplified nor keyed-up). Thanks to Mark Risjord and Jared Millson for encouraging us to be explicit about this.

15 For a discussion of the great-circle-distance, see Pearson (1990, Ch. 3).

16 See Frigg and Nguyen (2020) for a further discussion of similarity accounts of representation in general; see Winther (2020, Ch. 5) for a discussion of similarity in the context of maps.

17 Similar points can be made about the structuralist conception of representation, because the Mercator projection does not preserve certain structural features (e.g., distance ratios). This way of thinking about representation is discussed in Frigg and Nguyen (2020, Ch. 4).

18 We are not claiming here that the map user needs an explicit philosophical account of representation in order to use the map: rather it is that the map user should adopt the conventions associated with the map when using it, and that these conventions, which do not have to be understood in terms of similarity, are the subject of our account of representation.

19 An advocate of the idea that it is similarity (structural or otherwise) that establishes representation could argue that it is the "keyed-up" map (i.e., the map with the key applied to it) that is supposed to be similar to the target, not the "bare" map itself (thanks to Mark Risjord and Jared Millson for suggesting this possibility). But this would pull the rug from under such an account: the crucial move would be the application of the key, and then claiming that the results of such an application are "similar" to the target is just another way of describing how the outputs of the DEKI account of representation should be compared with the target.

20 For a detailed discussion of this point, see Frigg and Nguyen (2021).

21 Of course, as the example of Asbjörn's misuse of the Mercator projection shows, once the choices are made, they do constrain future uses of maps and models (at least if one aims at using them to generate useful knowledge about their targets). Conventions are freely chosen, but once chosen they are relatively fixed (thanks to Mark Risjord and Jared Millson for encouraging us to clarify this). But it should also be noted that there are occasions in the history of science where models can be fruitfully used by adopting new conventions that did not exist at the inception of the models. These include cases of model transfer, where scientists use models originally designed to represent some target system (e.g., a two-body celestial system) to represent an alternative kind of target system (e.g., the atomic nucleus as per Bohr's model), as well as cases where model behaviour is interpreted in novel ways (e.g., Dirac's observation that negative energy solutions to the Dirac equation, once understood as mathematical artefacts, could be interpreted in terms of positrons, cf. Bueno and Colyvan (2011, 365–365)).

22 For a vivid account of the vision of a final theory, see Weinberg (1993).

References

Aguilera, M. 2021. "Heterogeneous Inferences with Maps." *Synthese, Online First.* https://doi.org/10.1007/s11229-020-02957-w.

Boesch, B. 2019. "Scientific Representation and Dissimilarity." *Synthese, Online First.* https://doi.org/10.1007/s11229-019-02417-0.

Bolinska, A. 2013. "Epistemic Representation, Informativeness and the Aim of Faithful Representation." *Synthese, 190*(2), 219–234. https://doi.org/10.1007/s11229-012-0143-6.

Borges, J. L. 1999 "On Exactitude in Science." In J. L Borges (Eds.), A. Hurley (Trans.), *Collected Fictions* (p. 325). New York: Penguin.

Bueno, O., & Colyvan, M. 2011. "An Inferential Conception of the Application of Mathematics." *Nous, 45*(2), 345–374. https://doi.org/10.1111/j.1468-0068.2010.00772.x.

Contessa, G. 2007. "Scientific Representation, Interpretation, and Surrogate Reasoning." *Philosophy of Science, 74*(1), 48–68. https://doi.org/10.1086/519478.

Frigg, R., & Nguyen, J. 2016. "The Fiction View of Models Reloaded." *The Monist, 99*, 225–242. https://doi.org/10.1093/monist/onw002.

Frigg, R., & Nguyen, J. 2018. "The Turn of the Valve: Representing with Material Models." *European Journal for Philosophy of Science, 8*(2), 205–224. https://doi.org/10.1007/s13194-017-0182-4.

Frigg, R., & Nguyen, J. 2020. *Modelling Nature. An Opinionated Introduction to Scientific Representation.* Berlin and New York: Springer. https://doi.org/10.1007/978-3-030-45153-0.

Frigg, R., & Nguyen, J. 2021. "Mirrors without Warnings." *Synthese, 198*, 2427–2447. https://doi.org/10.1007/s11229-019-02222-9.

Giere, R. N. 1994. "Viewing Science." In *Proceedings of the Biennial Meeting of the Philosophy of Science Association, 1994, Volume Two: Symposia and Invited Papers* (pp. 3–16). https://doi.org/10.1086/psaprocbienmeetp.1994.2.192912.

Giere, R. N. 1999. "Using Models to Represent Reality." In L. Magnani, N. J. Nersessian, & P. Thagard (Eds.), *Model-based Reasoning in Scientific Discovery* (pp. 41–57). Dordrecht: Kluwer. https://doi.org/10.1007/978-1-4615-4813-3_3.

Goodman, N. 1976. *Languages of Art* (2nd ed.). Indianapolis and Cambridge: Hackett.

Kitcher, P. 2001. *Science, Truth, and Democracy.* New York: Oxford University Press. https://doi.org/10.1093/0195145836.001.0001.

Massimi, M. 2018. "Perspectival Modeling." *Philosophy of Science, 85*(3), 335–359. https://doi.org/10.1086/697745.

Monmonier, M. 1991. *How to Lie with Maps.* Chicago: University of Chicago Press.

Morrison, M. 2011. "One Phenomenon, Many Models: Inconsistency and Complementarity." *Studies in History and Philosophy of Science, 42*(2), 342–351. https://doi.org/10.1016/j.shpsa.2010.11.042.

Nguyen, J. 2020. "It's Not a Game: Accurate Representation with Toy Models." *The British Journal for the Philosophy of Science, 71*(3), 1013–1041. https://doi.org/10.1093/bjps/axz010.

Nguyen, J., & Frigg, R. 2020. "Unlocking Limits." *Argumenta*, 6(1), 31–45.

Nguyen, J., & Frigg, R. 2022. *Scientific Representation* (Cambridge Elements). Cambridge: Cambridge University Press.

Pearson, F. 1990. *Map Projections: Theory and Applications*. Boca Raton: CRC Press.

Robinson, A. H., & Petchenik, B. B. 1976. *Nature of Maps. Essays Toward Understanding Maps and Mapping Hardcover*. Chicago: University of Chicago Press.

Salis, F., Frigg, R., & Nguyen, J. 2020. Models and Denotation. In C. Martínez-Vidal, & J. L. Falguera (Eds.), *Abstract Objects: For and Against* (pp. 197–219). Cham: Springer. https://doi.org/10.1007/978-3-030-38242-1_10.

Shech, E. 2015. "Scientific Misrepresentation and Guides to Ontology: The Need for Representational Code and Contents." *Synthese, 192*, 3463–3485. https://doi.org/10.1007/s11229-014-0506-2.

Sismondo, S., & Chrisman, N. 2001. "Deflationary Metaphysics and the Nature of Maps." *Philosophy of Science (Proceedings), 68*, 38–49. https://doi.org/10.1086/392896.

Toulmin, S. 1953. *The Philosophy of Science*. London: Hutchinson's University Library.

Weinberg, S. 1993. *Dreams of a Final Theory: The Search for the Fundamental Laws of Nature*. New York: Vintage. https://doi.org/10.1063/1.2808871.

Winther, R. G. 2020. *When Maps Become the World* (Studies in History and Philosophy of Science). Chicago: University of Chicago Press. https://doi.org/10.7208/chicago/9780226674865.001.0001.

21 DEKI, Denotation, and the Fortuitous Misuse of Maps

Jared Millson and Mark Risjord

1 Introduction

In 1977, Erwin Kreuz, a 49-year-old Bavarian brewery worker, spent his life savings on a trip to the city he'd read so much about: San Francisco. After his transatlantic flight, Kreuz checked into a hotel, enjoyed some beers in a local tavern, and dined at a Chinese restaurant in what he presumed was the Bay City's famous Chinatown. It was only after 3 days that he began thinking that something was amiss. The Golden Gate Bridge looked much smaller, and greener, than it did in photographs. The hills surrounding the city were much less populated than the Bay Area suburbs that he had expected. Frustrated, Kreuz hailed a cab and asked to be taken to downtown San Francisco. The driver, dumbfounded, informed him that to do so, he would have to travel more than 3,000 miles—Kreuz was in Bangor, Maine. The man who came to be known as the "World's Last Lost Tourist" had mistaken the remote logging town for his final destination when his plane stopped there to refuel. Thanks in part to his limited English, a confusing remark from a departing flight attendant, and a habit of consuming 17 beers per day, Kreuz had wandered the streets of Bangor for 3 whole days, thinking he was in San Francisco (Remsen 1977). For a short period, he was a national celebrity (Chamings 2021).

Whether Kreuz had a map of San Francisco is unknown, but it may not have made a difference. The hills and waterways around Bangor fit Kreuz's knowledge of the Bay City's geography; the town's two Chinese restaurants reinforced his belief that he was in the world-renowned Chinatown district; and the rusted green bridge over the Penobscot River did not look like the Golden Gate. However, Kreuz reasoned that it must be one of the smaller bridges in the Bay Area. The same patterns of inference can be made on the basis of maps—and, with enough luck, they can yield true beliefs.

Consider a similarly confused traveler, Susan, who is attempting to navigate her way through Paris. Despite knowing that she is in Paris, she has mistakenly opened her map of Prague. Through a series of lucky mistakes and misreadings, she uses the map to successfully navigate

DOI: 10.4324/9781003202905-24

from *le Tour Eiffel* to *le Jardin des Tulieries*. That is to say, Susan draws inferences from beliefs about the figures on her map to beliefs about, for example, what turn to make and which bridge to cross. She draws *surrogative* inferences, resulting in a sequence of actions that leads her to *le Jardin*. Compare Susan with Zoe, who finds her way with a map of Paris by correctly reading it. Suppose that the conclusions of Zoe's surrogative inferences are the same as Susan's. However, Susan and Zoe differ significantly: Susan's conclusions were obtained by *luck* and unjustified, whereas Zoe's conclusions were justified. Unlike Zoe, Susan's misuse of the map made the true answers to her questions fortuitous.

Maps are paradigmatic examples of non-linguistic (or not-just-linguistic) representations. They are epistemic representations in the sense that we can learn something about the landscape by inspecting the map. The story of Susan and Zoe highlights the fact that not just any true belief formed from inspecting the map is the conclusion of a justified surrogative inference. We will call the challenge to account for this fact: *the problem of fortuitous misuse.*

In this chapter, we will explore the problem of fortuitous misuse with the aid of one of the more sophisticated accounts of epistemic representation currently on offer. Roman Frigg and James Nguyen's "DEKI" account, as recently presented in their book *Modeling Nature* (2020), has a natural account of surrogative inference based on a detailed and articulate analysis of epistemic representation. The D in DEKI stands for "denotation," and DEKI is one of the many accounts of epistemic representation that make denotation central to their analysis. One of the differences between Susan and Zoe is that their maps denote different cities. Part of Susan's mistake was to use a map that denoted Prague. Although this is not the only epistemically relevant difference between them, this chapter will primarily focus on denotation.[1] By virtue of what does Susan's map denote Prague, and not Paris, given that she is using it to successfully navigate around Paris?

In their discussion of denotation, Frigg and Nguyen gesture toward the extensive literature in the philosophy of language. They suggest that the various accounts are compatible with DEKI, and that they "remain agnostic" about the choice (Frigg and Nguyen 2020, 180). Although outsourcing the account of denotation to philosophers of language makes sense, doing so assumes that accounts of linguistic denotation can be applied *mutatis mutandis* to epistemic representations without difficulty. In truth, their promissory note that the reader can supply DEKI with her favorite story about denotation—"plug and play"—is quite difficult to redeem.

This chapter works through the prominent accounts of denotation in the philosophy of language to see whether they can provide DEKI with the resources needed to account for fortuitous misuse. Section 3 argues that none does so successfully. Searching for supplementary resources,

Section 4 dives into the ways in which maps are created. This section argues that mapmakers are subject to a variety of constraints, without which map users would not be in a position to draw justified surrogative inferences. The correct reading of a map depends on a relationship between the mapmakers and the users. DEKI epistemically decouples the users from the mapmakers, and the role that it reserves for denotation is not sufficient to reconnect them. DEKI's shortcomings have an interesting upshot: the constraints under which mapmakers labor mirror the conditions that justify map users' surrogative inferences. Indeed, the features that establish a map's denotation seem to prefigure the very inferential relationships that underwrite its proper use. Section 5 concludes with the suggestion that inferences play a hitherto neglected role in establishing the denotation of epistemic representations.

2 DEKI on Fortuitous Misuse: A First Pass

The distinctive feature of epistemic representations is that they are non-linguistic representations from which we can learn (cf. Frigg 2006, 51). By inspecting or reasoning about them, we come to have justified beliefs about the phenomena they represent. Any account of epistemic representation, thus, has to satisfy what Frigg and Nguyen call the "Surrogative Reasoning Condition." As they articulate it, an adequate account of epistemic representation must show how scientists "can generate claims about target systems by investigating models that represent them" (Frigg and Nguyen 2020, 3). The problem of fortuitous misuse shows that the "generate" must have epistemic force. Both Susan and Zoe are "generating" inferences, but only Zoe's are justified. We cannot learn from unjustified inferences. Therefore, to satisfy the Surrogative Reasoning Condition, an analysis should show how justified claims are generated by the epistemic representation, as analyzed. Minimally, an analysis should distinguish between justified and unjustified surrogative inferences. Something in the analysis must block unjustified surrogative inferences or otherwise show how they arise from a misuse or misunderstanding of the epistemic representation. A stronger demand would be that an analysis of epistemic representation should show how surrogative inferences are justified *in virtue of* satisfying the analysis.

The process of making surrogative inferences is built into the very structure of DEKI (Frigg and Nguyen 2020, 180). In its full regalia, it looks like this (cf. Frigg and Nguyen 2020, 176):

Let $M = < X,I >$ be a map. According to DEKI, M is an indirect epistemic representation of T if M represents T as Z, whereby M represents T as Z if all of the following conditions are satisfied:

1 **Denotation:** M denotes T (and, possibly, parts of M denote parts of T)

2 **Exemplification:** M I-exemplifies Z-features $Z1,...,Z_n$

3 **Keying-up:** M comes with a key, K, associating the set $Z1,...,Z_m$ with a set of features $Q1, ..., Q_l$: $K(<Z1,...,Z_m>) = <Q1,...,Q_l>$

4 **Imputation:** M imputes at least one of the properties $Q1,...,Q_l$ to T

The idea underlying DEKI's account of epistemic representation is that properties of the model carrier, X, are associated with properties of the target $(Q1,...,Q_l)$ via a set of properties exemplified by the model $(Z1,...,Z_m)$. Of course, any carrier will have many properties, only some of which are relevant to surrogative inference. The printed lines and shapes on Susan's map are relevant to its status as an epistemic representation of Prague in a way that the coffee stains are not. Interpretation plays the crucial role of determining which properties of the carrier will be "exemplified." Hence, models are identified as the order pair of the carrier, X, and an interpretation, I. Below, we will speak of "the map as interpreted" to mark this two-part conception of an epistemic representation. As illustrated in Figure 21.1, to interpret the carrier is to treat it as representing in a certain way (a Z-representation).

Each of the DEKI conditions provides a place where an unjustified surrogative inference might be blocked. For the purposes of this chapter, we will set aside Susan's potential abuses of conditions 2, 3, and 4, and focus on the denotation condition. Susan's use of her map of Prague to navigate Paris clearly fails the denotation condition, and for this reason it fails to be an epistemic representation of Paris. Her inferences are unjustified, because she is drawing conclusions that just happen to be true of a city by inspecting a map that does not denote that city.

There is something fundamental about denotation with respect to justifying surrogative inferences. Denotation functions somewhat differently than the other conditions, as is illustrated in Figure 21.1. The chain of arrows leading clockwise from the carrier X to the target T captures surrogative inference. The interpretation determines exemplification, which grounds the keying-up, which, in turn, grounds imputation. The denotation arrow, by contrast, leads directly from the model (which, again, is the map as interpreted) to the target. Should the denotation link be broken, presumably the rest is irrelevant and no justified surrogative inferences could be generated.

DEKI, thus, assumes the existence of a plausible generic account of denotation that, when married to the rest of the DEKI conditions, yields a plausible account of epistemic representation. And to yield a plausible account of epistemic representation, DEKI must satisfy the Surrogative Reasoning Condition. The central question of this chapter, then, is: by virtue of what is the denotation condition of DEKI satisfied? And can any available account of denotation serve to show why the denotation failure in cases of fortuitous misuse results in unjustified inferences?

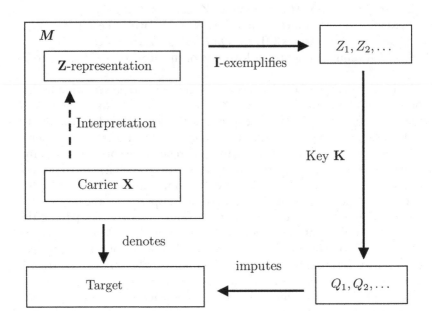

Figure 21.1 Authors' schematic representation of DEKI (based on Frigg and Nguyen 2020, 188).

3 Opening the Black Box of Denotation

Although other elements of DEKI get extended treatment, Frigg and Nguyen have little to say about denotation in *Modeling Nature.*[2] They direct their readers to the two prominent views on denotation in the philosophy of language: the descriptivist account and the causal-historical account. DEKI, thus, treats the denotation condition as a kind of black box to be filled in by philosophers of language. This section explores how the descriptivist and causal-historical accounts of denotation in language might apply to non-linguistic epistemic representations like maps.

3.1 *The Descriptive Account of Denotation*

According to descriptivist theories of singular terms,[3] a singular term is associated with descriptive content, even when it is grammatically simple (like the name "Kareem"). A singular term denotes a specific object when the descriptive content associated with that name suffices to uniquely pick out that object. Applying descriptivism to maps raises the question of how the descriptive content is "associated" with the map. In the philosophy of language, it is common to distinguish between speaker meaning and linguistic meaning. The association between the singular term and the description can be understood, then, either as occurring in the mind of the speaker (speaker denotation) or as somehow built into the semantics of the language (linguistic denotation). By analogy, the descriptive content that fixes the denotation of a map will be found either among the map user's beliefs or somehow "in" the map.

Suppose that unique descriptions of the target are associated with the map in the mind of the user. Susan associates her map with a unique definite description, for example, *the city in which the first man-made structure taller than 300 meters is located*. By descriptivist lights, her map now successfully denotes Paris. The problem with such a response is that it collapses use with fortuitous misuse. In the example, Susan is confused, and is associating her map of Prague with a description that is true only of Paris. Thus, in this version of a descriptivist account, her map *does* denote Paris, and her unjustified surrogative inferences would not be blocked by denotation failure.

Suppose the associated descriptions are "in" the map. The natural way to understand such a supposition is that the denotation of the map is determined by the markings on the map.[4] If the location has a unique conjunction of those features described by the map, it thereby denotes the location. The immediate problem, of course, is to determine which properties of the map determine the descriptive content. Intuitively, the printed lines and shapes count, but the coffee stains do not. We have already seen that DEKI solves this problem by taking the epistemic representation to be the map as interpreted. The user interprets the map by determining which properties to take as exemplified, then adopts a key relating these properties to those imputed to the location. The map denotes a location when the resulting imputations are together uniquely true of it. A descriptive account of denotation, thus, seems to fall right into the DEKI framework.

Although the foregoing story captures a plausible intuition about maps, it still fails. Recall that Susan succeeded in navigating Paris because of her mistaken, but fortuitous, reading of the map. This means that she created idiosyncratic exemplifications, keying-up, and imputations. Nothing in the DEKI conditions forbids such an interpretation. Frigg and Nguyen exacerbate the problem by writing that users are "free to choose *X*-features and *Z*-features as they please" (Frigg and Nguyen 2020, 170) and "enjoy complete freedom in choosing their keys" (Frigg and Nguyen 2020, 182). The combination of DEKI and the descriptive view entails that Susan's idiosyncratic epistemic representation *does* denote Paris, and we can no longer distinguish between her misuse of a map and Zoe's correct use of it.

In response, one might note that our discussion of DEKI has ignored the role of context (Frigg and Nguyen 2020, 148). Frigg and Nguyen say that exemplification is "highly sensitive" to context, and they require that the context "select" the exemplified features. In Section 4, we will explore the context of map making and how it might constrain correct use. For now, we only note that this ground must be tread carefully. Context has to be identified independently of the descriptive content that fixes the correct denotation. Appealing to the correct interpretation of a map to establish the descriptive content, and hence denotation, of the map would beg the question, since it is precisely the line between use and misuse that denotation is, in part, supposed to establish.

A descriptive account of denotation is, thus, an apparent dead-end for DEKI. On this family of views, descriptive content has to be associated with the map. Neither taking the user's beliefs nor taking the map's markings as the source of descriptive content will permit DEKI to distinguish between use and misuse. Both options foundered on the same difficulty: DEKI has no conditions that restrict the range of descriptive content. As a result, users are free to find any denotation at all, and the problem of fortuitous misuse stands unsolved. Appealing to standards of "correct use," while tempting, begs the question unless those standards can be shown to have some kind of epistemic grounding independent of the DEKI conditions.

3.2 *Two Causal Accounts of Denotation*

The alternative to descriptivist accounts of denotation is the so-called "causal-historical" family of views. These views share the idea that denotation requires some kind of causal link between the current use of the singular term and the object denoted. A speaker's capacity to use a term to denote is inherited from other speakers' use of the term. The causal chain thus created among users has its origin in the object denoted, and the term denotes *that* object because it is causally linked by the chain of uses. Causal views differ about whether the causal link alone is sufficient (the simple causal theory) or whether it needs to be supplemented by some descriptive content (the hybrid causal theory).

On the simple version of a causal theory, each speaker inherits their denoting use of the singular term from a previous speaker. That is, a speaker A uses a term N to denote whatever another speaker, S_i, uses N to denote. S_i inherited their use from a previous speaker, S_{i-1}, and the causal chain is traced back to an original use. At this "baptismal" event, the causal relation between the name and the named is established by an act of dubbing: "Call this one Freda!" says $S0$ while pointing at O. A's use of N refers to O, because it stands at the end of a causal chain leading back to the original naming of O.[5]

The simple causal theory could be applied to maps. The original user of a map is the mapmaker, who on finishing their creation says, "This is a map of Prague."[6] Subsequent map readers intend to use the map to denote whatever the mapmaker denoted with the map. The causal chain is somewhat more direct for maps than it is for singular terms insofar as the causal links are not passed through a series of map users. Rather, the causal link between the map reader and the mapmaker is established by the printing and re-printing of the original map. In our earlier example, then, Susan is using a map that is causally related to a mapmaker in Prague. It thereby denotes Prague, and Susan's use of it to navigate in Paris is a misuse. The DEKI conditions are not satisfied, because Susan is not imputing properties to the target denoted by the epistemic representation. So, according to DEKI, Susan's surrogative inferences are unjustified and we have a solution to the problem of fortuitous misuse.

Unfortunately, this tidy story will not stand up to scrutiny. Kripke presented the core idea of the simple causal view in *Naming and Necessity* (1980) as an alternative "picture" of denotation, but did not develop the details. It was quickly pointed out that the simple theory has some substantial defects, at least two of which are directly relevant to DEKI's potential assimilation of the simple causal theory of denotation.

First, the simple causal view requires that the named object exist, since the original dubber must stand in a causal relationship to it. This is also true for the more elaborate causal view (which we call a "hybrid" view) discussed below. No view requiring a causal link between representation and the represented can straightforwardly handle names like "Santa Claus." In the context of applying a causal account of denotation to epistemic representation, this problem is devastating. Accounting for targetless models is one of Frigg and Nguyen's criteria of adequacy (Frigg and Nguyen 2020, 13). One response would be to narrow the scope of "epistemic representation." Because a map of Middle Earth is not causally grounded in Middle Earth in the way that a map of Europe is causally tied to Europe, we might deny that the former is an epistemic representation. Developing an argument for such a demarcation is outside the scope of this essay. So that we might explore the consequences of causal theories of denotation, we will proceed on the assumption that targetless maps, like Tolkein's map of Middle Earth, are not (the right sort of?) epistemic representations.

The second standard objection to the simple causal theory motivates philosophers of language to look for more sophisticated alternatives. A simple causal account is subject to various kinds of "deviant" causal chains. These often concern mistakes made at or near the original baptismal event. Evans imagines this example:

> Two babies are born, and their mothers bestow names upon them. A nurse inadvertently switches them and the error is never discovered. It will henceforth undeniably be the case that the man universally known as "Jack" is so called because a woman dubbed some other baby with the name.
>
> (Evans 1973, 196)

Such troublesome examples are easy to transpose to epistemic representations. Consider a map made by an absent-minded, over-worked mapmaker. Having mapped several European capitals, he returns home to Paris. Holding the map he created in Prague, he says, "This is a map of my home" and gesturing broadly to the city around him, he prints the name "Paris" on the cover. The map now denotes Paris, rather than Prague, according to the simple causal theory. In the philosophy of language, the counterexample ends with the intuition that the name is somehow wrong. When transposed to epistemic representation, the objection cuts more deeply. The dubbing is irrelevant to whether subsequent users can draw justified surrogative inferences about Prague. Hence, the mapmakers' mistake does not disrupt the capacity of the map to be an epistemic representation of Prague.

Deviant causal chain examples motivate a family of views that combine elements of causal and descriptivist approaches. Evans, who was one of the innovators of such hybrids, expresses the core of the idea this way:

> a *necessary* (but not sufficient) condition for x's being the intended referent of S's use of a name is that x should be the source of causal origin of the body of information that S has associated with the name.
>
> (Evans 1973, 198)

The hybrid view generates the right intuition about Susan's map: it denotes Prague because Prague is the origin of the map's information. Similarly, the lines and shapes on the absent-minded mapmaker's map have their positions because of the mapmaker's activities in Prague. Again, the information in the map's markings is causally grounded in Prague, and hence it denotes Prague in spite of the mistaken baptism. A further attractive feature of this view would be that it explains why the absent-minded mapmaker's mistaken baptism does not disrupt the usefulness of the map. We can learn about Prague from the map, because the information in the map is causally linked to Prague.

We are immediately faced with a familiar problem: which of the many properties of the map constitute "the body of information" in the map? Suppose Susan's map of Prague sported a variety of stains caused in Paris. These link the map to Paris, but presumably stains are not the kind of property that constitutes information in the map. DEKI's solution, again, is to treat the epistemic representation as the combination of a paper map and interpretation, where the interpretation would specify the blue and red lines, but not the coffee stains, as exemplified properties and thereby information. Once again, the freedom that Frigg and Nguyen give to map users undermines DEKI's capacity to distinguish use from misuse. Suppose Susan interpreted the stains as on the map as exemplified properties, and keyed-up the map on that basis. Since these features are causally grounded in Paris and constitute information according to Susan's interpretation, the hybrid view treats the map as denoting Paris. Since Susan satisfies all the DEKI conditions, once again, DEKI is unable to distinguish between justified and fortuitously successful surrogative inferences.

Combining DEKI with either the descriptive or hybrid views of denotation runs into the same problem. By giving users free rein to choose interpretations, Frigg and Nguyen weaken the epistemic link between the features of the map and features of the target. To properly fix denotation, and thereby solve the problem of fortuitous misuse, we need to identify the epistemic relevance of certain features of the vehicle. As mentioned above, Frigg and Nguyen require that context "selects" the exemplifications, so that context identifies epistemic relevance. They construe context as "a certain set of problems and questions that are addressed by a group of scientists using certain methodologies while being committed to certain norms" (Frigg and Nguyen 2020, 148). The question—one not discussed in *Modeling Nature*—is how such problems, questions, norms, and methodologies can constrain the choice of interpretation so as to exclude unjustified surrogative inferences. To explore how this might work, we turn to a scientific context within which mapping is an important methodology.

4 Map Making and Measurement

Historically, the careful creation of site maps is one of the activities that distinguished archeology as a science. Reflection on the construction of site maps is integral to the systematic reflection on methodology in archaeology.[7] It, thus, provides an excellent example of the way that scientific questions, norms, and methods influence the interpretation of such maps.

Site maps represent the spatial relationships among the artifacts (e.g., pottery), structures (e.g., walls), and features (e.g., rivers) of the site. As a matter of good research design, a site mapmaker needs to determine from the outset what *kinds* of artifacts, structures, and features will be represented. Will the map depict large stones found at the site? The determination of kinds also requires identifying relevant distinctions to be represented. Some pottery shards, for example, may show signs of burning. Are burnt and unburnt shards two kinds of artifact or two instances of one kind (pottery shards)? The distinctions among kinds of artifacts, structures, and features to be represented requires criteria to determine whether a particular object is a member of the kind. By what criteria do we distinguish between burnt and unburnt pottery shards? A map's key reflects these decisions by the mapmaker and explicitly identifies the relevant kinds (and sometimes the associated criteria, e.g., "shards larger than 2cm") to be represented on the map.

Site mapmakers are constrained in their choice of kinds and criteria. Archaeological site maps serve two epistemic purposes: the research purposes of the site mapmaker and the purposes of future archaeologists who may use the map to answer different questions. The current archaeologists come to the site with questions motivated by current background knowledge. Whether there is an important distinction between, for example, burnt and unburnt pottery shards will depend on, for example, whether there was a known functional or cultural difference between burnt and unburnt pots among the site's original inhabitants. Site maps also record the dig for posterity, and since they cannot map everything, archaeologists must anticipate the purposes and questions of future map users. To do so, they use their current background knowledge and best guesses about the range of archaeological questions to determine what further artifacts, structures, and features should be represented.

Appropriate use of background knowledge and research questions to determine the kinds to be represented and their criteria plays a crucial role in justifying surrogative inferences drawn from the map. If it is known (or even just a guess) that burnt and unburnt pottery functioned differently in the culture, and if those differences were relevant to answering the archaeologists' research questions, then failure to record these as two kinds of artifact would be a scientific mistake. It would render the map useless for drawing a set of inferences that *should* be available. The justification of a particular surrogative inference, then, depends on the right kinds being identified with the right criteria, where "rightness" is determined by background knowledge and research questions.

With a set of kinds and criteria in hand, the archaeologist may proceed to actually make the map. Making any map requires a survey, and this has two elements: features of the terrain have to be *identified* and *measured*. Identification of the features involves using the criteria associated with the chosen kinds to determine whether a given part of the terrain needs to be recorded. Recording the features requires measuring them in some way. Typically, symbols on the map stand in the spatial relationships they do because artifacts, structures, and features corresponding to the kinds were measured as having a particular location relative to a point of reference (in archeological map making, this is traditionally called the "datum post," cf. Renfrew and Bahn 2016). But non-spatial measurements might be required as well, as when the population of cities on a map is represented by their color.

The foregoing brief survey of map making identifies several elements of the context relevant to the justification of surrogative inferences from a map. The map is produced so as to answer a certain range of questions, and this justifies the criteria for identifying and measuring features of the landscape. By identifying and measuring these features, the mapmaker puts information into the map. The justification of any surrogative inferences ultimately drawn by the users depends on the structure of justifications that are a part of the map production. Measurement, along with the justified selection of kinds and criteria, thus provides the precise sense in which the information "in the map" is grounded in the site. Since the surrogative inferences will be based on these features, the map user needs to be sensitive to them when making inferences from it. This is at least part of what it means to read the map correctly. The justification of the inferences, thus, depends on a fit between the kind of questions that the map has been created to answer and the kind of questions that the user is trying to answer.

These dimensions of the justification of surrogative inference distinguish between Susan's and Zoe's uses of their maps. These maps were designed (let us suppose) to help tourists navigate European cities. This justifies the choice of public thoroughfares (streets, bridges, etc.) and tourist attractions as features of the cities to be recorded. The mapmakers then surveyed the cities and marked up the maps accordingly. However, Susan's map is based on measurements of Prague, whereas Zoe's map is based on measurements of Paris. Susan is trying to answer questions about Paris, but none of the markings on her map is based on Paris. Hence, she cannot draw justified inferences about Paris from it. Zoe, of course, can do so, since her map is based on a survey of Paris. Note that this also makes it clear as to why the absent-minded mapmaker's map remains useful as an epistemic representation, independently of the (mis-)dubbing of the map. Although the name printed on the cover might help, practically, to orient the user, nothing in it is based on the survey of streets and landmarks. Hence, the printed name is largely irrelevant to the justification of the inferences drawn from the map.

Superficially, the account just given seems to fit the causal-hybrid account of denotation. Measurement establishes a causal relationship between the map and the terrain that it represents. The map denotes a particular place, because the information in the map is causally related to that place. There are important caveats, however, arising from the difference between proper names and epistemic representations like maps. Unlike maps, we learn little to nothing about an object by inspecting its name. Hence, in the causal-hybrid view, the information is contained in descriptions "associated" with the name. The contents of the descriptions associated with a name are independent of the causal chains connecting them to the baptismal event in the sense that the descriptions could all be false and the name still refer.

In contrast with names, the information in the map is more intimately related to its causal origin. The causal relations are caught up in the web of justifications described above. For example, suppose the archaeological site mapmaker represented burnt shards with a square, and unburnt shards with a triangle. However, because pens kept malfunctioning in the dry air, she had to use both black and blue pens. So, sometimes the squares and triangles are blue and sometimes they are black. The map instantiates four properties: square, triangle, blue, and black. Even though the properties of blue and black are causally connected to the site and carry information, it is not relevant information for making inferences from the map. Unlike the difference between squares and triangles, the difference between blue and black figures does not distinguish between kinds or measurements relevant to the purposes of the map. Despite being caused by a feature of the site, the properties of black and blue do not play a role in any of the inferences relevant to the map making and hence are epistemically irrelevant. By contrast, the difference between squares and triangles on the map is not only caused by differences in how burnt and unburnt shards are arranged on the site, but also the mapmaker's choice of these features is itself scientifically justified.

The causal origins of map features, then, establish denotation by being part of a process that has a crucial justificatory dimension. The choice of objects to measure and identify must be justified by the kind of questions that the map is designed to answer. The choice of measurement techniques needs justification too, since measurements are subject to different sorts of error, and the possible sources of error influence the kinds of questions that a map can answer. The justification of the surrogative inferences that might be made from a map depends on the intimate relationship between the causal grounding of the map and the web of justification between the questions, the criteria for identification of objects, and their measurement.

The relationship between causal grounding and justification found in epistemic representation is strikingly analogous to the inferentialist critique of reliabilist accounts of perceptual knowledge. A tradition stretching back to Wilfrid Sellars, with echoes of Kant and Hegel, and winding through the work of McDowell (1994), Brandom (1994), and Williams (2008), has charged such accounts with a failure to appreciate the normative character of knowing. A causal relationship between a subject's belief and its object is insufficient to entitle a subject to that state. The causal relationship between a believing subject and the object of their belief is made epistemically relevant by its uptake in a system of inferences, and it is through these inferences that the subject becomes accountable for their perceptual beliefs.

The foregoing account shows how context "selects" the epistemically relevant properties of a vehicle in a way that distinguishes between appropriate use and fortuitous misuse. The key ingredients are causal *and* normative. Criteria for the identification of features of the terrain and their measurement are justified by the questions that the map is designed to answer. The process of identification and measurement, thus, establishes the causal relationships between the map and the terrain that it denotes. The user is in a position to draw justified surrogative inferences about that terrain only if their reasoning is based on those features of the map that are themselves based on (previously justified) measurement and identification criteria.

5 Conclusion

Section 4 has provided a detailed account of how context "selects" properties of a map relevant to surrogative inference. At first pass, this seems like a positive result for DEKI. By adding details about the specific context of map making and map reading to DEKI, we get an account that distinguishes between justified and unjustified surrogative inference and resolves the problem of fortuitous misuse. The dark cloud within this silver lining, however, is that the details raise important problems for DEKI.

First, the mechanism by which contexts select relevant features of the map strongly constrains both mapmakers and map users, if justified surrogative inferences are to be generated by a map. This contradicts the idea that producers and users are free to choose exemplifications and keys.

Second, the causal grounding provided by measurement and identification is a crucial element of the context. Causal grounding runs contrary to the idea that maps can represent terrains that do not exist. A defender of DEKI must either let go of the idea that a "map" of the Middle Earth is an epistemic representation or provide criteria for distinguishing between real and imaginary maps.

Third, DEKI treats denotation as independent of the other conditions. DEKI would categorize the identification of kinds and measurement in terms of the exemplification and keying-up conditions. But Section 4 has shown that these are part of establishing denotation. Hence, the denotation condition cannot be satisfied unless the properties exemplified and keyed-up are also specified. At the very least, this means that DEKI does not capture some of the important dependencies of surrogative inference.

Finally, even if elements of the context could be neatly fit into the DEKI categories, Section 4 draws a straight line from the conditions of map production through the users' justified surrogative inference. No appeal to the DEKI conditions was necessary to distinguish between justified and unjustified inference. DEKI seems to miss the main action. Once we have told the story of how the inferences are justified, we have everything we need in order to say that *this* map represents Paris and *that* one represents Prague. This suggests that epistemic representation might be better understood in *inferentialist* terms.[8]

Acknowledgments

Dr. Risjord's work on this essay was supported by the Czech Science Foundation (GAČR), GX20–05180X "Inferentialism Naturalized: Norms, Meanings and Reasons in the Natural World."

Notes

1 Susan also misreads her map, whereas Zoe reads it correctly. These differences would be captured in other elements of the DEKI analysis. In our commentary on Frigg and Nguyen's main essay for this volume, we discuss these aspects of misreading.

2 However, they develop the denotation condition somewhat in Frigg and Salis (2019) and Salis et al. (2020).

3 The literature on denotation and reference concerns *both* the speech act of referring and the related uses of singular terms (e.g., the use of a definite description to direct the hearer's attention to a specific object), *and* the semantic meaning of a singular term or definite description, wherein it refers to an object. Since the problem of fortuitous misuse does not involve the use of the map to communicate, we set speech acts to the side and focus only on the semantic relation. In addition, since we are not concerned with potential differences between definite descriptions, proper names, natural kind terms, etc., we will simply speak of "singular terms."

4 In Salis et al. (2020) and Frigg and Salis (2019), they develop the idea that models "are constituted by model descriptions (linguistic and mathematical symbols) and propositional content" (Salis et al. 2020, 209). Although there is no space for a detailed discussion, this is susceptible to the same objection presented below.

5 One might think that the simple causal view runs afoul of Frigg and Nguyen's rejection of "Stipulative Fiat" as a means for fixing denotation (cf. Frigg and Nguyen 2020, 24–29). They argue that stipulation *alone* cannot determine denotation. However, note that the simple causal theory does not fix denotation by stipulation alone. The baptismal event uses the name in a speech act of baptism, not denotation. Subsequent denotations are constituted by the causal chain leading back through this event.

6 One has to be careful not to conflate the reference of the name with the reference of the map. We cannot require that the map acquires its reference from the name "Prague," which is cheerfully printed on the cover. There can be maps of unnamed places, so the baptismal event would involve the map, and a pre-existing name is not necessary.

7 See, for example, the extensive discussion mapping in the textbook *Archaeology Essentials: Theories, Methods, and Practice* (Renfrew and Bahn 2016).

8 We spell out the details of such an account in Khalifa et al. (2022).

References

Brandom, R. 1994. *Making It Explicit: Reasoning, Representing, and Discursive Commitment*. Cambridge, MA: Harvard University Press.

Chamings, A. February 11, 2021. "The Bizarre Tale of the World's Last Lost Tourist, Who Thought Maine Was San Francisco." *SF Gate*.

Evans, G. 1973. "The Causal Theory of Names." *Proceedings of the Aristotelian Society, Supplementary Volumes 47*, 187–225. https://doi.org/10.1093/aristoteliansupp/47.1.187.

Frigg, R. 2006. "Scientific Representation and the Semantic View of Theories." *Theoria. Revista de Teoría, Historia y Fundamentos de la Ciencia* 21(1), 49–65.

Frigg, R. and F. Salis. 2019. "Of Rabbits and Men." In B. Armour-Garb and F. Kroon (Eds.), *Fictionalism in Philosophy*. Oxford: Oxford University Press, pp. 187–206. https://doi.org/10.1093/oso/9780190689605.003.0010.

Frigg, R. and J. Nguyen. 2020. *Modelling Nature: An Opinionated Introduction to Scientific Representation*. New York: Springer International Publishing. https://doi.org/10.1007/978-3-030-45153-0.

Khalifa, K., J. Millson and M. Risjord. 2022. "Scientific Representation: An Inferentialist-Expressivist Manifesto." *Philosophical Topics 50*(1).

Kripke, S. A. 1980. *Naming and Necessity*. Cambridge, MA: Harvard University Press.

McDowell, J. 1994. *Mind and World*. Cambridge, MA: Harvard University Press.

Remsen, N. October 20, 1977. "Golden Gate-Bound German Visits Bangor by Mistake." *Bangor Daily News*.

Renfrew, C. and P. G. Bahn. 2016. *Archaeology Essentials: Theories, Methods, and Practice* (7th ed.). London: Thames & Hudson.

Salis, F., R. Frigg and J. Nguyen. 2020. "Models and Denotation." In J. Falguera and C. Martínez-Vidal (Eds.), *Abstract Objects*. New York: Springer, pp. 197–219. https://doi.org/10.1007/978-3-030-38242-1_10.

Williams, M. 2008. "Responsibility and Reliability." *Philosophical Papers* 37(1), 1–26. https://doi.org/10. 055686408094852121080/.

22 DEKI and the Mislocation of Justification

A Reply to Millson and Risjord

Roman Frigg and James Nguyen

In their "DEKI, Denotation, and the Fortuitous Misuse of Maps", Jared Millson and Mark Risjord (MR) take the DEKI account to task for being unable to "distinguish justified surrogative inferences from unjustified ones", something they take to be problematic because an analysis of representation "must block unjustified surrogative inferences" (this volume, 282). This, they say, means that DEKI fails to meet our own Surrogative Reasoning Condition (Frigg and Nguyen 2020, 3).

They reach this conclusion through a thought experiment that they take to show that the DEKI account is open to what they call the *fortuitous misuse* of a representation. Their protagonist is the absent-minded traveller Susan, who, when visiting Paris, by mistake uses a map of Prague to navigate. She uses that map to find her way from the Tour Eiffel to the Jardin des Tulieries, and due to a string of lucky coincidences, she does so successfully. However, despite reaching her destination, "Susan's conclusions were obtained by *luck* and unjustified" (this volume, 281). Hence, Susan's use of the map is fortuitous because she gets to where she wants to be, and it is a misuse because the conclusions she draws from the map lack justification. MR point out that Susan's use of the map is consistent with DEKI, and DEKI therefore lacks the means to block unjustified inferences, which MR take to be a red flag.

To assess this argument, it is important to be clear on what exactly requires justification, and there seem to be (at least) two different aspects of inferences from representations that might need to be justified. These are *very roughly* analogous to the validity and soundness of a logical argument, and we call them *derivational correctness* and *factual correctness*, respectively. An inference drawn from a representation is derivationally correct if the inferential steps that lead to the conclusion are correct with respect to the rules of the representation and only use premises that form part of the representation. The conclusion of an inference is factually correct if the conclusion is true of the representation's target. MR do not distinguish these two aspects clearly, but the context of their discussion suggests that they require a justification for both. In particular, they seem to require a justification for factual correctness when they sum up their conclusion of the thought experiment by saying that

DOI: 10.4324/9781003202905-25

it "highlights the fact that not just any true belief formed from inspecting the map is the conclusion of a justified surrogative inference" (this volume, 281).

We agree that when drawing inferences from representations we would ultimately like to have a justification of both their derivational and factual correctness. Our disagreement with MR is about where this justification has to come from. MR think that it must come from an account of representation, and they criticise DEKI because it fails to provide the required justifications (and hence fails to distinguish between justified and unjustified inferences). We disagree: although one should expect an account of representation to have something to say about derivational correctness, factual correctness lies squarely outside the remit of an account of representation, DEKI, or otherwise. Further, DEKI does provide a justification of derivational correctness, and so it successfully performs those tasks that it actually should perform.

Let us begin with derivational correctness. The locus of justification here is DEKI's key, which provides the rules associated with the representation. Drawing inferences from a representation has to be done using the key. Inferences not based on the key are unjustified. If someone looks at the map in our contribution to this volume and infers that everything in Stockholm is grey because the area on the map (as shown in Figure 20.1) denoting Stockholm is grey, then their inference is unjustified because it is not based on the key, which does not connect colours on the map with colours of the landscape.

Factual correctness is an altogether different matter. In our exposition of DEKI, we are careful to say that representations generate *claims* about their targets, but that these claims can be false (Frigg and Nguyen 2020, 178). Further, their truth-values are not something that the representation adjudicates, let alone justifies. Neither the truth-value of a claim nor the justification of why a certain truth-value is assigned to it is intrinsic to the representation. So, neither DEKI nor any other account or representation should be asked to account for either. Maps (and models) are not special in this regard. The same goes for other representations. Neither sentences, nor paintings, nor caricatures, contain their own truth-values along with justification for them; indeed, not even photographs do. "The cat is on the mat" says something about the cat; it neither assigns nor justifies, a truth-value to the sentence's content. Likewise, Constable's *Salisbury Cathedral from the Meadows* represents the Salisbury Cathedral as an early gothic building with one tower, but neither the truth-value of the claim that Salisbury Cathedral really is an early gothic building nor its justification is part of the painting. Truth-values and their justification are not part of a representation, and so they cannot be expected to appear in an analysis of representation. Contrary to what MR say, the absence of such specifications from DEKI is a feature, not a bug, of the account.

To be clear, we are not saying that conclusions of inferences drawn from representations should not be justified, let alone that there is no such (factual) justification. Of course we want justification! Our claim is that this justification is not part of a representation's content, and therefore it does not have to be analysed in a theory of representation. This sort of justification comes from the outside. We believe that "the cat is on the mat" is true because we see that the cat is on the mat, and we have reasons to trust our senses, or perhaps we have reliable testimony to that effect, not because of what the sentence means. Likewise, we can deem the claim, derived from Constable's painting, that Salisbury Cathedral is an early gothic building with one tower true because we stand on the meadow ourselves and see this to be the case. Or we can rely on testimony. This might involve Constable's own testimony, and justifying our belief that the testimony is reliable may involve appealing to Constable's artistic practices. In the context of scientific modelling, justification might come from the theoretical framework in which a model is embedded, the track record of the model as applied to systems of a certain kind, or the model's fit with data (obviously, the details of this will depend on the details of the model in question).[1]

The same goes for maps. MR go to a great length in explaining what justifies the conclusions of inferences drawn from maps (see Section 4 of their paper). They point out that to produce a map, map-makers first identify the relevant territory and then perform detailed measurements (this volume, 290). Based on these measurements, and possibly other relevant information that they have gathered (about the character of roads, the size of cities, and so on), they then draw up the map and put this information in it. The conclusions we draw from the map are justified because we know that map-makers have followed these procedures and carried them out correctly (this volume, 290–291).

We agree but insist that this is not part of the representational content of the map. If we wish to know whether the conclusions we draw from maps are true, and if we want to justify our views, we have to look outside the map. First, we have to make sure that we know what projection the map-makers have used, rather than using any projection that comes to mind. And we have to have reasons to believe that the map has been produced correctly. We believe that conclusions drawn from Ordinance Survey maps are true and justify this by pointing out that we know that these maps have been produced through careful and meticulous processes; we do not believe (or believe to a much lesser degree) conclusions drawn from a cheap tourist map because no such justification is available. But none of this is part of the content of the map itself. In fact, we can be misled. One of us has vivid memories of travelling through Yucatán with official government maps, which turned out to indicate non-existent roads (we later learned that the maps were drawn up based on the government's plans to build roads, which, alas, fell prey to budget cuts). The map simply represents the territory as being thus and so; it does not say whether the claims that follow from it are true, nor does it provide justifications.

Justifying claims is the province of epistemology and much can be said about how this is done. At this point, we are merely insisting that epistemology stands apart from semantics. A theory of representation tells us what the representational content of a map is; epistemology tells us under what circumstances are justified in thinking that the content (or parts of it) is (are) true. MR's criticism of DEKI is based on mingling the two.

Once this is recognised, it is clear that DEKI does not fail the Surrogative Reasoning Condition. The condition says that "[e]very acceptable theory of scientific representation has to account for how reasoning conducted on models can yield claims about their target systems" (Frigg and Nguyen 2020, 3). In the idiom previously introduced, the condition asks for a justification of derivational correctness, and for the reasons mentioned DEKI fits that bill.

This said, we would also like to highlight how special the circumstances are where the purported lack of justification becomes problematic. It is crucial to the thought experiment that Susan makes a *genuine* mistake when using a map of Prague to navigate Paris. MR confirm in personal communication that if Susan *deliberately* took a map of Prague to navigate Paris, then this would not be a fortuitous misuse of the map. At least in the context of scientific practice, this is the right thing to say. Repurposing maps about one city to navigate another is hardly a frequent occurrence, whereas repurposing models of one target to describe another is a common model building technique and it is good to see that MR do not recommend banning this technique.[2] On the other hand, the requirement of making a genuine mistake makes the case so unlikely that one wonders about its relevance in practice. The cases are so rare for maps, that someone who made a related mistake became a national celebrity (cf. the Erwin Kreuz case reported in MR's paper). Rarity turns to near-impossibility when we look at science. Scientific representations like models, theories, or diagrams are typically used by entire scientific communities. The chances that an entire community makes a genuine mistake and conflates a model of one target (e.g., the hydrogen atom) with a model of another target (e.g., Andromeda) are minute. Unless there are more realistic scenarios where the issue arises, we are not sure of how much of a problem this is.

Finally, MR suggest that inferentialist accounts of representation avoid this problem because they provide the required justifications. MR mention this in passing and do not provide details. This is an interesting claim and it would advance the debate to see it developed. But it is not immediately clear as to how inferentialism would do this. Consider Suárez's account, which is the most widely cited inferentialist account of scientific representation, according to which a model M is an epistemic representation of a target T only if (i) the representational force of M points towards T, and (ii) M allows competent and informed agents to draw specific inferences regarding T (Suárez 2004, 773). At least on the face of it, it is unclear how this account meets MR's justificatory requirements, but we are excited to see how they develop the inferential conception to do so.

Notes

1 Of course, given that science is a truth-directed enterprise (or, depending on your other philosophical views, an enterprise directed at empirical adequacy or understanding), scientists may *intend* to choose keys that yield conclusions justified in these ways. But these intentions are not necessarily satisfied (and so the intention alone does not provide such a justification), and the question of justification is still external to the representations themselves.
2 For a discussion of this way of constructing models, see, for instance, Ch. 11 of Frigg (2022).

References

Frigg, R. 2022. *Models and Theories*. London: Routledge.

Frigg, R., & Nguyen, J. 2020. *Modelling Nature. An Opinionated Introduction to Scientific Representation*. Berlin and New York: Springer. https://doi.org/10.1007/978-3-030-45153-0.

Millson, J., & Risjord, M. this volume. "DEKI, Denotation, and the Fortuitous Misuse of Maps."

Suárez, M. 2004. An Inferential Conception of Scientific Representation. *Philosophy of Science*, 71(5): 767–779. https://doi.org/10.1086/421415.

23 DEKI and the Justification of Surrogative Inference

A Reply to Nguyen and Frigg

Jared Millson and Mark Risjord

In their response to our essay, Frigg and Nguyen (henceforth F&N) propose that the problem of fortuitous misuse can be resolved by DEKI's key, since this is the "locus of justification" (this volume, 297). This is a fair reply, since our original essay only gestured toward the other DEKI conditions, hinting that they are vulnerable to parallel arguments. In this chapter, we will develop these arguments and show that keying-up does not save DEKI from the problem of fortuitous misuse.

Before doing so, it will be helpful to be precise about the character of the problem. We recognize F&N's distinction between "derivational" and "factual" correctness. When a map is misread, the surrogative *inferences* are unjustified, regardless of whether or not the conclusions are true. The problem of fortuitous misuse is a problem of misreading, and thus only concerns derivational correctness. As F&N rightly say, the Surrogative Reasoning Condition demands that an account of epistemic representation provide a "justification for derivational correctness" (this volume, 299). To do so, as argued in our main essay, an account must at least provide grounds for distinguishing derivationally correct from incorrect surrogative inferences. The problem of fortuitous misuse presents exactly this challenge: Zoe's inferences are derivationally[1] justified, whereas Susan's are not. An account of epistemic representation that could not differentiate Zoe's inferences from Susan's would, thus, fail the Surrogative Inference Condition.

In their lead essay, F&N set out to show what "conventions we must be aware of" when correctly reading a map and the way that "the mode of production of a map" matters to its proper reading (this volume, 266). Doing so would show that DEKI satisfies the Surrogative Reasoning Condition only if something in the DEKI conditions picked out, or at least circumscribed, the conventions and modes of production that informed the correct readings. The mere fact that the DEKI conditions fit the correct readings tells us nothing if they fit the incorrect readings equally well.

F&N highlight the difference between incorrect (naive) and correct (non-naive) readings by introducing the imaginary Swedish Prime Minister Asbjörn. He is tasked with dividing Sweden along its north-south axis. Asbjörn does not recognize that the map (Figures 20.1, 20.2) uses a Mercator projection. By simply dividing the map-distance between Sweden's northernmost and southernmost points in half, he reads the map as if the projection were flat, and thus mislocates the axis. Asbjörn thus misreads the map. A correct reading would use the inverse Mercator project to calculate distance. According to DEKI, Asbjörn's mistake was located in his key. Since his key was incorrect, his inference was unjustified.

Unfortunately, simply noting that Asbjörn's key is incorrect does not yet resolve the problem. Asbjörn is faced with a choice of two possible keys: one that uses the flat projection and one that uses the Mercator projection. One of these is the correct way to read the map in this context and the other is not. Both satisfy the DEKI definition of a key, since they associate features of the map with properties of the landscape. Satisfying the DEKI keying-up condition, then, is not sufficient to distinguish between justified and unjustified surrogative inferences, since incorrect keys satisfy them just as well as correct ones. Hence, the DEKI keying-up condition does not help distinguish between justified and unjustified surrogative inferences. Therefore, unless some other condition blocks unjustified surrogative inferences, DEKI fails the Surrogative Reasoning Condition.

In response, one might note that the DEKI conditions are "skeletal" (this volume, 262), as are all accounts of scientific representation. They need to be filled in by information from the particular context of use. In *Modeling Nature*, they distinguish between "functional" and "inherent" characterizations of scientific representation (Frigg and Nguyen 2020, p. 93). A functional characterization of a boiler, they write, will describe it as a "contraption that makes water hot" (Frigg and Nguyen 2020, p. 93), whereas the inherent characterization will show how different kinds of boiler do so. The DEKI conditions provide only a functional characterization of a justified surrogative inference. To show why Asbjörn's reading uses the wrong key, the inherent characterization will draw on details of the context to capture the differences between correct and incorrect readings. Note that the functional characterization must constrain the inherent characterization. If an architect's plan functionally specifies a "boiler," then no matter how they do it in context, the engineer needs to come up with something that heats water. Therefore, the DEKI conditions must circumscribe justified surrogative inference enough to determine which elements of context might count toward justification, even if it leaves the specific details open. The question, then, is whether a closer inspection of the conventions and modes of production for maps can be incorporated into the DEKI conditions so as to explain why some inferences are justified, and others, like Asbjörn's, are not.

We may begin with F&N's point that "maps are made for a particular purpose" (this volume, 273). Of course, as they would no doubt agree, not every purpose of the mapmaker is relevant to the representational capacity of a map. Situating mapmaking in the process of inquiry is one way to identify the epistemically relevant purposes. Inquiry begins with questions and ends with their answers. The conventions and conditions of production for a map determine what kinds of questions it can answer correctly. For instance, the Mercator projection preserves radial relationships among points. Maps made with this projection are designed to answer questions regarding the direction of travel between two points on the globe. Since distance is systematically distorted, it can be read from the map only with the inverse Mercator projection formula. On the other hand, the Mercator projection does not preserve geodesic distance over the globe; so if the user has questions about the shortest route between two points, a different map must be used. The justification of surrogative inference thus depends, in part, on the "fit" between users' questions and the questions the map was produced to answer. We might say that the map must be *relevant* to a user's questions. The DEKI conditions do not demand any fit between a map's conditions of production and the user's questions, but these are clearly part of why some readings are correct.

Map reading needs to be sensitive to those features of the map that are the result of measurement, and it should distinguish them from those that are arbitrary or accidental. The primary mode of production for maps is to survey the landscape. As our main essay describes in detail, mapmakers use the expected purposes of the map to determine which features will be represented. These features must then be identified in the landscape and their relationships (distance, radial angle) or properties (population, color) measured (or identified). Making correct inferences depends on knowing that not every feature of a map is the result of measurement. As F&N point out, inferring that buildings in Stockholm are pink from the pink dot on the map would be an unjustified surrogative inference because the key does not associate dot color with landscape color (this volume, 270). To key up the color of the dot with the color of the landscape would be a mistake because the map was not created to answer questions like that, and hence no measurements were made of the color of the landscape. However, the DEKI conditions do not require that keying-up has any relationship to the measurements determining the features of the map; hence, the conditions cannot specify which features are appropriate candidates for the key, and which are not.

Justified surrogative inferences also depend on the users reasoning properly with the resources of the map. Asbjörn's mistake in calculating distance was of this sort. A similar kind of error would have occurred if he had mis-identified which area was marked "Sweden" or if he had treated the blue lines as national boundaries rather than as waterways. Map users, thus, need to know the conventions with which a map was constructed. Broadly speaking, the user must correctly *derive* the premises of the surrogative inferences from the map. Clearly, Asbjörn's justification was deficient on this score. His mistake is treated by the DEKI conditions as an incorrect keying-up, since he associated the map distances and the terrain distances in the wrong way. But again, since the DEKI conditions do nothing to relate the conventions or conditions of production to the key, they do not account for the difference between correct and incorrect keys.

By construing maps as tools of inquiry and map reading as inferring answers from them, we have identified three general characteristics of mapping conventions and conditions of production. *Relevance*: the map must be produced in such a way as to provide potential answers to the user's questions. *Measurement*: the relationships among the elements of the map relevant to the user's questions must be based on measurements of the terrain. *Derivation*: the map user must correctly derive the premises of the surrogative inference from the characteristics of the map. Each of these can be used to show why a particular reading of a map is correct or incorrect, and hence are at least part of what supports the justification, the derivational correctness, of surrogative inference. However, as we have noticed along the way, these general characteristics are not circumscribed by the DEKI conditions in a way that distinguishes between justified and unjustified surrogative inferences. Asbjörn's unjustified inferences, thus, fit DEKI because the analysis underspecifies the function of justifying surrogative inferences. The "architectural plans" of DEKI's analysis do not constrain the "engineering" of surrogative inferences.

Although it is poor solace for DEKI, notice that similarity and structuralist accounts share DEKI's deficit. As F&N argue, similarity and structuralist accounts require the "right kind" of similarity or morphism, but they cannot antecedently specify it. As they say, "One first has to know the projection and all the conventions used, and only when everything has been spelled out one can turn around and say 'see, they are similar'" (this volume, 272). We argue that this applies to DEKI as well: keying-up only blocks unjustified surrogative inference if we begin with the conventions and conditions of production that justify a surrogative inference and then key up on that basis.

Inferentialism is distinct from DEKI, similarity, and structuralist views because it accounts for epistemic representation in terms of surrogative inference, not the other way around. The root of the problem for non-inferentialist views is that none of these analyses track the conditions of justification for surrogative inference. They are thus bound to capture surrogative inferences after the fact, so to speak. Inferentialism is attractive because the elements of the analysis track the justification of surrogative inference. The three elements of the justification of surrogative inference characterized above (relevance, measurement, derivation) are part of what we call the *inferential pedigree* of the inference

(Khalifa, Millson, and Risjord 2022). Once the inferential pedigree of the surrogative inferences about a target, *T*, from an epistemic representation, *M*, is in place, we suggest, nothing more is needed for *M* to represent *T*. Every good architect should first be a good engineer.

Acknowledgments

Dr. Risjord's work on this essay was supported by the Czech Science Foundation (GAČR), GX20–05180X "Inferentialism Naturalized: Norms, Meanings and Reasons in the Natural World."

Note

1 Since only the derivational sense of justification is in play, we will drop this modifier below.

References

Frigg, R., and J. Nguyen. 2020. *Modelling Nature: An Opinionated Introduction to Scientific Representation*. New York: Springer International Publishing. https://doi.org/10.1007/978-3-030-45153-0.
Frigg, R., and J. Nguyen, this volume. "DEKI and the Mislocation of Justification: A Response to Millson and Risjord."
Khalifa, K., J. Millson, and M. Risjord, 2022. Scientific Representation: An Inferentialist-Expressivist Manifesto. *Philosophical Topics*, 50 (1).
Nguyen, J., and R. Frigg. this volume. "Maps, Models, and Representation."

24 How Values Shape the Machine Learning Opacity Problem

Emily Sullivan

1 Introduction

Machine learning (ML) models have an opacity problem. At least this is the impression that one gets by the proliferation of papers in computer science developing explainable AI (XAI) methods and philosophers describing various conceptions of opacity (Creel 2020). However, to what extent is opacity really a problem for explaining and understanding phenomena with ML models? If we look to the built consensus that philosophers have taken on general issues of understanding and explanation, opacity is an insurmountable problem. Many agree that for a model to enable understanding, transparency (Dellsén 2020; Strevens 2013), simplicity (Bokulich 2008; Kuorikoski and Ylikoski 2015; Strevens 2008), and the ability to manipulate the model (De Regt 2017; Kelp 2015; Wilkenfeld 2013) are all necessary. But this cannot be the full story. While transparency is important, full-fledged transparency is not a sought-after goal. When we explain things, we often leave things out. Not all the details matter. Moreover, adding true but irrelevant or tangential details do not improve our understanding. Therefore, the important question is *what* should be transparent.

In this chapter, I argue that the problem of model opacity is entangled with non-epistemic values. I will look at three different stages of the scientific process surrounding ML models providing understanding of phenomena: model acceptance and linking the model to the phenomenon (§2), explanation (§3), and attributions of understanding (§4). I argue that at each of these stages non-epistemic values, in part, determine how much ML model opacity poses a problem. My aim is to provide a broad outline about how non-epistemic values impact ML model opacity with regards to understanding and explanation. In the end, much more could be said about the role of non-epistemic values in each stage of the ML model pipeline. However, I hope that what I provide here illuminates that ML opacity, explanation, and understanding can be entangled with non-epistemic values.

DOI: 10.4324/9781003202905-27

2 Epistemic Risk in ML Model Construction and Acceptance

The role that non-epistemic values have in scientific theorizing and modeling practices has a rich history in the philosophy of science. Although there are those who argue that science should be a value free enterprise, there is a general consensus that values are ineliminable.[1] I will not rehearse these arguments here. Instead, I will argue that assuming the arguments that non-epistemic values are inescapable in scientific practice are successful—which I very much think they are—then non-epistemic values provide boundaries to the ML opacity problem.

Hempel (1965) and Rudner (1953) originally discussed *inductive* risk as the paradigm case of values entering scientific practice: there is always a risk of error when accepting whether a given hypothesis is true or false. When a hypothesis has downstream societal consequences, societal values should be part of its acceptance considerations (Douglas 2000). Others have since argued that there are other types of epistemic risk that require value choices beyond the risk of error in accepting (or rejecting) a scientific hypothesis (Biddle and Kukla 2017). For example, Harvard and Winsberg (2022) have recently argued that there are two core types of epistemic risk: inductive risk and representational risk. The former narrowly concerns the risk of error from endorsing a false hypothesis or statement, whereas the latter is the risk that a given scientific representation is inadequate for a given purpose. These kinds of epistemic risk are ever present among ML models.

The ML modeling requires considering tradeoffs between type I and type II errors, a standard tradeoff for inductive risk.[2] Moreover, Biddle (2020) pinpoints several aspects of the model pipeline that involves tradeoffs closer to representational risk that must be resolved in non-epistemic ways, such as identifying the problem to be modeled, training and benchmarking, algorithm design, and model deployment decisions. However, there is one specific area in the ML model pipeline that Biddle overlooks that is especially important when considering how much model opacity and complexity threatens explanation and understanding: model acceptance and establishing the link between the model and the phenomenon.

2.1 Epistemic Risk in Connecting ML Models to Phenomena

In a previous work, I argued that the problem of model opacity should not be understood as simply an internal problem that requires greater transparency of how the model works. Model opacity *qua* opacity need not undermine explanation or understanding from complex ML models. Instead, the problem of model opacity is largely an *external* problem connecting the model to the target. Specifically, the problem of opacity is a function of how much *link uncertainty* (LU) the model has (Sullivan 2022a). It is not the inner details of how the model works, but the higher-level abstract features that the model relies on to make its decisions, and most importantly, how those features are externally supported in providing insight into the target phenomenon that matters for understanding. There are various interpretability techniques available for "black-box" ML models that provides us with the necessary details regarding how the model made its decision such that the problem of model opacity becomes an external problem of LU.[3]

The framework of LU fits nicely with various theories regarding the way in which models provide understanding of phenomena. For example, a common view is that models explain when the counterfactual inferences that the model makes are true of their target (Bokulich 2011). Thus, one central aspect of accepting whether a model could be used to explain and enable understanding is in linking the model's inferences to the target phenomenon. In my view, when this link is weak or involves several uncertainties, understanding is limited. Moreover, strengthening this link also dispels the problem of model opacity. On a different (yet arguably compatible) theory—the adequacy for purpose view (Parker 2020)—models are either adequate or inadequate for a very specific scientific purpose, such as answering a specific research question. Similarly, on such a view, the link between the model and the specific purpose needs to be established as adequate *enough* to provide insight into the research question identified. All said, one of the central features of accepting whether a model can provide understanding of phenomena is accepting whether the links between the model and the phenomenon are strong enough. Moreover, if the proponents of inductive risk are right, then deciding when we have reduced LU enough to understand can involve the consideration of non-epistemic values.

Consider an ML model that has a high degree of epistemic risk. Medical researchers sought to develop an accurate predictor model about the risk of death for patients presenting with pneumonia at a hospital. One goal of this model was to increase the efficiency of allocating medical resources to those who need it, while letting the others receive more comfort by recovering at home. Researchers found that an opaque neural network model achieved the highest accuracy rates. However, there is a clear epistemic risk in accepting whether the model should be used in practice, given the consequence of error.

It is my contention that there is also an epistemic risk in accepting that the model could *explain* or *provide understanding* of risk factors for patients with pneumonia. If scientists use the model to explain or represent the risks facing patients with pneumonia, but the model is inadequate for that purpose, then there are real non-epistemic consequences.[4]

Moreover, the model explores the tradeoff between recovering from home and staying in the hospital and using up hospital resources, which is not purely epistemic. Thus, accepting whether a particular model is adequate to explain or provide understanding involves the (implicit) weighing of these values. This is a traditional problem of inductive risk (or representational risk) in deciding what kinds of evidence and the level of evidential support that is necessary in the face of uncertainty, and how strong a connection linking the model and the target is necessary.

2.2 Epistemic Risk and Opacity

If I am right, that the problem of model opacity is a function of LU, then the questions of inductive risk and representational risk are relevant for model opacity. Thus, ML models face *link uncertainty risk*. Judgments about when there is enough evidence connecting a model to its target, such that model opacity is not an epistemic barrier, can involve epistemic risk entangled with non-epistemic values.[5] Representational questions regarding what data should be used to represent the target phenomena, specific ML architectures suited for the problem, and even the specific interpretability technique chosen to gain high-level insight into black-box ML models require judgments that reflect values.

In the pneumonia case discussed above, as it turns out, when researchers sought to reduce LU, it became clear that the model did not provide understanding of the intended target of assessing which patients should be admitted. The data that the model used relied on the following underlying assumption:

> If a hospital-treated pneumonia patient has a very low probability of death, then that patient would also have a very low probability of death if treated at home.
>
> (Cooper et al. 1997, p. 136)

An especially astute observer may be able to see that such an assumption faces a large risk for inadequately representing the target. And indeed, a more interpretable rule-based model, trained on the same data, found that someone having asthma had a very low risk of death (see Caruana et al. 2015). However, the reason for the low risk of death was precisely because of the hospital treatment intervention. Patients with asthma are immediately placed in ICU care. The above representational assumption behind the opaque ML model faces not only a high degree of LU, but also a high degree of LU-risk because of the non-epistemic consequences. To my knowledge, current explainability techniques were not applied to the opaque ML model directly; instead, researchers inferred that the ML model likely made similar inferences as the interpretable model (Caruana et al. 2015). And thus, because of the potential downstream social consequences and the risk of misrepresentation, the ML model was not placed into practice, and researchers disregarded the epistemic value of the model.

Since the LU between the original neural network model and the target was high, the opacity of the model created a greater epistemic barrier. Moreover, since the LU-risk was high, the need for more research into the external connection between the model and the target increases further. Therefore, if the extent to which model opacity undermines explanation and understanding is based on the degree to which there is an external connection between the model and the target, then the problem of opacity in ML is entangled with non-epistemic values, since the process of accepting whether there is sufficient connection between the model and its target is itself entangled with non-epistemic values.

3 Social Values and Explanation

Once researchers accept that a particular model is suitable for explaining phenomena, the next stage in the pipeline is actually constructing explanations. In this section, I argue that non-epistemic values, in part, determine the type and depth of the explanation that is required to adequately explain phenomena and the extent to which ML opacity poses an obstacle.

3.1 Non-Epistemic Explanatory Functions

Models on their own are not explanations; only when models help answer questions about some event or phenomenon do they explain (Bokulich 2011; Lawler and Sullivan 2021; Van Fraassen 1980). What I want to suggest here is that it is not just the specific question that we ask that matters for explanation, nor just the specific stakeholder or person who asks the question that matters (Zednik 2021). We must also consider the *function* or purpose of the explanation. Norms of explanation change depending on the function that an explanation has in a given context. Importantly, non-epistemic values are relevant when considering the functions that explanations should and do have, and the norms that follow.

The two explanatory functions that have gained the most attention in the epistemology and philosophy of science are the ontic and epistemic functions of explanation, namely, to discover relations in the world (Craver 2014; Illari 2013) and to enable understanding (Grimm 2010; Khalifa 2017). As a result, discussions about the norms of explanation are clustered around issues of representation (Frigg and Nguyen 2018), factivity (Elgin 2017), causality (Lange 2016; Sullivan 2019), and asymmetry (Reutlinger 2016). However, someone can explain for other purposes too. For example, an explanation is often sought to *justify* someone's actions. In this case, a reasons explanation is warranted instead of the type of causal explanation often required for scientific explanation, which comes with its own norms (Majors 2007).

One notable difference between the way "explanation" is used in the computer science (CS) context compared with the philosophy of science is that, in CS, explanations are generally understood as a product separate from whatever model was used to make a decision or classification. This means that there are aims of explanation that are divorced from how the model itself works. For example, Tintarev and Masthoff (2007) discuss several different aims of explanations found in CS literature that are not epistemic, such as trust, effectiveness, persuasiveness, and satisfaction, among others. More recently, Lipton (2018) also discusses the different aims of recent XAI techniques, such as trust. Here too, depending on the aim or function of the explanation, the norms of what makes an explanation a *good* explanation change.

I will focus on two non-epistemic aims: trust and persuasiveness. Consider the example of an ad explanation on Amazon. On Amazon when you are searching for products to buy, the platform often provides the user with recommendations of additional products to look at and consider purchasing. These recommendations are generated by various types of ML models. Amazon provides the user with simple "explanations" explaining why they are seeing the recommendations that they do. The explanations are usually along the lines of: "because people who bought this product also bought this other one" or "sponsored products related to this item." Such explanations are built seamlessly into the platform so that users may not even realize that there is a question that needs answering. If the purpose of these explanations is for users to *feel* more trust toward the platform, then Amazon can conduct user studies to see whether the feeling of trust is actually increased. Do users trust Amazon more when this explanation is provided over this other one, or over not having an explanation at all? On the other hand, if the purpose is to persuade users to buy more products or to buy a certain product, then again, Amazon can measure the difference in user buying behavior just by changing the explanation. This is exactly the type of thing that platforms, like Amazon, do. The best explanation that satisfies these functions is an explanation that increases user trust or purchases. It does not matter whether the explanation is faithful to how the model works or satisfies other important epistemic norms to fulfill these functions. This means that model opacity is not a barrier to explaining if the purpose of the explanation is to build the impression of trust or to persuade.[6]

Discussing the function that ad explanations *should* have is beyond the scope of this chapter. However, if we assume that one of these functions is an epistemic function, such as to provide users with an understanding of how Amazon's recommendation algorithm works, then the above ad explanations provide little insight and fail to explain. In this latter case, we need more detail about how the model works, and the explanation would need to be true or at least *true enough* (Elgin 2017).

Discussions concerning the various functions that explanations can have, and the norms needed to satisfy these functions, demand social considerations. Even in the context of science, there are social factors that can determine the various epistemic functions of interest and how to satisfy these functions. Thus, non-epistemic considerations play a role in the explanatory phase of ML research insofar as researchers need to decide what purpose their explanations have, and some of these purposes are non-epistemic. Further, model opacity does not prevent explaining for various non-epistemic purposes.

3.2 How Non-epistemic Values Influence Epistemic Explanatory Purposes

What about cases where the purpose of an explanation is clearly an epistemic purpose, such as enabling understanding? Here too, non-epistemic values can impact the type of explanation that is required, what information is relevant, and the extent to which opacity is a problem.

First, when we are explaining various scientific phenomena, we often need to idealize some aspect of the phenomena to explain. Some phenomena are too complex to explain fully in an understandable way. Further, some argue that idealization improves an explanation even in the absence of complexity, because idealizations highlight the *difference-makers* in a way that a complete explanation does not (Strevens 2008). There is considerable discussion about what idealization norms entail (Weisberg 2007), and sometimes these norms depend on non-epistemic considerations. For example, Potochnik (2015, p. 76) argues that social aspects influence when something is *true enough* for explanation, even if we restrict the purpose of explanation to understanding. The research focus and context determine the way in which models can be idealized. For instance, researchers interested in explaining cooperative behavior could use the same evolutionary game theory model while focusing on different aspects, such as genetic differences, or non-selective traits, such as learning. It is largely the interests of scientists, and often the interests of funding bodies, that determine these research foci. I want to take Potochnik's discussion of social influences impacting explanation further beyond the research focus and interests of scientists. I want to suggest that the various interests of those who are *receiving* the explanation and the *practical domain* that the phenomenon is situated in impact the type of explanation, what information is relevant, and the extent to which model opacity is a problem.

Consider the COMPAS model (Northpointe 2012). It is a risk assessment model that uses ML technologies in determining risk for prison recidivism. It was developed by Northpointe (now Equivant), a profit-seeking private company. The model is opaque both in the sense that Equivant will not disclose the algorithm and because it is based on ML technologies. The COMPAS algorithm has been used in decisions regarding sentencing and parole in the United States. COMPAS has been charged with racial bias and using features such as a zip code to make its decisions (Angwin et al. 2016; Larson et al. 2016).

Consider, on the other hand, a different ML model that seeks to give a risk assessment about whether someone is at risk of developing certain types of cancers. Call this model HRisk+. Suppose further that this algorithm is also developed by a profit-seeking company that will not release its algorithm and it is opaque in the same way as COMPAS, because it is based on ML technologies. Further, suppose that HRisk+ is being used by doctors to decide whether certain patients should be considered for new medical trials or for increased medical screenings. HRisk+ also uses features such as a zip code to make its decisions.

Suppose further that it is the case that living in a particular zip code increases the risk that someone is arrested for a crime because of policing methods and living in the same zip code increases the risk of someone developing a particular type of cancer because it is an old Superfund site. Further, in both cases, there is an authority figure (judge or doctor) explaining why they came to a decision they did via an ML model. How do the non-epistemic differences in the COMPAS case and HRisk+ case change the requirements for explanation and proper model transparency? First, the *type* of explanation that is appropriate for why someone was denied parole or why someone was chosen for a medical trial differs because both the practical domain and the interests of the person receiving the explanation differ.

Consider the practical domain. In the case of the COMPAS model, the domain of interest is a sociopolitical domain. COMPAS is used solely to help determine whether an individual is able to participate fully as a member of the larger social community. On the other hand, in the medical case, the domain of interest is the health sciences and clinical medicine. If a judge used the COMPAS algorithm to deny someone parole and gave an explanation simply citing the higher rates of crime and recidivism in the zip code in which that person lived, though perhaps true in the aggregate, this would not be satisfactory. The incarcerated person would rightly say, "it isn't relevant what someone *like me* in various respects might do, what matters is what *I* personally would do." However, in the medical case, the same type of statistical explanation that the judge provides seems completely appropriate, since medical decisions are a very different type of decision and are often based on aggregate patterns. In other words, in one case treating someone with, as King (2020) calls, the statistical stance is appropriate, but in another case it is not. To put this yet another way, it would not be surprising for a doctor to base a medical decision on what worked for your identical twin; however, it would be unjustified for a judge to make a sentencing determination based on what your identical twin did and not you. To be clear, it is not that the judge is wrong per se to use information regarding a zip code as evidence or even to use it in an explanation; however, given the sociopolitical domain in which the decision is situated, various non-epistemic considerations are equally or more salient, namely fairness and justice. Thus, what must be included in an explanation of why someone is being denied (or granted) parole is some connection between statistical trends and some *normatively* salient features that are relevant to the particular person under consideration and to larger norms of justice and fairness.[7] This is not the case with a medical explanation regarding who is a good fit for a medical trial or needs more screenings for a particular disease. Importantly, non-epistemic values are determining whether a statistical explanation over a reasons-based moral explanation is needed to explain the decisions of an ML model.

Further, the interests of those receiving an explanation can constrain the type of explanation that is required. For example, Zednik (2021) argues that various stakeholders in the ML modeling pipeline are interested in different epistemically relevant elements of how the model works. Zednik discusses this in terms of stakeholders asking different types of questions (e.g., where- vs why-questions) and that different question types require alternative levels of analysis. However, even if those receiving an explanation want an answer to the same broad question type—"why this decision?"—the specific interests of specific individuals can impact the explanation that is required. For example, someone who is a member of a group that is known to be subject to bias may be asking "why this decision?" specifically to find out whether bias was a part of the decision. The scope of such an explanation would include different epistemically relevant information compared with an explanation provided to someone who was not from such a group and where potential bias was not relevant. In the next section, I will discuss in more detail the role that different individual interests can have in impacting the scope of understanding.

The extent to which *model opacity* gets in the way of explanation also depends on the interests of the person receiving the explanation and its practical domain, and it thus depends on non-epistemic considerations. The level of detail needed concerning how the COMPAS model and the HRisk+ model work toward adequately explaining why a judge or doctor made a specific decision differs. Even though this is contested, it might be argued that in clinical settings, race, ethnicity, or gender are predictively useful (Vyas et al. 2020). For example, melanoma is a greater risk for white patients. However, in law, it is illegal in the United States and unjust for judges to make decisions regarding sentencing and parole based on race or to use various proxies for race in their decisions. Racial and gender bias in medical decisions looks different from criminal justice decisions. This suggests that the level of detail and transparency regarding how the model reaches a decision in the COMPAS case differs from the HRisk+ case, because the potential problems of bias differ. Again, this difference is due to a difference in non-epistemic considerations.

All said, when we are explaining using ML models non-epistemic values partly determine the content of what makes for an acceptable explanation, and how much ML opacity poses a problem.

4 Opacity, Non-epistemic Values, and Attributions of Understanding

I briefly discussed two stages in the scientific process surrounding the use of ML models in providing understanding of phenomena—model acceptance and explanation—and how non-epistemic values, in part, determine the extent to which model opacity is a problem. Lastly, in this section, I consider the next stage: attributions of understanding from ML models. I argue that non-epistemic values also, in part, determine when ML model opacity prevents attributing understanding of phenomena.

4.1 The Stakes of Understanding

In epistemology, *pragmatic encroachment* theories of knowledge suggest that the stakes of a situation influence attributions of knowledge (Fantl and McGrath 2009; Hannon 2017). For example, if it is really important for someone to get to a meeting on time, that person may need more evidence of the train schedule than someone who does not have any particular place to be. I want to suggest that understanding from ML models also depends on the stakes. Specifically, depending on the stakes, someone might need to know more about how a given ML model works to understand.

It is not a settled question as to what constitutes understanding. Some argue that understanding is just a kind of knowledge (Riaz 2015), whereas others argue that understanding is distinct from knowledge (Hills 2016; Lawler 2019). I will not touch on this debate here. Instead, I aim at simply motivating that there are interesting cases in the ML context that suggest a pragmatic encroachment view of understanding. I aim at motivating these cases using common shared touchpoints for understanding.

One shared touchpoint for understanding is that understanding comes in degrees. Someone can understand something more or less. The simplest way to motivate a pragmatic encroachment view of understanding is to consider the more what-if-things-had-been-different questions someone can answer the *more* they understand (Hu 2019). In addition to describing the degree to which someone understands, we can, and should, talk about understanding attributions in terms of some minimum threshold condition.[8] On the simple view I am suggesting here, attributing understanding, or what we can call full-fledged understanding, depends on the threshold of the number of what-if questions in the set of all possible what-if questions on a given topic in a particular context that is necessary to attribute understanding. In some cases, if someone can only answer a few questions, then we should not attribute them with full-fledged understanding. For example, answering the simple question that the house burned down because of faulty wiring does not seem to be enough for *really* understanding why the house burned down. In some contexts, it seems that simply knowing the cause is too minimal for understanding. A fire marshal who is responsible for investigating the cause of the fire would surely need to answer more what-if questions for a proper attribution of understanding why the house burned down, such as why the wiring was faulty or the cause that sparked the wire failure. My claim here is that if we accept that there is a minimum threshold for understanding attributions, then we can motivate a pragmatic encroachment view for when model opacity becomes a problem for understanding.[9]

There are two broad ways that non-epistemic values impact our attributions of understanding in the context of ML models: (*i*) the domain requires greater model transparency to attribute understanding, and (*ii*) the personal stakes in a given context can require greater model transparency to attribute understanding. I consider each in turn.

4.2 Varying Importance Concerning the Domain of Inquiry

First, consider how the domain of inquiry might demand greater model transparency for us to attribute an agent with understanding. A common way that recommendation systems work for various platforms is through a process called *collaborative filtering*. A collaborative filtering algorithm finds users that are similar to each other in various ways. It might be that they are in the same age group or that they tend to read the same news articles (e.g., sports and cryptocurrency). Various ML techniques are used to cluster similar users together.

Various domains use recommender systems based on collaborative filtering algorithms. However, different domains have more or less significance. For example, a news recommender system has lower stakes than a doctor recommender system. My claim is that there are different requirements for model transparency for understanding attributions based on the stakes of a given domain. In the news recommendation case, it is not important for the average user to know much at all about how users are clustered or how collaborative filtering works to attribute an understanding of why they are being shown a particular recommendation. An explanation along the lines of "users like you also enjoyed this article" seems sufficient.

However, in the case of a doctor recommender platform, given the importance that doctors have in someone's well-being (it could be a matter of life or death), a user would need to answer more what-if-things-had-been-different questions about how the algorithm works in order to understand. The consequences of error is greater in the doctor recommender case compared with the news recommender case. Because of the greater consequences of error, we need to be able to answer more what-if-things-had-been-different-questions regarding how the model works in order to attribute someone with understanding.

In some ways, a pragmatic encroachment view follows directly from the previous section that more model transparency is necessary in order to explain some phenomena compared with others. Given the close connection between explanation and understanding (Khalifa 2017; Strevens 2013), if explaining demands more details about how the model works, then it is necessary to know these more details to attribute understanding.

4.3 Greater Personal Stakes

Now, consider how the personal stakes of a practical situation could impact the demand for model transparency regarding the attributions of understanding. Imagine there are two people looking for a new apartment. Zoe needs an apartment quickly, whereas Thijs does not need one for at least 6 months or more. Zoe is also aware that recommendation platforms can have biases and make decisions based on race, gender, and nationality. Thijs, on the other hand, is not aware of such biases and does not fall into any of the concerned groups often impacted by bias. Both Zoe and Thijs are using a new housing platform that connects users to potential listings in their area. The recommendation system is primarily designed using collaborative filtering technologies. Both Thijs and Zoe get recommendations and explanations that often include the phrase "users like you."

In order for Zoe to understand from these explanations, given the urgency of her particular situation, she needs to know more about what "users like you" means. Is the system filtering listings based on her nationality or her race? She needs to understand more about how the ML clustering algorithm works to gain understanding of why she is seeing the listings that she does. Specifically, she needs to know more about what makes a specific user *like her*. Is the recommendation system just using her preference profile for a home office and a children's playroom? Or is the recommendation system filtering out listings in specific neighborhoods because she comes from a country that is considered "non-Western"?

In other words, given how important it is for Zoe to find an apartment, and the higher risk of potential bias, she needs to be able to answer more what-if-things-had-been-different questions compared with Thijs for her to understand why she is seeing the recommendations that she sees. Greater model transparency is necessary to attribute understanding due to various non-epistemic factors regarding personal stakes.

5 Conclusion

One of the main worries with ML model opacity is that we cannot know enough about how the model works to fully understand the decisions they make. Without fully understanding how decisions are made how can we possibly trust the system or act based on its decisions? Everything I have said so far in this chapter is consistent with the view that there are some cases where the function of explanation is such, or the domain is such, or the stakes for an individual are such that the opaque nature of ML models prevents understanding. However, I have not argued for such a skeptical outcome across the board. Instead, my aim was to argue that non-epistemic factors contribute to the question as to how much of a problem opacity really is. I also have only focused on a cluster of issues surrounding explanation and understanding. There could be other reasons that demand greater model transparency, for example, being able to maintain privacy or some other value (Müller 2021). However, in order to explain and gain understanding with an ML model the problem of opacity greatly depends on features external to the model instead of features internal to it (i.e., link uncertainty and empirical support, explanatory functions, and the social and personal significance of the model and its domain).

Acknowledgments

For helpful comments and conversations, I would like to thank Thomas Grote, Insa Lawler, Elay Shech, and Mike Tamir, with special thanks to John Mumm. I presented this paper at the University of Rochester and I am very grateful for the conversation that resulted. This work is supported by the Netherlands Organization for Scientific Research (NWO grant number VI.Veni.201F.051). This work is also part of the research program Ethics of Socially Disruptive Technologies, which is funded by the Gravitation program of the Dutch Ministry of Education, Culture, and Science and the Netherlands Organization for Scientific Research (NWO grant number 024.004.031).

Notes

1 See Betz (2013) for a defense of the value-free ideal and Johnson (forthcoming) for a defense of value-ladenness in science and algorithms.
2 See Karaca (2021) for a discussion of type I and type II errors in ML and a proposal for cost-sensitive error classification to minimize inductive risk during model construction.

3 See Gilpin (2018) for a review of various ML interpretability methods.
4 See Lusk and Elliott (2022) for an adequacy-for-purpose theory of values in science.
5 In Sullivan [2022b], I call this the external problem of model opacity. I also discuss how values impact a further internal problem of opacity.
6 One worry here is that not every response to a why-question should count as an explanation. So, in what sense does an Amazon ad explanation aimed at persuasion actually count as an explanation at all, if it does not satisfy any epistemic constraints or have any real relationship to the ML model that it is meaning to explain? This objection already assumes that explanation has a specific function and thus comes with specific normative constraints. The suggestion here is that given that other scientific fields use the concept of explanation in a very different sense, instead of correcting their use of the concept, it is better to identify different functions and norms of explanation. Various functions of explanation can also be seen throughout philosophy, for example, causal, moral, mathematical, and metaphysical explanation.
7 For a discussion on statistical evidence in law, see Enoch and Fisher (2015).
8 See Kelp (2015), Khalifa (2017), and Wilkenfeld (2013) for discussions about degrees and thresholds for understanding.
9 Where exactly this threshold lies for understanding is beyond the scope of this paper; however, my argument relies on the claim that practical stakes have some influence.

References

Angwin, J., L. Jeff, M. Surya, and L. Kirchner. 2016. "Machine Bias." *ProPublica*. Accessed May 2021. https://www. propublica.org/article/machine-bias-risk-assessments-in-criminal-sentencing.

Betz, Gregor. 2013. "In Defence of the Value Free Ideal." *European Journal for Philosophy of Science*, 3(2): 207–220. https://doi.org/10.1007/s13194-012-0062-x.

Biddle, Justin B. 2020. "On Predicting Recidivism: Epistemic Risk, Tradeoffs, and Values in Machine Learning." *Canadian Journal of Philosophy*, 1–21. https://doi.org/10.1017/can.2020.27.

Biddle, Justin B., and Rebecca Kukla. 2017. "The Geography of Epistemic Risk." *Exploring Inductive Risk: Case Studies of Values in Science*, 215–237. https://doi.org/10.1093/acprof:oso/9780190467715.003.0011.

Bokulich, Alisa. 2008. *Reexamining the Quantum-classical Relation*. Cambridge: Cambridge University Press. https://doi.org/10.1017/CBO9780511751813.

Bokulich, Alisa. 2011. "How Scientific Models Can Explain." *Synthese*, 180(1): 33–45. https://doi.org/10.1007/s11229-009-9565-1.

Caruana, Rich, Yin Lou, Johannes Gehrke, Paul Koch, Marc Sturm, and Noemie Elhadad. 2015. "Intelligible Models for Healthcare: Predicting Pneumonia Risk and Hospital 30-day Readmission." *In Proceedings of the 21th ACM SIGKDD International Conference on Knowledge Discovery and Data Mining*, pp. 1721–1730. https://doi.org/10.1145/2783258.2788613.

Cooper, Gregory F., Constantin F. Aliferis, Richard Ambrosino, John Aronis, Bruce G. Buchanan, Richard Caruana, Michael J. Fine, et al. 1997. "An Evaluation of Machine-learning Methods for Predicting Pneumonia Mortality." *Artificial Intelligence in Medicine*, 9(2): 107–138. https://doi.org/10.1016/S0933-3657(96)00367-3.

Craver, Carl F. 2014. "The Ontic Account of Scientific Explanation." In A. Hutteman, M. Kaiser (eds.), *Explanation in the Special Sciences*, pp. 27–52. Dordrecht: Springer. https://doi.org/10.1007/978-94-007-7563-3_2.

Creel, Kathleen A. 2020. "Transparency in Complex Computational Systems." *Philosophy of Science*, 87(4): 568–589. https://doi.org/10.1086/709729.

Dellsén, Finnur. 2020. "Beyond Explanation: Understanding as Dependency Modelling." *The British Journal for the Philosophy of Science*. https://doi.org/10.1093/bjps/axy058.

De Regt, Henk W. 2017. *Understanding Scientific Understanding*. New York: Oxford University Press. https://doi.org/10.1093/oso/9780190652913.001.0001.

Douglas, Heather. 2000. "Inductive Risk and Values in Science." *Philosophy of Science*, 67(4): 559–579. https://doi.org/10.1086/392855.

Elgin, Catherine Z. 2017. *True Enough*. Cambridge: MIT Press. https://doi.org/10.7551/mitpress/9780262036535.001.0001.

Enoch, David, and Talia Fisher. 2015. "Sense and Sensitivity: Epistemic and Instrumental Approaches to Statistical Evidence." *Stanford Law Review*, 67: 557.

Fantl, Jeremy, and Matthew McGrath. 2009. *Knowledge in an Uncertain World*. Oxford: Oxford University Press. https://doi.org/10.1093/acprof:oso/9780199550623.001.0001.

Frigg, Roman, and James Nguyen. 2018. "The Turn of the Valve: Representing With Material Models." *European Journal for Philosophy of Science*, 8(2): 205–224. https://doi.org/10.1007/s13194-017-0182-4.

Gilpin, Leilani H., David Bau, Ben Z. Yuan, Ayesha Bajwa, Michael Specter, and Lalana Kagal. 2018. "Explaining Explanations: An Overview of Interpretability of Machine Learning." In 2018 *IEEE 5th International Conference on Data Science and Advanced Analytics (DSAA)*, pp. 80–89. IEEE. https://doi.org/10.1109/DSAA.2018.00018.

Grimm, Stephen. 2010. "The Goal of Explanation." *Studies in History and Philosophy of Science Part A*, 41(4): 337–344. https://doi.org/10.1016/j.shpsa.2010.10.006.

Hannon, Michael. 2017. "A Solution to Knowledge's Threshold Problem." *Philosophical Studies*, 174(3): 607–629. https://doi.org/10.1007/s11098-016-0700-9.

Harvard, Stephanie, and Eric Winsberg. 2022. "The Epistemic Risk in Representation." *Kennedy Institute of Ethics Journal* 32(1): 1–31. doi:10.1353/ken.2022.0001.

Hempel, Carl. 1965. *Aspects of Scientific Explanation*. Vol. 1. New York: Free Press.

Hills, Alison. 2016. "Understanding Why." *Noûs*, 50(4): 661–688. https://doi.org/10.1111/nous.12092.

Hu, Xingming. 2019. "Is Knowledge of Causes Sufficient for Understanding?" *Canadian Journal of Philosophy*, 49(3): 291–313. https://doi.org/10.1080/00455091.2018.1497923.

Illari, Phyllis. 2013. "Mechanistic Explanation: Integrating the Ontic and Epistemic." *Erkenntnis*, 78(2): 237–255. https://doi.org/10.1007/s10670-013-9511-y.

Johnson, Gabbrielle. forthcoming. "Are Algorithms Value-free? Feminist Theoretical Virtues in Machine Learning." Journal Moral Philosophy.

Karaca, Koray. 2021. "Values and Inductive Risk in Machine Learning Modelling: The Case of Binary Classification Models." *European Journal for Philosophy of Science*, 11(4): 1–27. https://doi.org/10.1007/s13194-021-00405-1.

Kelp, Christoph. 2015. "Understanding Phenomena." *Synthese*, 192(12): 3799–3816. https://doi.org/10.1007/s11229-014-0616-x.

King, Owen C. 2020. "Presumptuous Aim Attribution, Conformity, and the Ethics of Artificial Social Cognition." *Ethics and Information Technology*, 22(1): 25–37. https://doi.org/10.1007/s10676-019-09512-3.

Khalifa, Kareem. 2017. *Understanding, Explanation, and Scientific Knowledge*. Cambridge University Press. https://doi.org/10.1017/9781108164276.

Kuorikoski, Jaakko, and Petri Ylikoski. 2015. "External Representations and Scientific Understanding." *Synthese*, 192(12): 3817–3837. https://doi.org/10.1007/s11229-014-0591-2.

Lange, Marc. 2016. *Because Without Cause: Non-Casual Explanations in Science and Mathematics*. New York: Oxford University Press. https://doi.org/10.1093/acprof:oso/9780190269487.001.0001.

Larson, J., M. Surya, L. Kirchner, and J. Angwin. 2016. "How We Analyzed the COMPAS Recidivism Algorithm." *ProPublica*. Accessed May 2021. https://www.propublica.org/article/how-we-analyzed-the-compas-recidivism-algorithm.

Lawler, Insa. 2019. "Understanding Why, Knowing Why, and Cognitive Achievements." *Synthese*, 196(11): 4583–4603. https://doi.org/10.1007/s11229-017-1672-9.

Lawler, Insa, and Emily Sullivan. 2021. "Model Explanation Versus Model-induced Explanation." *Foundations of Science*, 26(4): 1049–1074. https://doi.org/10.1007/s10699-020-09649-1.

Lipton, Zachary C. 2018. "The Mythos of Model Interpretability: In Machine Learning, the Concept of Interpretability is Both Important and Slippery." *Queue* 16(3): 31–57.

Lusk, G. and Elliott, K.C. 2022. Non-epistemic Values and Scientific Assessment: An Adequacy-for-purpose View. *European Journal for Philosophy of Science*, 12(2): 1–22.

Majors, Brad. 2007. "Moral Explanation." *Philosophy Compass*, 2(1): 1–15. https://doi.org/10.1111/j.1747-9991.2006.00049.x.

Müller, Vincent C. 2021. "Deep Opacity Undermines Data Protection and Explainable Artificial Intelligence." *Overcoming Opacity in Machine Learning* 18.

Northpointe. 2012. COMPAS Risk and Need Assessment System: Selected Questions Posed by Inquiring Agencies. http://www.northpointeinc.com/files/downloads/FAQ_Document.pdf.

Parker, Wendy. 2020. "Model Evaluation: An Adequacy-for-Purpose View." *Philosophy of Science*, 87(3): 457–477. https://doi.org/10.1086/708691.

Potochnik, Angela. 2015. "The Diverse Aims of Science." *Studies in History and Philosophy of Science Part A*, 53: 71–80. https://doi.org/10.1016/j.shpsa.2015.05.008.

Reutlinger, Alexander. 2016. "Is There a Monist Theory of Causal and Non-causal Explanations? The Counterfactual Theory of Scientific Explanation." *Philosophy of Science*, 83(5): 733–745. https://doi.org/10.1086/687859.

Riaz, Amber. 2015. "Moral Understanding and Knowledge." *Philosophical Studies*, 172(1): 113–128. https://doi.org/10.1007/s11098-014-0328-6.

Rudner, Richard. 1953. "The Scientist Qua Scientist Makes Value Judgments." *Philosophy of Science*, 20(1): 1–6. https://doi.org/10.1086/287231.

Strevens, Michael. 2008. *Depth: An Account of Scientific Explanation*. Harvard University Press.

Strevens, Michael. 2013. "No Understanding Without Explanation." *Studies in History and Philosophy of Science Part A*, 44(3): 510–515. https://doi.org/10.1016/j.shpsa.2012.12.005.

Sullivan, Emily. 2019. "Universality Caused: The Case of Renormalization Group Explanation." *European Journal for Philosophy of Science*, 9(3): 1–21. https://doi.org/10.1007/s13194-019-0260-x.

Sullivan, Emily. 2022a. "Understanding from Machine Learning Models." *The British Journal for the Philosophy of Science*, 73(1): 109–133. https://doi.org/10.1093/bjps/axz035.

Sullivan, Emily. 2022b. "Inductive Risk, Understanding, and Opaque Machine Learning Models." *Philosophy of Science*, 1–13. doi:10.1017/psa.2022.62

Tintarev, Nava, and Judith Masthoff. 2007. "A Survey of Explanations in Recommender Systems." In *2007 IEEE 23rd International Conference on Data Engineering Workshop*, pp. 801–810. IEEE. https://doi.org/10.1109/ICDEW.2007.4401070.

Van Fraassen, Bas C. 1980. *The Scientific Image*. Oxford University Press. https://doi.org/10.1093/0198244274.001.0001.

Vyas, Darshali A., Leo G. Eisenstein, and David S. Jones. 2020. "Hidden in Plain Sight-Reconsidering the Use of Race Correction in Clinical Algorithms." *New England Journal of Medicine*, 383(9): 874–882. https://doi.org/10.1056/NEJMms2004740.

Weisberg, Michael. 2007. "Three Kinds of Idealization." *The Journal of Philosophy*, 104(12): 639–659. https://doi.org/10.5840/jphil20071041240.

Wilkenfeld, Daniel A. 2013. "Understanding as Representation Manipulability." *Synthese*, 190(6): 997–1016. https://doi.org/10.1007/s11229-011-0055-x.

Zednik, Carlos. 2021. "Solving the Black Box Problem: A Normative Framework for Explainable Artificial Intelligence." *Philosophy & Technology*, 34(2): 265–288. https://doi.org/10.1007/s13347-019-00382-7.

25 Understanding from Deep Learning Models in Context

Michael Tamir and Elay Shech

1 Introduction

Advances in machine learning (ML) techniques, deep learning (DL) especially, are drawing increased philosophical consideration. ML-trained algorithms are often called *models*, encouraging questions about how such automated effective estimation techniques fit in with existing accounts of scientific modeling and representation for understanding. In contrast to how simple idealized models arguably enable understanding by reducing complexity (Bokulich 2008; Khalifa 2017; Potochnik 2017; Strevens 2008), Sullivan (2022) rejects the possible claim that ML models enable understanding by reducing complexity. In particular, Sullivan (2022, 110) claims that "model simplicity and transparency are not needed for understanding phenomena," arguing that DL models can provide understanding despite the ostensible opaqueness or "blackbox" nature of how particular estimations are generated. Instead, she suggests that understanding with a DL model depends on what she calls *link uncertainty*, or "the extent to which the model fails to be empirically supported and adequately linked to the target phenomena."

In this chapter, we place into context how the term *model* in ML contrasts with traditional usages of scientific models for understanding, resolving core ambiguities involving representational links to the target phenomenon. We explore standard techniques involving direct analysis of an estimator's learned transformations (viz., the hidden layers of a DL model). Next, we show that such direct analysis can improve understanding of the target phenomenon and reveal how the model organizes relevant information. In Section 2, we lay the groundwork for contrasting non-ML scientific models with ML models, and set the stage for our interaction with Sullivan's take on understanding with ML models. Section 3 provides a brief overview of ML and DL models and considers a candidate for framing the proper target of ML understanding leveraged in later sections. Section 4 identifies three modes of understanding given the proposed target of ML models. We then disambiguate what we describe as the difference between implementation irrelevance and functionally approximate irrelevance, and explore

DOI: 10.4324/9781003202905-28

how this distinction impacts potential understanding with these models. Section 5 addresses an ambiguity in the concept of link uncertainty, arguing that distinguishing empirical link failures from representational ones clarifies the role played by scientific background knowledge in enabling understanding with ML. In Section 6, we conclude with a brief summary.

2 C-schema Models and Sullivan on Understanding from ML Models

In one of her paradigm examples of non-ML models, Sullivan (2022) considers Thomas Schelling's model for explaining and understanding why human populations tend to be segregated. Schelling wanted to investigate "some of the individual incentives and individual perceptions of difference that can lead collectively to segregation" (Schelling 1971, 138). This means that segregation behavior among populations is the *target phenomenon* of interest, that is, the object of study that we want to understand. He constructed a simple model: A checkerboard represents spatial locations with two types of individual households, represented by dimes and nickels. Each household, or "actor," is stipulated to prefer that at least 30% of its neighbors be of the same type (similarity preference parameter). If this condition is met, the actor remains in place; if not, one moves the actor to the closest unoccupied space. It turns out that for most initial configurations, the equilibrium state of the board results in segregation. Such results may suggest that racial segregation can occur without an organized institutionally racist influence. But how do models like Schelling's afford understanding? Sullivan (2022, 3) adopts the view (found in, e.g., Khalifa 2017; Strevens 2008) that explanation aims at understanding whereby "explaining why helps us understand why." Schelling's model succeeds in affording understanding of actual segregation behavior found in some populations if it links faithfully with causal factors that explain such segregation. Similarly, Schelling's model fails at providing understanding and explanations of actual populations when, for instance, the similarity preference parameter is inaccurate.

It is helpful to view a paradigmatic non-ML model like Schelling's model, along with the explanations and understanding that it may provide, within a framework that we will call the "C"-shaped modeling schema, or *C-schema* (Figure 25.1).

There are many accounts of scientific modeling, representation, and inference that fit the C-schema. For example, similarity accounts (e.g., Weisberg 2013), mapping isomorphism or structuralist accounts (e.g., da Costa and French 2003), inferentialist accounts (e.g., Suárez 2004), and more recent views (e.g., Frigg and Nguyen 2016) can all be viewed within a C-schema. Such models M are used or interpreted to explain, understand, or investigate some target T, which is a phenomenon

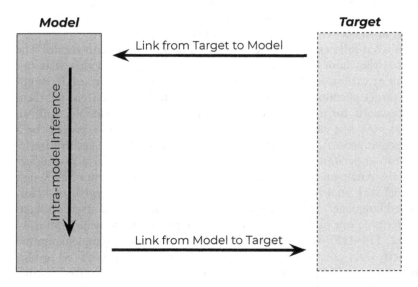

Figure 25.1 "C"-shaped modeling schema.

in the world, with specified audiences and goals in mind (e.g., epistemic or pragmatic). The horizontal top of the C-schema concerns the modeling or representational relationship between M and T. For example, some accounts note that a scientist must stipulate that M denotes T (e.g., Frigg and Nguyen's (2016) DEKI account). On the vertical side of the C-schema, one interacts with M directly to draw inferences, for example, by running iterations of Schelling's model and identifying specific relationships between similarity preference parameters and state equilibria, or by performing mathematical calculations given certain conditions. Last, on the horizontal bottom side, M is used to relevantly draw inferences and answer questions about, or impute properties to, T.

Models that fit into the C-schema can range in kinds such as concrete models like the San Francisco Bay-Delta model, mathematical models such as the Lotka-Volterra predator-prey model, or simulation/computational models like Schelling's model (Weisberg 2013, Ch. 2). Such models tend to facilitate inferences about systems and phenomena of interest, limited to the scope of representation, and afford understanding in a number of ways. For instance, in a typical map (or map-like) model, we can not only use M to draw inferences about T but we can also see how such inferences are drawn, for example, because of relevant similarities or isomorphic relations associated with M and T (say, a map and a city). There are many varied accounts as to how the C-schema works and facilitates successful inference, such as those accounts mentioned above.

In one prominent family of C-schema accounts, to which we return, understanding is powered by some relevant structural similarity or isomorphism between M and an abstraction of features of T.

In what follows, we will focus on two of Sullivan's main claims. The first is the claim that opacity and complexity of ML models that can occur at various levels is not an in-principle impediment to understanding target phenomena. Models do not have to be transparent for them to be useful for understanding and explanation. For example, Sullivan notes how one can compute factorials using an iterative process or a recursive process, but that such implementation differences are irrelevant to understanding with, for example, a climate model that uses factorials. Analogously, "one does not need to know whether Schelling's model was implemented using a functional, object-oriented, or actor-based language" and so "implementation blackboxing in itself does not undermine our ability to explain or understand phenomena" (Sullivan 2022, 114–115). We agree with Sullivan on this point—implementation opacity does not matter anymore than drawing a map with red or blue ink matters—but we think that there is an important overlooked distinction between this type of implementation irrelevance and (what we call) functionally approximate irrelevance.

Second, Sullivan holds that a fundamental impediment to understanding in the context of both ML and non-ML models concerns the link between model and target (horizontal sides of the C-schema), what she calls "link uncertainty," which "constitutes a lack of scientific and empirical evidence supporting the link connecting the model to the target phenomenon" (2022, 124). For instance, in discussing Schelling's model, she notes that without "empirical evidence validating that the possible causes identified by Schelling's model are actual causes, there is no link connecting the model to the phenomenon" (2022, 125). Sullivan (2022, 128) then continues to argue for the same point in the context of ML models, holding that "lack of understanding is not due to implementation or model illegibility." Instead, it is due to high "link uncertainty (the amount, kind, and quality of scientific and empirical evidence supporting the link connecting the model to the target-phenomenon) that is present." Generally, we are in agreement but we will highlight an important contrast between empirical link uncertainty (e.g., found in Schelling's model due to an inaccurate similarity preference parameter, and other potential empirical questions concerning actual populations) and what we describe as data-misrepresentation link uncertainty.

3 The Target of ML Models

ML classification[1] is an algorithmic process for generating an estimator $f : x_i \rightarrow p_i$ that, for given "input elements" $\{x_i\}_{i \in N} \subset \Omega_X$ in a data set, estimates a scoring function p_i defined over the set of "output elements" or *y-targets*[2] $y_i \in \Omega_Y$. Typically, f is parameterized by some set of parameters $\theta \in \Theta$ establishing a family of estimators $\{f_\theta\}_{\theta \in \Theta}$ for the given estimation procedure. Training an ML estimator f can be distinguished from non-ML rule-based estimation algorithms in that the parameter values θ determining the trained f_θ are optimized so as to best "fit the data" according to a prescribed loss function. Namely, the parameters defining the estimator f_θ are optimized by taking a set of sample pairs $\{(x_i, y_i)\}_{i=0}^{N} \in \Omega_X \times \Omega_Y$,[3] called the *training set*, and, for a given loss function $L : (p_i, y_i) \to l_i \in \mathbb{R}$, finding the optimal parameterization θ^* such that:

$$\theta^* = \arg\min_{\theta \in \Theta} \sum_{i=0}^{N} L\big(f_\theta(x_i), y_i\big)$$

The process of finding the optimal f_{θ^*} for a given training set is called *training the model*.

DL is a family of ML techniques in which estimators are constructed through the composition of multiple linear and nonlinear transformations, called a *neural network*. Neural networks can consist of multiple (hidden) layers of learned parameterized transformations of the data, where each *hidden layer* (parameterized transformation) is learned through ML optimization. We argue below that while DL models do not necessarily provide understanding in the same manner as (say) map representations, learned representation layers may be leveraged to improve understanding by providing insight into how a DL model learns to organize raw input data to optimally estimate y-targets.

As an example, consider an ML model trained to solve "complete the sentence" tasks on text sampled from a given corpus. An x_i might be "In the morning, I enjoy _____" where the answer y_i for a particular data point is "coffee." An ML model f_θ is trained to best estimate how to fill in this blank given the x_i input. Other (similar) data points may have the same input for $x_{i'}$ but a different $y_{i'}$. Underlying how English tends to be written (for our samples) is some distribution over the entire vocabulary: "coffee" may be likely, but terms like "tea," "eggs," "sunshine," etc. also get non-negligible probability mass, whereas other arbitrary terms like "economic," "about," or "transcendental" do not. We can imagine there is some "true" distribution $p(Y = y_i | X = x_i)$ underlying the conditional probability for the (x_i, y_i) data generated by a given sampling process. An ML algorithm learns from actually sampled examples in the training set to best estimate this distribution. That is, the result of an ML training process is an estimator f_θ for the random variable Y given X, such that the model can estimate conditional probabilities $p_\theta(Y = y | X = x_i)$

induced by the scoring function for given x_i values.[4] If, for appropriate metrics, the induced $p_\theta(Y|X)$ matches the actually sampled y_i values well on data not used for training, called *test data*, we can start making judgments about how reliably it can relate such data more generally, but, strictly speaking, we are still just talking about the model and its (potential) data.

We can view this within a generalized C-schema framework: Inferences about how well the model can estimate y-targets given x-data, exploration of what particular features of x-data tend to play important roles in (correct) estimations of y-targets, and the study of how network parameters and hidden layer transformations organize and restructure x-data to effectively estimate y-targets are all examples of "vertical" inferences in a C-schema. As Sullivan (2022) observes, we need to "link" to the external phenomena intended to be understood. What is missing are the *horizontal* links of a C-schema connecting facts about the model and the data considered in isolation of a target phenomenon.

In order to horizontally link an ML model to a target of understanding (and hence evaluate the links' level of uncertainty which Sullivan correctly argues is vital), we must be specific about the target. Our hypothesis is that the appropriate target of understanding with ML models closely relates to how a learned distribution $p_\theta(Y|X)$ estimates the "actual" distribution $p(Y|X)$ describing a phenomenon's studied features. Specifically, we propose the following:

> (TML) **Target of ML Hypothesis:** The target phenomenon of understanding with ML models is the *relationship(s) of features represented by the data.*

The *target phenomenon* is not a particular object or sampling instance, but relationships of the properties or features found potentially in individual or multiple objects or object types.[5] The target is the relationship patterns of these represented features. These relationship(s) are typically those described (indirectly or directly) by some underlying actual distribution $p(Y|X)$ estimated by the model. Although the targeted relationships are described by such a $p(Y|X)$, we emphasize that it is the real-world relationships between features implied by the description $p(Y|X)$ that are the target phenomenon and not $p(Y|X)$ itself. The concept of *represented features* intuitively can be thought of as measurable properties associated with the phenomenon,[6] but as we elaborate in Section 5 the specific features represented by data are intimately tied to measurement methodology and sampling practice. We argue that focusing on precisely what features sampled data do *and do not* represent is paramount to evaluating ML model link uncertainty.

4 Implementation Irrelevance vs. Functionally Approximate Irrelevance

Relationships of features in the TML hypothesis play an important role for interpreting the target of ML understanding. Assume that with proper methodology one rules out the sort of link uncertainty (discussed in Section 5 below) associated with how data represent features of the target phenomenon (horizontal top side in the C-schema). Assume further that an ML model is well trained with appropriate metrics, data, and research practices for testing generalizability to support the claim that $p_\theta(Y|X)$ reliably estimates similarly sampled data (vertical side in the C-schema). In order to study trained ML models, especially complex DL models, for insight into external targets, we must first clarify how a link from ML models to TML targets (horizontal bottom side of a C-schema) may work.

To frame this challenge, let us contrast understanding from ML models and their targets with examples of understanding from subway maps and their targets. In the latter case, relevant facts about the target are abstracted and then represented typically as some graph-like visualization, a map. The map can then be used to understand how to navigate the represented subway system so long as these navigation insights are circumscribed by what is faithfully supported through the captured facts. For instance, one can infer which paths are available from point A to point B, but not (always) details about the decor or physical distance associated with taking the paths. The map represents an abstraction of the right details, enabling judgments based directly on the map's topology to be linked back to things like understanding how to navigate the represented subway system.

There are multiple details about the map that are also irrelevant to understanding navigation. For instance, color choices, whether it is physical or digital, and so on, are irrelevant details that fall under what Sullivan (2022) describes as "implementation" details. However, topological facts depicted by a map are relevant and cannot be ruled out as mere "implementation" details of the visualization, ostensibly because one must use those details directly to generate judgments like "this path from A leads to B." What about complex DL trained neural networks? After all, an ML model's estimations of $p_\theta(Y|X)$ depend directly on the parameter values θ and network architecture. Sullivan (2022, 114–115) argues that when the "[d]etails regarding the implementation are unnecessary for explaining and understanding," opacity of these details ("implementation blackboxing") is "not in principle problematic for explaining or understanding phenomena." Sullivan's argument refutes that DL models cannot be used for understanding merely due to *some* opacity, but the reliance on the *implementation irrelevance* of certain details, particularly their irrelevance to the target of understanding, is essential for this defense.

We have a dilemma: To understand feature relationships with an ML model despite detail opacity with respect to learned parameter instantiations, such details ostensibly cannot be used for insight into the relationship, but if estimations of $p_\theta(Y|X)$ directly depend on the instantiation of these parameter values how can they be used to understand the relationship between X and Y without such details? Parameter instantiation details are not purely irrelevant implementation details like coding language, color choices, material constitution, or even provably equivalent algorithmic techniques. The dilemma resolves by disambiguating implementation irrelevance, where the variations in question make no difference to the target of understanding, from a second sort of *functionally approximate irrelevance*, active in the case of DL understanding. What distinguishes functionally approximate irrelevance from implementation irrelevance is that in the former *varied details matter to the studied target*, but they are varied only in ways that approximately preserve the relevant aspects of the phenomenon to be understood.[7]

When the target is well specified by the TML hypothesis, we argue that by also accounting for the role and degree of functionally approximate irrelevance, DL may help us understand the following aspects and relationships of features represented by the data:[8]

> **IR** (Informative Relationship): If and to what extent mutual information between the features of a target phenomenon exists.
> **FI** (Feature Importance): Which features associated with the input x-data either individually or in combination are more/less important to such an informative relationship.
> **LR** (Learned Representation): How transforming the input x-data to better enable estimation of the y-target reveals informative ways of organizing the x-data and the features it represents for this or other estimation purposes.

To illustrate how functionally approximate irrelevance plays a role, consider the sentence completion task discussed in Section 3. Using the same neural network architectures and the same (or similar sampling of) training data, we could train two different models. Random differences in how the parameters are initialized before training, order differences in how the models see the training data, or differences in the hyper-parameters used to define how the models are trained can result in two substantively different learned parameterizations θ and θ', even though the probability estimates $p_\theta(Y|X)$ and $p_{\theta'}(Y|X)$ generated by the respective DL models approximate the feature relationships described by $p(Y|X)$ equivalently. Sullivan defines "highest-level" ML opacity (blackboxing) as cases where one merely has access to model inputs and outputs but not execution details. IR understanding is compatible with Sullivan's highest-level opacity. If both models are successfully trained, performing sufficiently and equivalently across different strata of the data, we can infer that since both $p_\theta(Y|X)$ and $p_{\theta'}(Y|X)$ approximate $p(Y|X)$, which, in turn, describes features of the target phenomenon, significant parameter detail differences (i.e., $\|\theta - \theta'\| \geq \epsilon$) are irrelevant to understanding that there is an informative relationship. With the ML model's target clearly defined as the relationships of features represented by X and Y, the fact that $\|\theta - \theta'\| \geq \epsilon$ has approximate irrelevance becomes clear: $p_\theta(Y|X)$ approximates $p(Y|X)$ well enough, so although the details matter to *how the estimates are made*, they can still be (approximately) irrelevant for the target of understanding. With the functionally approximate irrelevance of particular parameter details ensured, IR understanding even with "highest-level" opacity is possible. At minimum, the reliable generalizability of an ML model to similarly sampled data increases our understanding that there is *some* signal in the x-data that is useful for y-target estimation. The existence of even an opaque but a sufficiently reliable estimator $p_\theta(Y|X)$ can entail that there is mutual information between the features represented by X and Y.

Turning to FI, assigning feature importance in DL is an active research area. Certain "permutation" style techniques treat the DL model with highest-level opacity where x-features are manipulated (e.g., occluding part of an image as in (Seiler and Ferguson 2014)) and the impact on estimation performance is then analyzed. Other methods attribute importance from internal network properties, tracing individual contributions (typically involving gradients) from neuron to neuron (Shrikumar et al. 2017; Simonyan et al. 2013). Arguments for FI-based understanding for either permutation methods, or methods that meticulously trace contributions through the network are similar to the arguments for IR above. If two similarly trained models irreconcilably diverge in their feature importance implications, approximate irrelevance comes into question and the researcher should doubt whether such FI attribution yields very much genuine understanding of which features are predictive of the target. In contrast, if two similarly trained models tend to agree on which features are important, or better yet, multiple feature attribution techniques agree on which features matter to model predictions for certain sorts of input, then the functionally approximate irrelevance of the specific layer-by-layer calculations is evident, supporting FI understanding of the phenomenon. Again, the details matter in the sense that changing them has a direct impact; however, by establishing that certain detail variations approximately preserve resulting insights into the target phenomenon's feature relationships (viz. FI relationships), understanding is possible.

In their influential discussion of representation learning, Bengio et al. (2013) describe how DL models must "learn to identify and disentangle the underlying explanatory factors hidden in the observed milieu of low-level sensory data." Research into how DL models "disentangle" and organize the "low-level" input data focuses not only on studying the informative relationship between input and output substantiated by the possibility of training a DL model to detect these relationships successfully (IR), but also on how the data are represented via transformations from one hidden layer to the next hidden layer of the network (LR), revealing what kind of information is preserved and how it is represented in the associated vector (tensor) spaces of these layers.[9] Sullivan (2022, 119) echoes Bengio et al., describing how DL models learn to "tease out the relevant features from the irrelevant," as in image classification where each hidden "layer gradually picks out higher and higher-level abstractions until it reaches a classification of the image." How hidden layer representations interact and combine for the ultimate classification goal is studied not just for a deeper understanding of the neural networks themselves, but also to understand how they organize and incorporate relevant information content in the "low-level sensory data" as they mathematically transform it into new representations for optimal estimation (Olah et al. 2020).

A simple example of such informative hidden layer representations is traditional word vectorization.[10] The early efforts of Mikolov et al. (2013) used shallow neural networks to map individual terms to vector representations, which were optimized for "fill in the blank" style tasks similar to the example in Section 3. Such vectorized *word-embeddings* are not merely useful for the original task. The representations could be reused as *pre-trained* representations for novel text-based tasks.[11] Embeddings were widely used to study and leverage ostensible semantic relationships (e.g., analogies, synonymy clusters) manifested by their usage patterns for practical applications.

In contrast to IR and FI, LR must engage *directly* with the learned representations of data (like word-embeddings) associated with hidden layers or neurons. As described above, the individual parameterizations of the learned representations are opaque, but in the case of LR, understanding is achieved through a direct engagement with these representations.[12] The ostensible target phenomenon of the original Mikolov et al. (2013) task is the relationship between the terms "filled in" (y-targets) and their surrounding context terms (x-data). This relationship is described by some distribution capturing (some of) this frequency in context information. Study of these vectorizations, such as the relative position of vector differences, can capture analogical ($\vec{man} - \vec{king}$ vs $\vec{woman} - \vec{queen}$) or morphological ($\vec{smart} - \vec{smarter}$ vs $\vec{hard} - \vec{harder}$) term usage in (sampled) text.[13] Similarly, projecting word-embeddings onto lower-dimensional visualizations, or studying the clustering patterns of embeddings can inform our understanding of term usage as a surrogate for synonymy.[14] As with the discussion of IR, different embeddings likely have non-identical parameterizations due to differences in the training data, the way that the models were initialized, neural architecture, training process, etc. However, if these representations can be used for LR understanding of the represented features, properties such as the relative positions of word-embeddings used to complete analogies should be evident in the respective representations despite these differences. Dev et al. (2019) explain that "rotation or scaling of the entire dataset will not affect synonyms (nearest neighbors), linear substructures (dot products), analogies, or linear classifiers" because "there is nothing extrinsic about any of these properties." For example, in studying the impact of basis rotations to align GloVe (Pennington et al. 2014) and Word2Vec (Mikolov et al. 2013) embeddings, they confirm that using a vector from Word2Vec to complete an analogy using GloVe embeddings "is very poor, close to 0; that is, extrinsically there is very little information carried over" by the (basis dependent) parameter values themselves. However, when the learned rotations were used to align the embeddings first, near equivalent performance was recovered.

Studies like these illustrate how individual parameterization details can differ but still have functionally approximate irrelevance to the specific method of studying the properties of the DL model. In the simple case of word-embeddings, we see that properties such as relative angles or positions that are invariant to certain changes of coordinate values allow for a direct engagement with hidden layer representations to gain LR understanding.[15] This suggests a path to reconciling Sullivan's contrast of DL models with idealized models. Namely, although DL models have an overwhelming "blackbox level" number of parameter details, focusing on how scientists and researchers leverage these models for LR (FI and MI) understanding reveals the approximate irrelevance of these particular details. By attending to which details are (approximately) irrelevant to these more illuminating relationships and properties, we can see how such opacity (at an approximately irrelevant individual parameter detail level) need not prohibit improved understanding.

5 Disambiguating Link Uncertainty

Our above discussion explored the TML hypothesis that understanding with ML models targets the relationships of features represented by the data and described by some underlying distribution directly or indirectly estimated by the model. Further, if varying certain "blackbox" details impact the target relationships of an ML model only approximately, then DL understanding is still possible at least for MI, FI, and even LR when we "open up the black box" to acquire this understanding. In this section, we consider examples discussed by Sullivan to disambiguate two kinds of link uncertainty that we submit are conflated in Sullivan (2022) so that the understanding gained from FI and LR can potentially be leveraged to prevent such uncertainty.

Sullivan (2022, 126) presents Esteva et al.'s (2017) melanoma classifier as an example of low link uncertainty. Esteva et al. (2017) use the Inception-v3 (Szegedy et al. 2016) model and then fine-tune (further train) it for dermatology images. The authors tested their model using data points with biopsy verified labels, inferring that it could outperform most expert dermatologists evaluated on "the same data." What links justify the conclusion that their model was able to outperform expert dermatologists tested on "the same data?" First, although the model's x-data are processed from images that the dermatologists looked at, even these data are not identical: Inception-v3 takes as input 2-dimensional arrays of RGB values, whereas human dermatologists view images. Issues of proper lighting at test time, etc. might affect human performance but not the model. Further, issues of *data leakage* (Kuehlkamp et al. 2017; McCoy et al. 2019; Torralba and Efros 2011), where unintended signal correlated to the y-target is inadvertently left in the x-data, pose a risk to the validity of this horizontal link. For instance, Haenssle et al. (2018) reported that their DL trained algorithm with the similar Inception-v4 convolutional neural network (CNN) architecture performed competitively with expert dermatologists. However, in a follow-up study, Winkler et al. (2019) note that "[i]n clinical routine, suspicious lesions are frequently marked before being excised or photographed" and that benign images were "frequently labeled as being malignant by the CNN when ink markers were visible at the periphery of the dermoscopic image." In subsequent tests of the CNN on new data taken before and after marking, they found that adding the marking significantly increased false positivity, suggesting that human markings made before the original study leaked unintended human expert information about the y-target into the x-data.[16] Since ML models primarily "learn" from the provided data, ruling out data leakage and confounding bias in sampling methodology, data preparation, and y-target labeling are fundamental to preventing link uncertainty with ML models. In data leakage cases, the actual features represented by the data were not just the intended features (specifically, how skin looks (x-data) and dermatological state (y-target)). The data also included influential features, namely, experts selectively added markings in a way that correlated with suspicions of dermatological states. Conclusions drawn from a successfully trained DL model are about these features (also), because that is what the actually sampled data represent, rendering the link with intended features, namely, unmarked skin and their dermatological state uncertain. Let us call this kind of link uncertainty, resulting from relevant and unintended misrepresentation of target features by the data, *misrepresentation uncertainty*.

Misrepresentation uncertainty can clearly corrupt the horizontal links between the model and target (in both directions), confounding the intended features of the target phenomenon to be understood. This is different from the kind of empirical link uncertainty introduced in Section 2 that occurred with Schelling's model. In that case, the uncertainty did not arise from an *unintended* mismatch between what elements of the simulation model (coins as residents, squares in the grid as houses) are supposed to represent. Rather, the alleged uncertainty concerned a mismatch between empirical facts about homeowner reactions and intended model parameters (namely, the 30% similarity preference threshold), and it is not clear that these model assumptions veridically correspond to the target. In misrepresentation uncertainty such as leakage cases, however, the data still veridically correspond to *some feature* actually measured and encoded in the data, but substantively *not the intended features* to be understood.

Sullivan (2022, 126) argues that "[i]mplementation black boxes do not get in the way of understanding phenomena in the melanoma case because the model is operating within a background of existing scientific understanding." She highlights that "[t]he level of scientific justification and background knowledge linking the appearance of moles to instances of melanoma is extensive," noting that it is a "leading deciding factor for medical intervention" and biopsies. This supports claims that there may be meaningful relationships between features of the phenomenon, but it does not provide clear horizontal links for understanding. Certainly, background scientific knowledge can inform the kinds of features to target with an ML model, but in order to establish a link between the model and the target, more is needed. Disambiguating misrepresentation link uncertainty from empirical link uncertainty helps clarify what is needed. The data used by an ML model must *represent* the targeted features of the phenomenon as intended. If not, a background of scientific evidence does not prevent misrepresentation uncertainty. Haenssle et al. (2018) rely on the same "background knowledge" for a similar use case and even DL model but demonstrably fail to link the model to the target. Complementary to the important role of background scientific knowledge for informing how ML models link with their target phenomenon, misrepresentation uncertainty also highlights that appropriate sampling methodology and data preparation are necessary to establish a representation link between the ML model's data and the intended features of the target phenomenon.

Moreover, FI and LR, discussed in Section 4, can be instrumental not just as examples of understanding, but also in helping to rule out misrepresentation uncertainty. Consider Sullivan's (2022, 127) discussion of Wang and Kosinski's (2018) DL model. She explains that "researchers built the model . . . to see whether it was possible to identify an individual's sexual orientation based on facial features alone." Their model is trained on profile pictures taken from dating websites (x-data) and self-reported orientation for said websites (y-target). Although the reported estimator quality metrics are high enough to suggest mutual information between *these features* (profile pictures and self-reports), Sullivan (2022, 127) observes that the "link uncertainty is vast. . . . As the researchers themselves note, many of the features that the model tracks are cultural features, such as certain grooming patterns and dating-profile picture conventions." According to our account, this is a case of misrepresentation uncertainty. Instead of using data that represents "facial features alone" and "sexual orientation" they allow for data leakage in the form of "grooming patterns and profile conventions." The extent to which this data leakage was influential could be achieved by alternative data sampling methods. It might also be better understood through an analysis of FI using permutation importance techniques such as visual occlusion of grooming features, etc., which are believed to have leaked confounding information,[17] or more gradient-based techniques used to identify which parts of an image have a greater impact on model estimation. Similarly, studying how the DL model transforms and represents higher-level features may also inform greater LR understanding of the actually represented features.

Although the three modes of ML understanding explored in Section 4 are not intended to be comprehensive, we emphasize that a predictable relationship between the represented features is not necessarily causal. Background scientific theory is vital for inferring causal claims from information-theoretic claims about features. Sullivan argues effectively against the existence of any background scientific knowledge supporting parental hormone theory (PHT), an origin theory for sexual orientation. Even if the model were reconciled of its misrepresentation uncertainty using improved data sampling methods, etc., further inference from such an IR (between the features veridically representing facial structure and appropriately defined orientation features) cannot be made based on an ML model alone. Though PHT causal hypotheses may be loosely related to the features targeted by such a methodologically rectified ML model, far more scientific work (in the context of an appropriate background theory) is required before it could be said to even partially provide supporting evidence. It is a question of what kind of evidentiary patterns do (and do not) support causal claims within a scientific domain. Origin theories about causal links such as PHT are not merely suspect; they fall outside the scope of target features to be understood by ML models according to the TML hypothesis.

6 Conclusion

Our account can be taken in part as a development of Sullivan (2022), adding further distinctions. Specifically, (1) we have explored the TML hypothesis as an account of the appropriate kind of target for ML models, and (2) we have identified MI, FI, and LR as (at least) three modes of understanding such targets with ML models. (3) We have argued that functionally approximate irrelevance be distinguished from implementation irrelevance, and we have suggested that this distinction helps illuminate why parameter detail proliferation does not necessarily render the level of questions answered by MI, FI, and LR opaque. Last, (4) we have argued that the difference between empirical and misrepresentation-based link uncertainty brings more clarity to the role of background scientific knowledge in supporting ML model based understanding.

Notes

1 In the context of classification, scoring functions p_i are commonly probability distributions on the state space Ω_Y. In the special case where a single output is desired, p_i may also be an indicator function returning 1 for a single element $\hat{y}_i \in \Omega_Y$ and 0 for other elements.

2 The scored "output elements" are commonly referred to as the *target* or *target feature* of an ML model, which in our present contexts risks a misleading conflation of y-targets with the target phenomena of understanding of models as in the C-schema. To disambiguate, we shall use "y-" to prefix "y-targets" whenever the former is intended.

3 In the context of sampling, Ω_X and Ω_Y can here be thought of as state spaces for (marginalized) random variables X and Y, respectively, which are described by some joint distribution corresponding to the data sampling process.

4 In practice, the standard construction for DL classifiers includes a final "softmax" (multinomial logistic) layer estimating $p(Y|X)$, by reducing cross-entropy between the estimated distribution of scores over all y values and the actually sampled y-target in the data ensuring that $p(Y|X)$ satisfies the formal constraints of probability distributions and that (with enough data) learns to assign probability mass to non-peak values. The "hard" max of this learned distribution is returned when a single $\hat{y}$ value must be given.

5 We use the phenomena/data distinction in a similar manner to Bogen and Woodward (1988).

6 Note that in practice data scientists and ML researchers refer to each of the dimensions represented by x-data as a *feature*, and the y-target as the *label*, or the (y-)*target feature*, in the context of supervised learning. To be explicit, for our usage here both x and y data represent features of a target phenomenon.

7 Sullivan's view may be that what we distinguish as functionally approximate irrelevance and implementation irrelevance here both fall under some more general concept of implementation irrelevance. Our account may be interpreted as a refinement of Sullivan, making explicit that accounting for the role and degree of approximation matters to understanding.

8 We make no claims that this list is comprehensive.

9 See, for example, (Kingma and Welling 2013; Chen et al. 2018; Achille and Soatto 2018; Olah et al. 2020; and Tamir and Shech 2022), for a discussion of the various DL and information theoretic techniques for studying such representations.

10 More complex contemporary Transformer techniques use much deeper pre-training of text embedding methods bearing some similarities to

shallower early word vector pre-training representations but have been adapted to embed both chunks of text and individual terms in the context of the surrounding text in which they are written (Devlin et al. 2018; Vaswani et al. 2017).

11 Because information about frequency in contexts learned by these models can be useful in more general text-based tasks, these representations and more advanced techniques are commonly also used as pre-trained representations for new models. Such pre-training is standard for both novel text tasks and image classification tasks. As we see in Section 5, both Esteva et al. (2017) and Haenssle et al. (2018) use versions of pre-trained Inception models (v3 and v4 respectively) for image classification.

12 There are numerous methods of visualizing and analyzing learned representations directly, including projecting to a lower-dimensional space, inspecting particular weights indicating strength of importance to the task, developing generative models, sweeps in latent feature spaces (Chen et al. 2018; Kingma and Welling 2013), and more.

13 When learned representations are particularly effective as pre-trained representations, it is tempting for additional applications to infer that they capture relationships about language (or vision, etc.) in general. However, there are epistemic risks associated with adopting patterns in the data on which a pre-trained model was trained, leading, for example, to gender bias in word-vectorizations (Bolukbasi et al. 2016).

14 Esteva et al. (2017) discussed below does a similar low-dimensional projection of learned representation vectors of dermatology photos to study how clustering patterns of these images relates to their respective diagnosis labels.

15 For a more complex example, see Olah et al.'s (2020) universality hypothesis.

16 Esteva et al. (2017) have a similar potential source of data bias, where dermatologists selectively used a ruler when capturing images of lesions; one of the authors is quoted in popular media as recognizing that "[i]n our data set, dermatologists tended to do this only for lesions that were a cause for concern" (Patel 2017).

17 FI might similarly be used to detect the above identified potential dermatology data leakage risks, by understanding the importance of markings and the ruler usage included in the x-data.

References

Achille, A., & Soatto, S. 2018. "Emergence of Invariance and Disentanglement in Deep Representations." *The Journal of Machine Learning Research*, 19(1), 1947–1980. https://doi.org/10.1109/ITA.2018.8503149.

Bengio, Y., Courville, A., & Vincent, P. 2013. "Representation Learning: A Review and New Perspectives." *IEEE Transactions on Pattern Analysis and Machine Intelligence*, 35(8), 1798–1828. https://doi.org/10.1109/TPAMI.2013.50.

Bogen, J., & Woodward, J. 1988. "Saving the Phenomena." *Philosophical Review*, XCVII (3): 303–352. https://doi.org/10.2307/2185445.

Bokulich, A. 2008. *Reexamining the Quantum–Classical Relation*. Cambridge: Cambridge University Press. https://doi.org/10.1017/CBO9780511751813.

Bolukbasi, T., Chang, K. W., Zou, J., Saligrama, V., & Kalai, A. 2016. "Man Is to Computer Programmer as Woman is to Homemaker? Debiasing Word Embeddings." arXiv:1607.06520. https://doi.org/10.48550/arXiv.1607.06520.

Chen, R. T., Li, X., Grosse, R. B., & Duvenaud, D. K. 2018. "Isolating Sources of Disentanglement in Variational Autoencoders." *NeuroIPS*, 2610–2620. https://doi.org/10.48550/arXiv.1802.04942.

da Costa, N. C. A., & French, S. 2003. *Science and Partial Truth: A Unitary Approach to Models and Scientific Reasoning.* Oxford: Oxford University Press. https://doi.org/10.1093/019515651X.001.0001.

Dev, S., Hassan, S., & Phillips, J. M. 2019. "Closed Form Word Embedding Alignment." *IEEE ICDM*, 130–139. https://doi.org/10.1109/ICDM.2019.00023.

Devlin, J., Chang, M. W., Lee, K., & Toutanova, K. 2018. Bert: "Pretraining of Deep Bidirectional Transformers for Language Understanding." arXiv:1810.04805. https://doi.org/10.48550/arXiv.1810.04805.

Esteva, A., Kuprel, B., Novoa, R. A., Ko, J., Swetter, S. M., Blau, H. M., & Thrun, S. 2017. "Dermatologist-level Classification of Skin Cancer with Deep Neural Networks." *Nature*, 542(7639), 115–118. https://doi.org/10.1038/nature21056.

Frigg, R., & Nguyen, J. 2016. "The Fiction View of Models Reloaded." *The Monist*, 99(3): 225–242. https://doi.org/10.1093/monist/onw002.

Haenssle, H. A., Fink, C., Schneiderbauer, R., Toberer, F., Buhl, T., Blum, A., Kalloo, A., Hassen, A. B. H., Thomas, L., Enk, A., & Uhlmann, L. 2018. "Man against Machine." *Annals of Oncology*, 29(8): 1836–1842. https://doi.org/10.1093/annonc/mdy166.

Khalifa, K. 2017. *Understanding, Explanation, and Scientific Knowledge.* Cambridge: Cambridge University Press. https://doi.org/10.1017/9781108164276.

Kingma, D. P., & Welling, M. 2013. "Auto-encoding Variational Bayes." *ICLR*. https://doi.org/10.48550/arXiv.1312.6114.

Kuehlkamp, A., Becker, B., & Bowyer, K. 2017. "Gender-from-iris or Gender-from-mascara?" *IEEE Winter Conference on Applications of Computer Vision*, 1151–1159. https://doi.org/10.1109/WACV.2017.133.

McCoy, R. T., Pavlick, E., & Linzen, T. 2019. "Right for the Wrong Reasons: Diagnosing Syntactic Heuristics in Natural Language Inference." arXiv:1902.01007. https://doi.org/10.18653/v1/P19-1334.

Mikolov, T., Sutskever, I., Chen, K., Corrado, G. S., & Dean, J. 2013. "Distributed Representations of Words and Phrases and Their Compositionality." *Advances in Neural Information Processing Systems*, 26. https://dl.acm.org/doi/10.5555/2999792.2999959.

Olah, C., Cammarata, N., Schubert, L., Goh, G., Petrov, M., & Carter, S. 2020. "Zoom In: An Introduction to Circuits." *Distill, 2020.* https://doi.org/10.23915/distill.00024.001.

Patel, N. 2017. "Why Doctors Aren't Afraid of Better More Efficient AI Diagnosing Cancer." *The Daily Beast.* https://www.thedailybeast.com/why-doctors-arent-afraid-of-better-more-efficient-ai-diagnosing-cancer.

Pennington, J., Socher, R., & Manning, C. D. 2014. "Glove: Global Vectors for Word Representation." *EMNLP*, 1532–1543. https://doi.org/10.3115/v1/D14-1162.

Potochnik, A. 2017. *Idealization and the Aims of Science.* Chicago: University of Chicago Press. https://doi.org/10.7208/chicago/9780226507194.001.0001.

Schelling, T. C. 1971. "Dynamic Models of Segregation." *The Journal of Mathematical Sociology*, 1, 143–186. https://doi.org/10.1080/0022250X.1971.9989794.

Shrikumar, A., Greenside, P., & Kundaje, A. 2017. "Learning Important Features through Propagating Activation Differences." In *International Conference on Machine Learning* (pp. 3145–3153). PMLR. https://doi.org/10.48550/arXiv.1704.02685.

Simonyan, K., Vedaldi, A., & Zisserman, A. 2013. "Deep Inside Convolutional Networks." arXiv:1312.6034. https://doi.org/10.48550/arXiv.1312.6034.

Strevens, M. 2008. *Depth: An Account of Scientific Explanation*. Cambridge, MA: Harvard University Press.

Suárez, M. 2004. "An Inferential Conception of Scientific Representation." *Philosophy of Science*, 71(5): 767–779. https://doi.org/10.1086/421415.

Sullivan, E. 2022. "Understanding from Machine Learning Models." *The British Journal for the Philosophy of Science*, 73(1): 109–133. https://doi.org/10.1093/bjps/axz035.

Szegedy, C., Vanhoucke, V., Ioffe, S., Shlens, J., & Wojna, Z. 2016. "Rethinking the Inception Architecture for Computer Vision." In *Proceedings of the IEEE Conference on Computer Vision and Pattern Recognition*, 2818–2826. https://doi.org/10.1109/CVPR.2016.308.

Tamir, M., & Shech, E. 2022. Manuscript. "Understanding and Deep Learning Representations."

Torralba, A., & Efros, A. A. 2011. "Unbiased Look at Dataset Bias." In *IEEE CVPR 2011*, 1521–1528. https://doi.org/10.1109/CVPR.2011.5995347.

Vaswani, A., Shazeer, N., Parmar, N., Uszkoreit, J., Jones, L., Gomez, A. N., Kaiser, L., & Polosukhin, I. 2017. "Attention is All You Need." arXiv:1706.03762. https://doi.org/10.48550/arXiv.1706.03762.

Wang, Y., & Kosinski, M. 2018. "Deep Neural Networks Are More Accurate than Humans at Detecting Sexual Orientation from Facial Images." *Journal of Personality and Social Psychology*, 114(2), 246. https://doi.org/10.1037/pspa0000098.

Weisberg, M. (2013). *Simulation and Similarity: Using Models to Understand the World*. Oxford University Press.

Winkler, J. K., Fink, C., Toberer, F., Enk, A., Deinlein, T., Hofmann-Wellenhof, R., Thomas, L., Lallas, A., Blum, A., Stolz, W., & Haenssle, H. A. 2019. "Association between Surgical Skin Markings in Dermoscopic Images and Diagnostic Performance of a Deep Learning Convolutional Neural Network for Melanoma Recognition." *JAMA dermatology*, 155(10), 1135–1141. https://doi.org/10.1001/jamadermatol.2019.1735.

Zeiler, M. D., & Fergus, R. 2014. "Visualizing and Understanding Convolutional Networks." In *European Conference on Computer Vision* (pp. 818–833). Cham: Springer. https://doi.org/10.1007/978-3-319-10590-1_53.

26 Link Uncertainty, Implementation, and ML Opacity

A Reply to Tamir and Shech

Emily Sullivan

Scientific modeling is an avenue for understanding. Machine learning (ML) models are no exception. However, there are notable differences between ML methods and more "traditional" modeling methods. The ML models are data-driven instead of theory- or hypothesis-driven. They are complex and often opaque. Do these differences have implications for how models enable understanding? In a previous work, I argued that at least ML complexity and opacity does not get in the way of understanding phenomena so long as the link between the model and the target phenomena does not have a high degree of *link uncertainty* (LU) (Sullivan 2022).

In "Understanding from Deep Learning Models in Context," Tamir and Shech (this volume, ch. 25) seek to disambiguate aspects of my work on implementation irrelevance and LU. I generally welcome Tamir and Shech's thought-provoking distinctions; however, I will touch on three areas of disagreement: (1) the limits of implementation irrelevance, (2) empirical versus representational LU, and (3) the target of understanding with ML models.

1 The Limits of Implementation Opacity and Irrelevance

In a previous work, I highlighted that ML models are opaque due to *implementation* opacity (i.e., how ML algorithms and trained models implement functions), and that such opacity is not, in principle, a barrier to understanding phenomena with ML models. Tamir and Shech agree that implementation opacity can be irrelevant and does not matter "any more than drawing a map with red or blue ink matters" (this volume, 326). But ultimately, they argue that there is an important distinction between implementation irrelevance and what they call *functionally approximate irrelevance* (FAI). Unlike the colors of a map, other features, like topological facts, cannot be ruled out as mere implementation details. FAI is proposed to mark the difference between mere color choices and the more important topological-like features. The distinction seems to be that there are just some features of models that are *in principle* irrelevant, and other features that are intrinsic difference-makers

DOI: 10.4324/9781003202905-29

where the "varied details matter to the studied target" (this volume, 330). However, here too, instead of claiming that functional approximation opacity undermines understanding, they do a nice job showing that functional approximation can also be irrelevant for understanding in various contexts.

First, I want to clarify implementation irrelevance and how it relates to the problem of ML opacity, before discussing FAI. Consider a computational system that draws a subway map. Computational systems execute algorithms. These algorithms can be discussed at varying levels of abstraction. We can talk about the algorithm at the high(est) level of abstraction by referring to the draw_subway_map() function, or the various sub-steps of the algorithm, or the sub-steps of its sub-steps. Undoubtedly, some of these sub-steps involve computing various topological features. I am not suggesting that the topological features of the map are irrelevant for adequately representing, say, NYC's subway system. On the contrary, for the draw_subway_map() algorithm to be successful, the topological features of the map output must share the relevant topological features of NYC's subway system. How these topological features are expressed (or computed)—either by a visual depiction of nodes and edges or in mathematical notation—does not affect the general goal of representing NYC's subway system.

We must make a distinction between the *goal* or task of the algorithm and *how* the algorithm achieves this goal or task. My claim in Sullivan (2022) was that the way algorithms achieve their goals can be opaque—suffer from implementation opacity—and that this opacity is often irrelevant for assessing whether a model can enable understanding of the phenomenon it bears on. However, the general goal of the algorithm must be known for understanding to be possible. That said, this distinction is not so simple. Due to the layers of abstraction in computational systems, lower-level algorithmic goals, such as some topological calculation, become subsumed under a higher-level algorithm. The lower-level algorithms *become* the way of implementing higher-level algorithms. Thus, lower-level algorithm goals are also a matter of implementation.

In my view, higher-level goals of algorithms are necessarily important for evaluating LU and the scope of possible understanding. Concerning ML models, the trained model executes an algorithm that has a specific classification or predictive task / goal. How the algorithm achieves the goal is an implementation question of the various steps that the learned model takes to reach an output, such as a set of learned weights. Which specific learned weights are necessary to know—and their level of description—for understanding phenomena with ML models is *the* question. I have argued that the answer to this question partly depends on the epistemic risk facing the model and is largely an external question regarding LU.

The umbrella term "implementation irrelevance" does not make a distinction in kind between a model implementing color choices, coding language, or feature importance. I sympathize with the inclination for drawing distinctions here. However, I am not sure as to whether there really is a notable difference in kind when discussing understanding. Understanding requires a specific target. This could be the model itself, some phenomena it bears on, or something else.

Distinguishing between the features of a model that are *in principle* irrelevant and those that are intrinsic difference-makers—from a view from nowhere—is untenable. Relevance is related to a target. Perhaps someone is interested in the differences that coding language makes to building and running ML models in real-world scenarios. Some coding languages provide guarantees against certain runtime problems that other coding languages do not. Maybe this makes a difference to the way an ML model functions in a deployed scenario. In this case, the LU that would need to be reduced to provide understanding is the LU between the coding language, the model, and the deployed scenario. Coding language becomes centrally relevant.

In the cases I am generally interested in—understanding real-world phenomena through modeling—the link between the phenomenon and the model regarding color is certainly irrelevant. The link regarding the phenomenon and the coding language is certainly irrelevant. However, feature importance becomes more salient for determining *how* to reduce LU. It is not a difference in a kind of irrelevance or opacity that is operating here, but rather a difference in the target for understanding. There is no space to say that there are "varied details [that] matter to the studied target," but those details are also irrelevant for understanding the target. Difference-making is entwined with the target. The main disagreement seems to be that Tamir and Shech have a model-centric view of understanding phenomena, whereas in my view models are merely a means—not the target—for understanding phenomena (see Section 3). Different kinds of opacity or irrelevance are a red herring. The central question is how to reduce LU for various types of targets, and which aspects of how the model works need to be known to reduce LU for specific cases.

Reducing LU may require a comparison between different models, different interpretability methods, or even between a "black-box" model and a "white-box" model. As Tamir and Shech discuss, if applying different interpretability methods to the same ML model reveals that the same set of features are identified as being most important, the less we need to know about how the ML model makes its estimates. I agree. Various robustness checks are undoubtedly important and something I discuss in my (2022). However, in my (2022), I did not provide a roadmap for which levels of abstraction are necessary for understanding phenomena. My goals were more modest. I sought to show that the target of understanding is important. And that in actual cases of ML models, we generally know enough about the high-level decision points of a model that the real epistemic problem of opacity becomes an *external* problem, where we look to validating models not by looking inside to what a model is doing, but by looking at external connections or robustness checks with other models.

2 Link Uncertainty and Representation

Tamir and Shech introduce two types of link uncertainty: unintended representational uncertainty and empirical justification uncertainty. The former concerns how the data and the model may inadequately represent targets; the latter concerns whether the ML model has an empirical connection to its targets. I largely welcome this discussion. Answering representational questions is a necessary step for understanding, whether representational mismatches are intentional or not. Moreover, for various targets, issues of representation may go beyond empirical validation, such as the pneumonia risk assessment case I discuss in Sullivan (this volume, ch. 24). Medical researchers developed an ML model exploring the risk of death for patients with pneumonia. The neural network model was highly accurate at the *technical* task of optimizing for risk of death under a certain set of assumptions. However, the model also had a different goal of helping medical researchers predict and devise treatment plans. The model failed at this secondary goal; patients with asthma were classified as low risk. But this low risk was precisely because of existing hospital treatment plans. So, although the model and data uncovered a real empirical link, they did not represent the target of interest.

Does this distinction change the overall relationship between model opacity and understanding? Although I cannot address this question here, I suspect that it does not. There are unresolved disagreements regarding how similar representations need to be of their targets. Issues of data leakage and confounding bias seem like clear candidates for increasing LU. However, other aspects of data manipulation or normalization, like RGB color changes, may not be a representational worry or a case of LU in the way Tamir and Shech seem to indicate. Moreover, if similarity is not a direct aim of representation, then opacity can still be compatible with adequate representation.

3 The Target of ML Models

Tamir and Shech propose the following hypothesis about the target of understanding with ML models:

> *TML Hypothesis*: The target phenomenon of understanding with ML models is the relationship(s) of features represented by the data. (this volume, 328)

They say further that:

> If the TML hypothesis is correct...ML models help us understand relationships between features represented by the data, but expecting such models to further provide causal explanations (e.g. for why a feature is predictive) is inappropriate... (this volume, 349 (see also 335–336)).

I do not deny that in some contexts the TML hypothesis is true. However, notice that the TML hypothesis is *model* centric: the target consists of understanding the model. A model-centric view of ML overlooks an important way in which scientific modeling can provide understanding of the world. When models enable understanding of real-world phenomena (or possible real-world phenomena), the target of understanding is not the relationships of features represented by the data. Data relationships are used as a means to some further end, or, as I have argued elsewhere with Insa Lawler, models provide understanding by *inducing* explanations of phenomena instead of the model itself being an explanation (Lawler and Sullivan 2021). It is not the ML model alone that provides understanding; the model induces an explanation paired with external links connecting the model to the phenomena that enables understanding. This is why I argued that most ML models merely provide how-possibly explanations. The ML models do not provide causal support themselves, but rather indicate possible (causal) hypotheses that additional research must justify through reducing LU.

The fundamental question that I started with remains: how much detail about the model needs to be known to understand phenomena with ML models? Again, I submit that is largely an external problem of LU. Restricting ML models to the model-centric view of the TML hypothesis goes against the goals that many ML researchers themselves postulate, as well as keeping ML models apart from the rest of model-based science, which unnecessarily constrains our scientific toolbox.

References

Lawler, Insa and Emily Sullivan. 2021. "Model Explanation Versus Model-induced Explanation." *Foundations of Science*, 26(4): 1049–1074. https://doi.org/10.1007/s10699-020-09649-1.

Sullivan, Emily. 2022. "Understanding from Machine Learning Models." *The British Journal for the Philosophy of Science*, 73(1): 109–133. https://doi.org/10.1093/bjps/axz035.

Sullivan, Emily. this volume. "How Values Shape the Machine Learning Opacity Problem."

Tamir, Michael and Elay Shech. this volume. "Understanding from Deep Learning Models in Context."

27 Expecting Too Much from Our Machine Learning Models

A Reply to Sullivan

Elay Shech and Michael Tamir

In "How values shape the machine learning opacity problem" (this volume, ch 24), Sullivan identifies three stages at which non-epistemic values in part determine the extent to which machine learning (ML) and deep learning (DL) models have an opacity problem. Due to limitations in space, we concentrate here on the first stage that she discusses, namely, model acceptance and establishing the link between model and phenomenon, but we will also briefly interact with some of her comments on explanation.

Generally, we understand Sullivan's argument as follows. First, in a previous work, Sullivan (2022) argues that the extent to which model opacity and complexity is a hindrance to understanding and explanation depends on how much "link uncertainty" the model has. Link uncertainty seems to correspond to the external evidential support connecting the model to its target phenomenon. If link uncertainty is high, opacity is a hindrance to explanation and understanding. If it is low, then opacity is not a hindrance.

Next, when making non-deductive inferences that are ampliative-inductive in character, such as when making inferences about targets using models, there is a corresponding epistemic-inductive risk associated with such inferences. This is also true for inferences based on ML models, so that high link uncertainty will imply a corresponding high epistemic-inductive risk and lower link uncertainty does not imply as much epistemic risk. Is the risk worth taking? Namely, when is the risk low enough, and the evidence strong enough, to warrant making an inductive inference such as when we accept a hypothesis? The answer, we are told, will depend on values, including non-epistemic values. But if non-epistemic values, in part, determine matters of epistemic-inductive risk, then Sullivan argues that they will be "entangled" (this volume, 306) with link uncertainty since it is a source of epistemic-inductive risk, as well as opacity because link uncertainty decides whether opacity is a hindrance to explanation and understanding.

For example, Sullivan considers an opaque model that predicts the risk of death for patients with pneumonia where link uncertainty is

DOI: 10.4324/9781003202905-30

high since the data the model used relied on a false assumption. On her account, high link uncertainty and opacity hinder the model's ability to afford understanding: "If the researchers used the model to gain some insight into what factors increase the baseline risk for complications, the researchers would be led astray" (this volume). So far, so good,[1] but how do matters of non-epistemic values enter the mix? Sullivan suggests that since the model was intended to assist in deciding how to allocate medical resources—life and death decisions—then there is "clear epistemic risk in accepting whether the model should be used in practice, given the consequence of error" (this volume, 308). However, we worry about the idea that inferences based on models—including ML and non-ML models, opaque and transparent models, complex and simple models, etc.—necessitate extraordinary appeals to values, let alone non-epistemic values. Epistemic risk is not exclusively problematic for ML models. The sorts of risks that Sullivan highlights, though important in the sense that there is a possibility of error, extend generally to non-deductive inferences with both non-ML and ML model based reasoning.

Centrally, there also seems to be a conflation of two different concepts associated with the notion of risk. The first concerns the *epistemic risk* associated with making inferences based on an ML model, namely, the conclusions we draw may turn out to be false. The second concerns the *practical risk* associated with taking (or withholding) actions based on the inferences made using these models. Such actions clearly depend on context, including preferences, goals, dispositions, and non-epistemic values. Say, for instance, that a model effectively predicts aspects of stock market behavior and that you have to decide whether or not to invest a substantial amount of someone else's funds on the model's predictions. Should you risk it? Clearly, such a decision will depend on various subjective, context-dependent matters that are not related to the epistemic risk associated with making inferences by using the model (including the link uncertainty associated with a ML model). If, for example, such funds are needed for someone's cancer treatment, then even minimal risk may be too high of a practical risk to take in such a scenario regardless of how low the associated epistemic risk may be.

Consider, for instance, Sullivan's discussion of how non-epistemic differences in the COMPAS case and HRisk+ case change the requirements for explanation and proper model transparency. Recall that both involve opaque ML models, one determining the risk for prison recidivism and the other the risk of developing various cancers. We are to assume that living in a particular zip code raises both types of risk. Sullivan then notes that appealing to the ML model's determination of risk is not appropriate as an explanation of why someone was denied/granted parole but it is appropriate as an explanation for why someone was chosen for a medical trial. What is the source of this asymmetry in explaining the appropriateness of these respective potential actions? According to Sullivan, non-epistemic values determine whether a statistical explanation over a reasons-based moral explanation is needed to explain the decisions of an ML model.

To see what goes wrong with Sullivan's entanglement argument, consider the assumption about zip code. Sullivan says:

> Suppose further that it is the case that living in a particular zip code increases the risk that someone is arrested for a crime because of policing methods and living in the same zip code increases the risk of someone developing a particular type of cancer because it is an old Superfund site.
>
> (this volume, 313)

In other words, living in a particular zip code increases the risk that someone is convicted of a crime, but it does not increase the likelihood that a specific individual is disposed to committing a crime. This is as if you were more likely to get diagnosed with cancer only because you live in an area where people are more likely to be tested for particular cancers. It is quite different from, for example, living in an old Superfund site, which increases one's risk of cancer because of the presumed *causal* relation between living in a polluted area and cancer. The two cases are asymmetrical (in part) because of the potential causal relationships between the predictive features and what the ML model predicts.

This is where the distinction between epistemic risks and practical risks is critical. The ability (or lack thereof) of an ML model to detect causal relationships, so far, is an epistemic question. Drawing the conclusion of how to act based on implied relationships (e.g., medical intervention, parole decisions) belongs to the practical domain. In order to make sound decisions on what action to take, the causal relationships and better yet the causal consequences of interventions are relevant. However, these ML models do not and should not be expected (in isolation) to answer the core questions of causation and causal intervention at issue here. Doing so is expecting too much of ML models.

Of course, predictive relationships and other forms of understanding provided by ML models, including Informative Relationships (IR), Feature Importance (FI), and Learned Representation (LR), can help support investigations of causal relationships in the context of existing background theory and further study. Indeed, medical studies, including residents of a zip code as discussed above, can be a part of such further investigation.

Reflecting again on Sullivan (2022) and our analysis of it (in this collection), we specifically disregarded the how-possibly/how-actually distinction that Sullivan uses to frame much of her analysis. According to our account, this distinction distracts from the core issues. For instance, in arguing that the deep-patient model fails to give how-actually explanations, Sullivan (2022, 18) notes that "[t]he model does not speak to why a certain marker is linked to a disease." Again, if our Target of ML (TML) hypothesis is correct, such expectations are too high for ML models. ML models help us understand relationships between features represented (via sampling) by the data, but expecting such models to further provide causal explanations (e.g., for why a feature is predictive) is inappropriate. Such questions are the domain of subject matter specific research, and although DL models might help researchers understand that certain features are informative to the prediction (IR) and specifically to varying degrees (FI) (perhaps especially when that information is distilled with other relevant facts (LR)), such understanding still requires further subject-specific study into (say) pathologies and causes associated with a condition.

Taking a broader view, the position we argue for diverges from Sullivan in two key areas. First, we push for finer grained distinctions in terms of both why DL opacity is irrelevant and the nature of Sullivan's link uncertainty. It is easy to dismiss the possibility that, for example, choice of color, software library, or coding language have any relevance to a target. Things are more conceptually treacherous for functionally approximate irrelevance examples, because they have a direct impact on the ostensible target of understanding. Trained ML parameter values would seem to have relevance to the target, but in highly parameterized DL cases these individual values can be (and often are) approximately irrelevant at the individual parameter level. Sullivan argues that any model internals are uninformative, whereas we highlight this sort of (functionally approximate) irrelevance to illustrate how the "level" at which values become relevant directly informs what understanding can be gained. Second, we moderate expectations of the role that ML models play. ML models can be excellent estimators of predictive (and other) feature relationships, but drawing causal inferences (in isolation) from such models expects too much: *Prediction does not failsafe causation any more than correlation.* Emphasizing this epistemic standard for scientists, practitioners, and even philosophers prevents naïve strawman attacks when ML models fail as a panacea for detecting causation in isolation of further theory, scientific background, and well-designed experimentation.

Note

1 However, see our clarifications concerning link uncertainty and the appropriate target of ML models in our contribution for qualifications.

References

Sullivan, E. 2022. "Understanding from Machine Learning Models." *The British Journal for the Philosophy of Science* 73(1): 109–133. https://doi.org/10.1093/bjps/axz035.

Sullivan, E. this volume. "How Values Shape the Machine Learning Opacity Problem."

Part IV

Understanding and Scientific Progress

28 Understanding the Progress of Science

C. D. McCoy

1 Introduction

Much of the recent explosion of literature on scientific progress was ignited by Bird's provocative article "What Is Scientific Progress?" (Bird 2007), where he advocates an "epistemic" account of scientific progress, a kind of view that he alleges is lately overlooked despite its venerable history. In making room for his account, Bird criticizes what he takes to have been the most prominent accounts of scientific progress advanced in the latter half of the twentieth century. He claims that realist philosophers of science have often taken truth to be the ultimate goal of science and its progress to be properly characterized by an accumulation of truths, or at least an approximating approach to them. By contrast, historically-minded philosophers of science and anti-realists have frequently rejected the realists' notion of truth as a goal of science, instead preferring to characterize progress in terms of, among other things, success in problem-solving, as in the well-known views of Kuhn (1996) and Laudan (1977). Bird criticizes both camps: the realists for overlooking the importance of justification to progress and the anti-realists for giving up on truth—both regarded as being essential elements of the generally received conception of knowledge in analytic epistemology. According to Bird, it is knowledge that should properly be regarded as the principal aim of science: science makes progress precisely when it realizes the accumulation of scientific knowledge.[1]

The ensuing discussion has generated many responses to Bird's view (Dellsén 2018), many of which have come from supporters of (some version of) the truth-based realist account. There have also been some supporters of Bird's view that knowledge is the aim of science, as well as others that have promoted accounts based on the idea that understanding makes better sense as a goal of science than knowledge. For the most part, the authors involved in this debate are avowed scientific realists, so they typically dismiss anti-realism quickly, and along with it the problem-solving approach to scientific progress, which they take to be untenable (or irrelevant) due to its assumed commitment to anti-realism. In my view, this attitude toward the problem-solving approach to scientific

DOI: 10.4324/9781003202905-32

progress is unfortunate, not only because I have found it insightful for understanding important episodes in science in my own work (McCoy 2015, 2019) but also, more generally, because much of what is of value in the problem-solving approach is, in fact, independent of the essentially metaphysical issue of scientific realism, based as the approach is on the importance of scientific problems and problem-solving within scientific practice (Nickles 1981).

Nevertheless, regardless of the merits of the problem-solving approach, it is fair to say, as Bird does, that it lacks a clearly identified epistemic (or, as Dellsén prefers to say, cognitive) goal. Indeed, Kuhn and Laudan explicitly reject characterizing it in familiar epistemological terms, for they famously reject the presuppositions of mainstream epistemology, especially those concerning the nature of knowledge, truth, and justification.[2] As for the prospects of developing an alternative approach to epistemology, one which takes its cue from the problem-solving practices of science, one will find relatively little encouragement in the principal works of Kuhn and Laudan, instead only scattered remarks such as that problem-solving is "the single most general cognitive aim of science" (Laudan 1977, 124) (which leaves it completely opaque as to how problem-solving in itself should be a cognitive aim at all) or that scientists solve puzzles because they simply like the challenge (Kuhn 1996, xx).[3]

The course of epistemology has, however, broadened in the intervening years (i.e., since the heyday of debates over scientific progress), such that it now seems possible to assimilate the problem-solving approach to scientific progress and its distinctive characteristics into the currents of epistemological inquiry. My aim in this chapter is to furnish a problem-solving-motivated characterization of the progress of science that addresses the shortcoming identified by Bird, and I do this by taking understanding rather than knowledge as the principal epistemic aim of science. Although it is intuitive that the point of solving problems is to achieve a higher degree of understanding (and not just to come to know their solutions), this intuitive idea is able to receive a deeper philosophical explication thanks to comparatively recent developments in epistemology regarding the concept of understanding. Many of these developments have been in response to the value problem of knowledge, in the face of which philosophers have increasingly concluded that knowledge lacks a sufficiently distinctive value to deserve pride of place in epistemology, with several philosophers (Elgin 1996; Kvanvig 2003; Pritchard 2010; Riggs 2003; Zagzebski 1996) making compelling cases that it is, in fact, understanding that has the needed kind of distinctive value to make sense of our epistemic practices. Although I will not adopt any of their specific proposals here, I believe that the considerations that lead them to focus on understanding are importantly relevant to the proper characterization of progress in science and, in particular, have a strong affinity with the motivations of the problem-solving approach to progress.

As noted above, some responses to Bird's arguments have already made the move to base their accounts of scientific progress on understanding rather than knowledge (e.g., Bangu 2017; Dellsén 2016). Despite basic agreement on the significance of understanding to science, advocates of these views have embraced diverse analyses of the concept of understanding. To my mind, though, none of these analyses is fully adequate to capturing the essentials of problem-solving, which I regard as integral to the practice of science. Therefore, my approach in this chapter will be to consider the role and significance of problem-solving in science from the very beginning, in order to bring out its intuitive relation to our pre-theoretic concept of understanding (Section 2). This relation will then feature centrally in explicating my own account of what (scientific) understanding is and why I believe it is the principal epistemic aim of science (Section 3). In the final section, I defend the account against the threat of its reduction to (scientific) knowledge (Section 4).

2 Problem-Solving as a Measure of Scientific Understanding

Rather than survey and generalize from historical cases of problem-solving in science, or draw on recent scientific work on problem-solving, I will use Kuhn's well-known analysis of scientific change into recurring stages of normal and revolutionary science as my source for identifying the role and significance of problem-solving in science. I do this not because I believe Kuhn's views give the overall best characterization of how science proceeds but simply because his analysis of scientific change is widely known and sufficiently accurate to highlight the key aspects of problem-solving practices in science needed to develop my account. My principal aim in this section is to draw out an intuitive connection between these practices, as seen in Kuhn's stages of scientific change, and our ordinary, pre-theoretic conception of understanding.

Recall that for Kuhn science is roughly split into two periodic phases: normal science and revolutionary science. Normal science is carried out by practitioners working within a paradigm, which, besides the scientific theories relevant to the phenomena studied under that paradigm, includes such things as exemplars, shared values, a material culture of instrumentation, and also metaphysical views. Whatever else these elements contribute to a paradigm's role in science, during normal science they are meant to legitimate "the puzzles and problems that the community works on" (Kuhn 1996, xxiii). The paradigm sets out, broadly speaking, which problems practitioners should aim at solving and which methods and techniques should be used to solve them. According to Kuhn's picture, we should, nevertheless, expect that anomalies inevitably appear. These anomalies eventually come to threaten the paradigm, at last pushing scientists to seek a replacement paradigm, which then ushers in the phase of revolutionary science. Once a new paradigm is adopted, the cycle then begins anew.

Going beyond Kuhn's own particular descriptive aims, we may ask, "what is the primary epistemic aim of puzzle-solving during normal science?" To a philosopher tutored in mainstream epistemology (generally regarded as the "theory of knowledge"), the natural answer would be: "knowledge of puzzle-solutions." And indeed, normal science does have a highly cumulative character characteristic of ordinary knowledge acquisition. Kuhn himself mentions three ways that scientific practitioners operating within normal science contribute to the accumulation of knowledge (Kuhn 1996, 24): (1) by extending the knowledge of the facts that established the paradigm; (2) by increasing the extent of match between predictions and those facts; and (3) by further articulating the paradigm itself. It is, therefore, inviting, given the salience of normal science in the overall course of science (according to Kuhn anyway) and this highly cumulative creation of knowledge, to regard knowledge as the principal epistemic aim of normal science's puzzle-solving.

Certainly, extended and increased knowledge of facts, explanations, articulations, etc. are of significant epistemic value. However, no pure axiology of knowledge can explain why scientists seek certain facts, certain explanations, certain articulations, etc. in the context of normal science. Items of knowledge are all alike in being items of knowledge; to distinguish some items of knowledge as more valuable than others, therefore, demands that additional considerations be brought to bear (incidentally, a point that Bird (2007, 84) himself makes). For some philosophers, these additional considerations are pragmatic, psychological, sociological, etc.—in short, not properly epistemic. Is there another, better way to connect puzzle-solving and epistemology, one that does not leave much of the practice outside the latter's scope?

Indeed, there is, for there is another important function of solving puzzles and problems set by a paradigm, one that is primarily evaluative in character, namely, to determine how effective the paradigm is at solving the problems it sets. To be sure, an individual puzzle-solver need not *personally* care about assessing his or her paradigm's viability, explanatory power, or whatever else is taken to be epistemically valuable. Even so, solving puzzles can be interpreted as having this evaluative function for both the individual scientist and the community of scientists alike. Intuitively, whether a scientist solves a puzzle depends not only on whether it can be solved given the resources of the paradigm by the community of scientists but also on whether the individual scientist possesses the necessary skill and knowledge to solve the puzzles that are attempted. If a scientist succeeds at solving many and various puzzles, then, as Kuhn says, he or she "proves him[- or her]self an expert puzzle-solver" Kuhn (1996, 36). Although such expertise at puzzle-solving may certainly have its social merits, from an epistemological point of view, individual or collective success at solving puzzles is also naturally read as evidence (defeasible evidence, of course) of the paradigm's effectiveness at solving the problems that it sets.

To the typical realist philosopher of science, scientists' puzzle-solving practices would be understood as intending to produce evidence in order to assess the degree to which the paradigm (theory, model, etc.) is true or truthlike (or known or knowledge-like). However, even if scientists think that that is what they are doing, they are rarely in a position to actually make such an assessment (due, e.g., to the problem of unconceived alternatives, etc.). The realist reading of problem-solving thus being inadequate to practice, is there a way to read the aim of puzzle-solving practices in an epistemological way while still maintaining a clear connection to scientific practice?

I propose that there is such a way, one rooted in our intuitive, pre-theoretic concept of understanding: scientists are, whether aware of it or not, principally (individually or collectively) evaluating their degree of understanding, not knowledge, by solving problems. In the first place, it makes intuitive sense to say that a successful puzzle-solving scientist in normal science is both demonstrating the understanding afforded by the epistemic resources of the paradigm and also demonstrating his or her own understanding achieved through this paradigm as well. Scientists, of course, may be seen as demonstrating both their knowledge and the knowledge inherent in their paradigm too. Attention to the practice of puzzle-solving, however, suggests that this kind of understanding sought by puzzle-solving has a somewhat different character than knowledge. It emerges in particular from thinking about what makes a normal science practitioner skilled, what someone who has an expert understanding (of his or her paradigm) can do. Intuitively, a skilled scientific practitioner (in normal science) is someone who is able to produce facts and explanations, ones that solve empirical and conceptual problems set by his or her paradigm—in particular, by using the conceptual means at his or her disposal (i.e., as given in the paradigm). Thus, puzzle-solving is (1) a test of ability (and not of knowledge, at least not directly), which (2) does not generally involve the prior possession (knowledge) of puzzle-solutions (indeed, if the puzzle-solver had the solution already, then we could hardly regard his or her activity as puzzle-solving, much less him or her as a skilled puzzle-solver). Understanding, unlike knowledge, aptly describes the aim of puzzle-solving given these two conditions on it, for it is intuitively correct to say that puzzle-solving tests one's ability to understand (and not to know), and it is clear that one can understand how to solve puzzles without knowing their solutions in advance.

Further aspects of a problem-solving-oriented conception of scientific understanding are suggested by the second phase of Kuhn's scheme, revolutionary science. Let us briefly recall how Kuhn describes revolutionary science. The phase first emerges after persistent anomalies eventually overcome the resistance to change of scientists bound within a paradigm, which leads to the breakdown of the normal science puzzle-solving activity of scientists. As anomalies accumulate, Kuhn describes science as being "in a state of growing crisis" (Kuhn 1996, 67). Scientists do not respond to crisis by abandoning their paradigm right away, although they do begin considering alternatives that might solve the prevailing paradigm's mounting problems. However, if a newly developing paradigm's supporters are competent (and fortunate), then, as Kuhn says, "they will improve it, explore its possibilities, and show what it would be like to belong to the community guided by it" (Kuhn 1996, 159).

From this brief description of revolutionary science, it is evident that problem-solving is also an essential activity of this phase, although here it takes on a slightly different character than it does in normal science (i.e., it is not mere "puzzle-solving" anymore, for the puzzles are no longer set in advance by the dominant paradigm). An expert revolutionary science problem-solver may not be an expert normal science puzzle-solver, and vice versa, but both are distinguished by their ability to produce solutions to contextually salient problems. Thus, the epistemic rationale behind "revolutionary" activities is intuitively characterizable in terms of understanding as well, to wit, to develop a novel understanding that is lacking in the present paradigm. Of course, for Kuhn himself there are no rules to assess whether one such understanding is objectively better than another, since it is crucial to his thesis that there is no rationally compelling (i.e., rule-based) reason that practitioners adopt a particular paradigm (Kuhn 1977). Nevertheless, Kuhn does allow that scientists must be convinced on two important points: (i) that the new paradigm must promise to resolve some outstanding problem that can seemingly be solved in no other way; (ii) that the new paradigm must also preserve a large part of the prior paradigm's problem-solving ability. Both of these pertain to the "problem-solving effectiveness" (borrowing Laudan's term) of the new paradigm in comparison to the old. It is appealing, therefore, to interpret problem-solving in revolutionary science in similar epistemological terms, namely, as evaluative of the potential understanding afforded by a newly developing paradigm.

By examining both normal science and revolutionary science, we see that the problem-solving activities found in Kuhn's account of science suggestively point to a single, unified epistemic goal of understanding rather than knowledge. There are, to be sure, important differences between the problem-solving of the two phases. Normal science problem-solving activity is "backward-looking" and assessive, for it is primarily intended to evaluate the degree of understanding so far achieved in the current, established paradigm. Importantly, the production of cumulative knowledge in the form of new problem-solutions (facts and explanations) is secondary and instrumental to this more essential function. In revolutionary science, by contrast, problem-solving is "forward-looking" and diagnostic, for it is primarily intended to evaluate the potential for understanding that may be achieved through solving problems with the new resources furnished by the future paradigm under development. The production of individual items of knowledge is, again, secondary and instrumental to the more essential function of establishing a new understanding.

3 Understanding as the Principal Epistemic Aim of Science

The previous section presented the intuitive connection between problem-solving and understanding by reference to the problem-solving activities of science. In this section, I will now proceed more systematically in the task of developing an account of scientific progress that is based on the concept of understanding as understood through its connection with problem-solving. I argue that it is understanding and not knowledge that is the principal epistemic aim of science, hence the epistemic concept against which scientific progress should be primarily measured.

Observers of both the older and newer debates about scientific explanation and understanding will, no doubt, be aware that I am far from alone in suggesting the relevance of understanding to the issue of scientific progress. Philosophers of science have frequently mooted the idea that understanding is an important product of science, especially in motivating their own favored accounts of scientific explanation (e.g., Friedman 1974; Strevens 2008). However, this was often done in the past without explicating precisely how explanation is related to understanding—or even explaining what understanding is (as, e.g., Kim (1994) and de Regt and Dieks (2005) have remarked). This persistent uncertainty over the nature of understanding has made room for another long-standing tradition in the philosophy of science according to which understanding is merely "psychological" and hence lacking in any objective epistemic significance at all, starting with Hempel's (1965) influential studies of explanation, continuing through van Fraassen's (1980) pragmatic account of explanation, and finding more recent expression in the work of Trout (2002).

The position taken in this chapter is that understanding can and should be taken as a central concept of (scientific) epistemology. The recent tide of support in epistemology for a genuinely objective, epistemic notion of understanding (e.g., Elgin 1996; Kvanvig 2003; Pritchard 2010; Riggs 2003; Zagzebski 1996)—one of the main motivations for this chapter— is certainly not without its critics though. Against this tide stand some who, though accepting that there is such a notion, believe that the new epistemology of understanding has added little to the long-established epistemology of knowledge. Indeed, several philosophers have argued that understanding reduces to (or merely is a species of) knowledge, especially explanatory knowledge (e.g., Grimm 2006; Khalifa 2012; Strevens 2013). If it were true that understanding really is essentially knowledge (at least for all epistemological purposes), then the import of understanding would be quite minimal indeed. However, I will defer my discussion of this important issue to the final section of this chapter in favor of first arguing for the priority of understanding as the epistemic aim of science, relying for now on the intuition (and the indications of the previous section) that knowledge and understanding are sufficiently distinct concepts.

To begin elaborating my account of scientific understanding, I wish to draw attention first to the epistemic significance of (theoretical) explanations in science. That they are so significant clearly indicates that the distinctive locus of epistemic value in science is not the bare scientific fact—it is rather scientific theory. By theory, I mean not only theory in the sense of general, abstract laws, etc. but also other explanatory and descriptive resources, from scientific frameworks or paradigms to concrete models. Empiricist scruples may incline some toward the view that theory is too ephemeral to have real, substantial epistemic significance, which leads on further to the position that the sole epistemic content of science should not be identified with scientific theory but with the empirical facts. Nevertheless, even the most hard-nosed empiricist must admit that scientific theory is indispensable for securing the greater part of those empirical facts that we do possess. Thus, in the case where scientific explanations are accepted as genuinely epistemic in character, theory is clearly the fundamental seat of epistemic value in science; when not, theory remains the indispensable vehicle of epistemic value in science, and that will suffice for present purposes.

Since one can, linguistically at any rate, treat scientific theories as objects in epistemic locutions attributing understanding to an agent, one might, following Kvanvig (2003), characterize our understanding of scientific theories as "objectual." Although many philosophers have been willing to entertain a category of objectual understanding, it remains in dispute as to how this kind of understanding should be properly analyzed. For example, some authors, including Kvanvig, talk of the objects of objectual understanding as bodies of beliefs, while highlighting the coherence, holism, or integration of these beliefs as crucial for their status as objects. Although I agree with proponents of objectual understanding that it is a distinct category among kinds of understanding and that understanding a theory can be described as a kind of objectual understanding, I will not be following those who hold that theories are best thought of as bodies of propositions that collectively admit of belief (for reasons to be explained shortly and also further in the final section). I will also avoid making any explicit use of the notion of objectual understanding, since for the purposes of scientific epistemology the category of objectual understanding is far too broad. Phenomena can be understood "objectually" just as well as theory, yet theories and phenomena must be distinguished into epistemologically distinct categories in order to make sense of scientific methodology, particularly as it is theories that explain phenomena and not vice versa. Our understanding of phenomena is asymmetrically dependent on understanding theory. Thus, I will prefer to speak of theoretical understanding (Saatsi 2019) and understanding phenomena with theories (Strevens 2013).

One might think that there is, nevertheless, not much of epistemological significance in marking a distinction between scientific facts (or propositions, if you like) and scientific theories, since it might seem that theories are nothing more than collections of scientific facts (or propositions). Certainly, if scientific theories were merely collections of propositions, then "believing a theory" would be nothing more than believing that collection of propositions. I allow, at least for some logical and epistemological purposes, that treating theories as if they were such is sometimes fruitful and relatively harmless. Actual theories in scientific practice, however, the ones that scientists use, are surely not collections of facts, propositions, or whatever else. In the first place, scientific theories are clearly not learned as collections of facts and are not communicated as such either (which is not to say that learning and communicating facts are not an essential part of the learning and communication of theory). If someone memorized the entirety of Weinberg's *Quantum Theory of Fields* (in all of its three volumes), they would hardly thereby come to know quantum field theory, let alone understand it. That is because understanding (or knowing) a theory is a matter of practical engagement with that theory, applying it to problems set by the descriptive and explanatory scope of the theory. Thus, it is essential to distinguish between scientific facts and scientific theories in order to understand the practice of science.

Based on the foregoing remarks, I claim that theoretical understanding is essentially practical and constructive. To practicing scientists, scientific theories (or models) are tools to be applied, specifically in order to understand (explain and describe) phenomena. Much as the nature of a hammer is not its mere being as an object but primarily as a piece of equipment, a means of hammering, the nature of theories is not their mere being as objects (bodies of knowledge) but primarily as a means of theorizing about phenomena. They are the tools of science. Although to some this perspective on theories might suggest an unpalatable "instrumentalism," highlighting the practical dimension of theories in connection to understanding surely does not entail anything untoward about their factual status, much less about the descriptions and explanations produced by their application.[4]

Instead, the practical and constructive nature of understanding should draw attention rather to the central role of the agent's abilities to use such conceptual tools to solve problems. In the literature on scientific understanding, de Regt, in particular, has emphasized agent abilities in this way, making them central to his "pragmatic" account of scientific understanding (de Regt 2009, 2017; de Regt and Dieks 2005). De Regt's (and my) insistence on the essential role of the agent in an account of understanding is not a concession of the subjectivity of understanding but rather the recognition of an essential cognitive component to understanding, one that involves the ability to produce knowledge, not just possess it.

The basic statement of my account of understanding and progress with respect to it can now be given. First, possessing a (scientific) understanding of some subject matter is mediated by one or more (scientific) theories of that subject matter—this is the requirement of "understanding phenomena with theories." Second, understanding some subject matter with a theory depends on having the ability to produce explanations and descriptions pertaining to that subject matter, that is, solving explanatory and descriptive problems within the scope of that theory. Hence, understanding a theory means being able to understand phenomena with that theory. Third, and finally, progress in understanding simply involves improvement in theoretical understanding, which is a matter of improved ability to explain and describe phenomena, the degree of which is something evaluable by a problem-solving standard.[5]

To argue for the priority of this kind of understanding among epistemic aims of science, I will first introduce two broad, archetypal conceptions of epistemology and accounts of theoretical progress based on them. According to what I will call the knowledge-centered epistemology, scientific theories are collections of propositions, so that knowing a theory is simply knowing a collection of propositions. Epistemic progress according to the knowledge-centered epistemology is identified with the accumulation of knowledge (as in Bird's epistemic account), or the increasing verisimilitude of the epistemic content of scientific theories (according to the garden-variety realist). According to what I will call the understanding-centered epistemology, by contrast, scientific theories are conceptual tools, not collections of facts. Understanding a theory is having the ability to use that theory to produce knowledge (of facts and explanations), especially in response to scientific problems. Epistemic progress according to the understanding-centered epistemology is identified with the increase of understanding, an increase in the ability to solve scientific problems by producing good descriptions and explanations of phenomena that pertain to the subject matters of the relevant theories.

The understanding-centered epistemology acknowledges the epistemic significance of productive abilities in the epistemology of science. The knowledge-centered epistemology focuses only on the value and significance of certain products of science: items of knowledge. From the perspective of the understanding-centered epistemology, however, these items of knowledge mainly have instrumental value and significance, insofar as they contribute to increased understanding.

I maintain that, in leaving no place for the productive abilities necessary for explaining and grounding the growth of knowledge, the knowledge-centered epistemology can only give a static and skeletal picture of science and its epistemology. Certainly, such an epistemology is able to account for the storage and sharing of items of knowledge, as propositional knowledge may be recorded and transmitted (assuming some basic abilities of comprehension and of judging the reliability of testifiers). It can also give a simple standard of progress, since static bodies of propositional knowledge can be compared both synchronically and diachronically. But how can such items of knowledge by themselves license increases in knowledge? Epistemologists of science have traditionally avoided this crucial issue by shunting the means of increasing knowledge off into "psychology" or "pragmatics" (as in empiricist or hypothetico-deductivist approaches). However, this maneuver, though attractive for its simplification of the problems of epistemology, illegitimately limits the scope of epistemology and limits our philosophical understanding of science. There must be a place made for the scientist, and not just the "science."

In sum, it is only the understanding-centered epistemology that provides the conceptual framework needed for elucidating the epistemic progress of science. It is centered on the idea that growth is a matter of production, production that depends on having the right tools for the job and the ability to use them. Epistemically speaking, what is intrinsically valuable is cognitive ability, not its product alone. This is because knowledge's epistemic value is dependent on the understanding to which it is related. To be sure, having an understanding does depend on the possession of some knowledge, but this knowledge is only partly constitutive of the understanding possessed. Although the growth of knowledge is an essential concomitant of the growth of understanding, it is the growth of understanding that is the principal epistemic goal of scientific practice. Scientific progress should accordingly be characterized, then, as an increase in understanding.

4 Understanding, Explanation, and Knowledge

At the end of the previous section, I contrasted two broad epistemological approaches to characterizing progress in science: the knowledge-centered and the understanding-centered, arguing in favor of the priority of the latter, especially for how it captures the dynamic growth of science. The existence of this contrast, however, depends on my conception of understanding not being reducible to knowledge (or some arrangement of knowledge).[6] If it is, then my claims are illusory, for then the growth of understanding can be accounted for in traditional, knowledge-based epistemological terms. Although I cannot give an elaborate defense of the distinction on which my argument rests in the remaining space available here, I can offer some further considerations for why assimilating understanding to knowledge is a mistake.[7]

Above, I argued that scientific progress should be characterized in terms of theoretical progress, since the epistemic content and value of science resides predominantly in its theories. Of course, it is consistent with this claim that scientific theories have epistemic value because of the propositional knowledge that constitutes them (or because they approximate propositional knowledge in some way). This is a fundamental idea of what I have called the knowledge-centered account of scientific progress: improving our theories means improving our knowledge (whether that be restricted to empiricist-sanctioned knowledge or else approximate knowledge with respect to the world-as-everything-that-is-the-case). It is also consistent, though, with my preferred view, that scientific theories have epistemic value because with them one is able to understand. This is the core idea of what I have called the understanding-centered account of scientific progress: improving our theories means increasing our understanding. At this level of description, then, the two approaches are not obviously opposed, for it is conceivable that the cognitive abilities that are distinctive of understanding may be rendered in terms of propositional knowledge or that propositional knowledge may be rendered in terms of cognitive abilities.

However, any attempt to reduce one concept to the other would be ill-advised, for each plays an important and distinctive role in a comprehensive scientific epistemology. To see how, consider the activity of solving explanatory problems. When one is able to deliver an explanation, prompted by an explanatory problem, it is intuitive to say that it is because one has the relevant theoretical understanding (understanding "with," as Strevens (2013) puts it). Owing to one's theoretical understanding, one has additionally an explanatory understanding of the explanandum (understanding "why") and, further, propositional understanding of the propositions involved in that particular explanation (understanding "what"). Understanding, then, "flows down" from theory to explanation to individual propositions. In other words, it is in virtue of the relevant objectual (or theoretical) understanding that one has an explanatory understanding in the specific problem-context, and by virtue of these understandings together that one has the relevant propositional understanding of the facts involved in the explanation. Knowledge, as it is usually conceived in contemporary epistemology, works differently than understanding. It "flows up" from proposition to theory. This is because knowledge of propositions is the basic form of knowledge. In order to know an explanation (of a phenomenon), one must know the propositions involved in that explanation and know that they are related explanatorily; in order to know a theory, one must know certain explanations and descriptions that are important concomitants of that theory, and know that they are related together theoretically. Both cognitive schemes—the flow of understanding and the flow of knowledge—are important and intimately related aspects of scientific activity. The failure to acknowledge either one would leave some epistemically valuable elements of science unaccounted for.

The conceit of the knowledge-centered epistemology is that the latter scheme is entirely sufficient to account for the epistemic value of our theories, explanations, and propositions. This aggrandizement of propositional knowledge, the root of the knowledge-centered epistemology, is, of course, a deeply-rooted and long-standing dogma of analytic epistemology. The many legitimate successes of this dogma to account for much involved in our epistemic practices have, however, obscured essential facets of epistemology, such as learning, conceptual change, and inference. Insofar as these are handled at all in epistemology, they are treated as items to be baked into the propositional mold. Static propositions, however, are the wrong tools for the job, and they necessarily force dynamic concepts (like learning, change, and inference) to become mere trivialities (e.g., "learning is the acquisition of new propositions," "conceptual change happens when the intension associated with a concept term changes," "inference is a pattern of propositions where one is related to the others by valid rules") while failing to account for (or even acknowledge) the process itself. It is a fine epistemology for machine intelligence, perhaps, but it leaves out much that makes knowledge and other epistemic concepts humanly significant. These aspects, rooted in human experience, should not be obscured or eliminated for the sake of easy analysis.

To illustrate and underscore the point, I conclude with a well-worn example. The example, Carroll's famous story of the tortoise and Achilles (Carroll 1895), is meant to show (as I interpret it anyway) that no amount of knowledge can stand in for the ability to infer. The tortoise in the tale is willing to assent to as many premises as Achilles provides him: that P, that $P \supset Q$, that $\{P; P \supset Q\} \vdash Q$, etc. However, the tortoise refuses to assent to the conclusion that Q. That is, the tortoise will not infer that Q, despite the apparent logical compulsion to do so. What lesson is there in this parable? To my mind, it shows that propositional knowledge (including additional theorizing at meta-levels of analysis) by itself cannot fully capture the nature of inference, for the basic reason that inference is fundamentally a practical, processual, skilled activity.[8] If the tortoise fails to understand anything in Carroll's story, it is that he fails to understand how the theory of inference is supposed to be related to the practical activity of making inferences (although, then again, one might suppose that he actually understands it all too well, to the detriment of poor Achilles . . .).

Acknowledgments

I am grateful to Finnur Dellsén, Kareem Khalifa, Insa Lawler, Vera Matarese, and Juha Saatsi for commenting on various versions of this chapter, and to Adam Carter and Emma Gordon for valuable conversations during its genesis. Some of this material was presented at the 2017 meeting of the British Society for the Philosophy of Science in Edinburgh, "Yet Another Great Workshop on Idealization, Causation, and Explanation" at the Munich Center for Mathematical Philosophy, also in 2017, and the 2nd Scientific Understanding and Representation Workshop in 2020, all of whose audiences I thank as well. This work was supported by a New Faculty Seed Funding Grant from Yonsei University, which I gratefully acknowledge.

Notes

1 Bird calls his account "epistemic," in line with the usual assumption that epistemology is fundamentally concerned with knowledge (Bird 2010). This is a point that will be under dispute in this paper, since I (as well as many contemporary epistemologists) hold that the subject matter of epistemology has a broader scope than just knowledge, particularly the mainstream conception of it in analytic epistemology.

2 As Laudan says, "the initially plausible, and broadly held, view that theory testing and theory evaluation are at root . . . epistemological activities" is "fundamentally mistaken and systematically misleading" (Laudan 2000, 165). Kuhn, for his part, did think that truth was an important standard, even a necessary one, within the context of a paradigm—just not the "utopian" truth of the realist, a "goal set by nature in advance" of our inquiry into it (Kuhn 1996, 171).

3 That said, Laudan does promote a sort of meta-epistemology of his own, which he calls "normative naturalism" (Laudan 1984, 1987). According to normative naturalism, the aims of science shift over time in a way that requires re-conceiving the remit of "epistemology" as scientific values change.

4 Such a generally practical and constructive point of view is widely adopted already in the literature on scientific models. Some do use it to press for a more specifically anti-realist stance toward theories, for example Cartwright et al. (1995), but not all—cf. the different perspectives on offer in Morgan and Morrison (1999).

5 As Insa Lawler has pointed out to me, Strevens (2013, 515) describes a very similar sounding account of understanding, for example, when he states that "to understand a theory is to have the ability to use the theory to construct, or at least to comprehend, internally correct scientific explanations of a range of phenomena." Where my view departs most significantly from Strevens' is that he links understanding to the grasping of explanations, the latter construed as a structured set of propositions, whereas I believe that understanding involves the ability to provide explanations and descriptions that respond to problems set by the theoretical context.

6 It is worth mentioning that some relate understanding not to propositional knowledge ("know that") but to a different kind of knowledge, some kind of practical knowledge or "know-how," such as Zagzebski (2001) and Hills

(2016). Although relating understanding to knowledge in this way has a strong affinity with the account I offer here, I hesitate to make too much of the connection, because the, as it were, "cognitive know-how" of understanding is not perfectly analogous to conventional examples of know-how (like riding a bike or wielding a hammer). Cf. (Sullivan 2018).

7 Many other philosophers have argued that (objectual or explanatory) understanding is distinct from knowledge. However, they generally maintain the propositional nature of theories (i.e., regarding them as bodies of information) and appeal variously to a certain holism, coherence, integration, or unification inherent in those bodies of information as the mark of understanding (e.g., Cooper 1994; Elgin 2007; Friedman 1974; Gardiner 2012; Kvanvig 2003; Zagzebski 2001). Although I cannot argue it here, I believe that these suggestions cannot account for the epistemic value of understanding as understood in this chapter. Instead, it seems to me that these features have mainly to do with distinguishing theories from one another, that is, with epistemic compartmentalization. Some philosophers (e.g., de Regt and Dieks 2005; Elgin 2007; Grimm 2006; Hills 2016; Kvanvig 2003; Riggs 2003) appeal to the notion of "grasping" to distinguish between understanding and knowledge. Insofar as the metaphor of "grasping" is elaborated at all, grasping is usually described as either having in mind some appropriate body of information with the marks mentioned above (holism, coherence, etc.) or else some kind of ability, often as the "know-how" mentioned in note 6.

8 Khalifa claims that the possession of information concerning inferential connections is all that is required to complete an explanatory story, much like Achilles supposes in the story. Although he states that he is willing to concede that the ability to infer may not be fully reducible to knowledge, he dismisses the role of inference in explanatory understanding as "so thin as to trivialize understanding" (Khalifa 2012, 28). However, it is hardly thin and trivial. That inference only appears as such is because it is so familiar, even becoming invisible to us . . . if a computer were able to contemplate it, I should think that the human power to infer would hardly seem to it thin or trivial!

References

Bangu, Sorin. 2017. "Scientific Explanation and Understanding: Unificationism Reconsidered." *European Journal for Philosophy of Science* 7: 103–126. doi: 10.1007/s13194-016-0148-y.

Bird, Alexander. 2007. "What Is Scientific Progress?" *Noûs* 41: 64–89. doi: 10.1111/j.1468-0068.2007.00638.x.

Bird, Alexander. 2010. "The Epistemology of Science—A Bird's-Eye View." *Synthese* 175: 5–16. doi: 10.1007/s11229-010-9740-4.

Carroll, Lewis. 1895. "What the Tortoise Said to Achilles." *Mind* 4: 278–280. doi: 10.1093/mind/IV.14.278.

Cartwright, Nancy, Towfic Shomar, and Mauricio Suárez. 1995. "The Tool Box of Science." In *Theories and Models in Scientific Processes*, eds. William E. Herfel, Władyłslaw Krajewski, Ilkka Niiniluoto, and Ryszard Wójcicki, 137–149. Amsterdam: Rodopi.

Cooper, Neil. 1994. "Understanding." *Proceedings of the Aristotelian Society, Supplementary Volumes* 68: 1–26. doi: 10.1093/aristoteliansupp/68.1.1.

de Regt, Henk. 2009. "The Epistemic Value of Understanding." *Philosophy of Science* 76: 585–597. doi: 10.1086/605795.

de Regt, Henk. 2017. *Understanding Scientific Understanding.* New York: Oxford University Press.

de Regt, Henk, and Dennis Dieks. 2005. "A Contextual Approach to Scientific Understanding." *Synthese* 144: 137–170. doi: 10.1007/s11229-005-5000-4.

Dellsén, Finnur. 2016. "Scientific Progress: Knowledge Versus Understanding." *Studies in History and Philosophy of Science Part A* 56: 72–83. doi: 10.1016/j.shpsa.2016.01.003.

Dellsén, Finnur. 2018. "Scientific Progress: Four Accounts." *Philosophy Compass* 13: e12525. doi: 10.1111/phc3.12525.

Elgin, Catherine. 1996. *Considered Judgment.* Princeton: Princeton University Press.

Elgin, Catherine. 2007. "Understanding and the Facts." *Philosophical Studies* 132: 33–42. doi: 10.1007/s11098-006-9054-z.

Friedman, Michael. 1974. "Explanation and Scientific Understanding." *The Journal of Philosophy* 71: 5–19. doi: 10.2307/2024924.

Gardiner, Georgi. 2012. "Understanding, Integration, and Epistemic Value." *Acta Analytica* 27: 163–181. doi: 10.1007/s12136-012-0152-6.

Grimm, Stephen. 2006. "Is Understanding a Species of Knowledge?" *The British Journal for the Philosophy of Science* 57: 515–535. doi: 10.1093/bjps/axl015.

Hempel, Carl Gustav. 1965. *Aspects of Scientific Explanation.* New York: The Free Press.

Hills, Alison. 2016. "Understanding Why." *Noûs* 50: 661–688. doi: 10.1111/nous.12092.

Khalifa, Kareem. 2012. "Inaugurating Understanding or Repackaging Explanation?" *Philosophy of Science* 79: 15–37. doi: 10.1086/663235.

Kim, Jaegwon. 1994. "Explanatory Knowledge and Metaphysical Dependence." *Philosophical Issues* 5: 51–69. doi: 10.2307/1522873.

Kuhn, Thomas. 1977. "Objectivity, Value Judgment, and Theory Choice." In *The Essential Tension*, 320–339. Chicago: University of Chicago Press.

Kuhn, Thomas. 1996. *The Structure of Scientific Revolutions*, 3rd ed. Chicago: University of Chicago Press.

Kvanvig, Jonathan. 2003. *The Value of Knowledge and the Pursuit of Understanding.* Cambridge: Cambridge University Press.

Laudan, Larry. 1977. *Progress and Its Problems.* Berkeley: University of California Press.

Laudan, Larry. 1984. *Science and Values.* Berkeley: University of California Press.

Laudan, Larry. 1987. "Progress or Rationality? The Prospects for Normative Naturalism." *American Philosophical Quarterly* 24: 19–31.

Laudan, Larry. 2000. "Is Epistemology Adequate to the Task of Rational Theory Evaluation?" In *After Popper, Kuhn and Feyerabend*, eds. Robert Nola and Howard Sankey, 165–175. Dordrecht: Springer.

McCoy, C. D. 2015. "Does Inflation Solve the Hot Big Bang Model's Fine-Tuning Problems?" *Studies in History and Philosophy of Modern Physics* 51: 23–36. doi: 10.1016/j.shpsb.2015.06.002.

McCoy, C. D. 2019. "Epistemic Justification and Methodological Luck in Inflationary Cosmology." *The British Journal for the Philosophy of Science* 70: 1003–1028. doi: 10.1093/bjps/axy014.

Morgan, Mary S., and Margaret Morrison, eds. 1999. *Models as Mediators.* Cambridge: Cambridge University Press.

Nickles, Thomas. 1981. "What Is a Problem That We May Solve It?" *Synthese* 47: 85–118. doi: 10.1007/BF01064267.

Pritchard, Duncan. 2010. "Knowledge and Understanding." In *The Nature and Value of Knowledge*, eds. Duncan Pritchard, Alan Millar, and Adrian Haddock, 3–90. Oxford: Oxford University Press.

Riggs, Wayne. 2003. "Understanding 'Virtue' and the Virtue of Understanding." In *Intellectual Virtue*, eds. Michael DePaul and Linda Zagzebski, 203–226. Oxford: Oxford University Press.

Saatsi, Juha. 2019. "What Is Theoretical Progress of Science?" *Synthese* 196: 611–631. doi: 10.1007/s11229-016-1118-9.

Strevens, Michael. 2008. *Depth.* Cambridge: Harvard University Press.

Strevens, Michael. 2013. "No Understanding Without Explanation." *Studies in History and Philosophy of Science Part A* 44: 510–515. doi: 10.1016/j.shpsa.2012.12.005.

Sullivan, Emily. 2018. "Understanding: Not Know-how." *Philosophical Studies* 175: 221–240. doi: 10.1007/s11098-017-0863-z.

Trout, J. D. 2002. "Scientific Explanation and the Sense of Understanding." *Philosophy of Science* 69: 212–233. doi: 10.1086/341050.

van Fraassen, Bas. 1980. *The Scientific Image.* Oxford: Oxford University Press.

Zagzebski, Linda. 1996. *Virtues of the Mind.* Cambridge: Cambridge University Press.

Zagzebski, Linda. 2001. "Recovering Understanding." In *Knowledge, Truth, and Duty*, ed. Matthias Steup, 235–254. Oxford: Oxford University Press.

29 Scientific Progress without Justification

Finnur Dellsén

1 Introduction

Most people would agree that science is one of the human endeavors in which we have made the most progress in the past few centuries. But what makes us so sure that the various changes that have occurred in science are genuinely progressive? *What is required for scientific progress?*

Philosophical accounts of scientific progress contain different answers to this question. These accounts purport to tell us both what is, and what is not, required for scientific progress. In this chapter, I will be concerned with a particular requirement that is implied by some such accounts and not others. Roughly, this requirement holds that the acceptance of a new theory counts as progressive only if the theory is epistemically justified. Such a requirement is explicitly and enthusiastically endorsed by some in the debate about scientific progress (e.g., Bird 2007; Needham 2020; Park 2017). Others have rejected such a justification requirement as unnecessary or unmotivated (e.g., Cevolani and Tambolo 2013; Dellsén 2016; Niiniluoto 2014; Rowbottom 2008).

This chapter argues in favor of the latter position. My arguments will not rely heavily on our pre-theoretic intuitions about whether we would be inclined to use the term 'progress' to describe some hypothetical episode in science. Rather, I will attempt to identify a wide range of paradigmatic cases of scientific progress in which epistemic justification is lacking. Denying progress in these cases would not just be counterintuitive, but also go against various truisms about scientific methodology and imply that there is much less progress in science than most of us have previously thought. Thus, in brief, I hope to be moving beyond clashing intuitions about merely hypothetical cases, and instead provide a different type of argument against the justification requirement on scientific progress.

2 Scientific Progress and Epistemic Justification

What are accounts of scientific progress accounts of? What is the question to which an account of scientific progress is supposed to be the

DOI: 10.4324/9781003202905-33

answer? In what follows, I take the central question of scientific progress to be this:

> What type of cognitive change with respect to a given phenomenon *X* constitutes a (greater or lesser degree of) scientific improvement with respect to *X*?

For reasons of space, I will not defend my focus on this question here. I will, however, clarify three key phrases therein.

The first clarification concerns the phrase 'type of cognitive change'. This is meant to cover, *inter alia*, the process of *adopting* a new theory about some previously untheorized phenomenon, and the process of *replacing* one theory with another.[1] Different accounts of scientific progress disagree on exactly what types of adoption/replacement are required for scientific progress, such as whether the adoption of a new theory requires coming to *know* the contents of the new theory. Here and in what follows, the term 'theory' should be interpreted very broadly, so as to count any type of scientific representation – including hypotheses, models, and natural laws – as 'theories'.

A second clarification concerns the term 'constitutes'. It is a truism that there are many ways for science to make progress. For example, there is surely a sense in which scientists make progress as they collect more evidence for their theories or develop new formalisms in which to couch their theories. These are both forms of scientific progress... in a sense. But I say that we should distinguish them from the type of progress that occurred, for example, when J. J. Thomson's plum pudding model of the atom was replaced with Ernest Rutherford's nuclear model. The difference is that the former type of progress (e.g., collecting evidence) is progress *because and in so far as* it helps to achieve the latter type (e.g., improving atomic models). By contrast, the latter type of progress counts as progress regardless of whether it leads to some other instances of progress. To mark this distinction, I say (following Bird 2008, 280) that some improvements *promote progress*, whereas others *constitute progress*. It is the latter that accounts of scientific progress are accounts of.

A final clarification. In the question above, I use the term 'improvement' instead of 'progress' to emphasize two related points about our topic. The first is that accounts of scientific progress are not attempts to analyze the *term* 'scientific progress' as it is used in either common parlance or scientific practice. Even if there was no such term in our languages, there would remain the question of what types of cognitive changes we should count as imrpoving on what came before. So, the underlying philosophical question is not about language, and certainly not about the word 'progress'. A second reason to use 'improvement' rather that 'progress' is to emphasize that accounts of scientific progress are unmistakably *normative* – not mere descriptions of scientific practice or the history of science, but also prescriptive claims about how science ought to proceed (cf. Niiniluoto 2019, §2.2). In evaluating accounts of scientific progress, we thus need to think long and hard about whether the normative implications of each account are, all things considered, desirable.

Now, there are many accounts of scientific progress – many accounts, that is, of what type of cognitive change constitutes a scientific improvement on what came before. But four such accounts have been most prominent in the recent literature (see Dellsén 2018b; Niiniluoto 2019). *The truthlikeness account* holds that scientific progress consists in increasing the truthlikeness (i.e, the verisimilitude) of accepted theories (Niiniluoto 1984, 2014; Popper 1963). *The functional account* holds that progress consists in decreasing the number and/or importance of unsolved problems, by solving or eliminating existing problems without generating new ones (Kuhn 1970; Laudan 1977). *The epistemic account* holds that progress consists in accumulating knowledge in science, that is, in adding to the stock of scientific knowledge (Bird 2007, 2016). Finally, *the noetic account* holds that scientific progress consists in increasing our capacity to understand natural phenomena (Dellsén 2016, 2021). For each of these accounts, there is an achievement that lies at its heart (respectively: truthlikeness, problem-solving, knowledge, understanding). It is possible, of course, to develop each of these accounts in different ways depending on precisely how one defines each type of achievement.

In this chapter, I will focus on a feature of the epistemic account that distinguishes it from most versions of the other three accounts. According to epistemological orthodoxy – endorsed by proponents of the epistemic account (e.g., Bird 2007; Park 2017) – a known proposition must have a certain kind and degree of normative support, called *epistemic justification*. Roughly, epistemic justification (or simply *justification*) is what must be added to a true belief in order for it to constitute knowledge.[2] So, roughly speaking, the epistemic account implies that adopting or replacing a theory cannot count as a scientifically progressive unless the new theory is epistemically justified; otherwise, the new theory would fail to count as knowledge.[3] Call this *the justification requirement*.

Is epistemic justification also a requirement for progress on other accounts? Some have argued or assumed that understanding should be taken to entail epistemic justification or even knowledge (e.g., Khalifa 2017; Sliwa 2015). In my view, this is a mistake (Dellsén 2017, 2018a; see also Hills 2016; Lawler 2016). Hence, in contrast to Bangu (2015), I take it that the most plausible version of an understanding-based account does *not* require justification for progress. Similarly, although proponents of the truthlikeness account sometimes write as if justification is essential to science as a whole and therefore to scientific progress (e.g., Niiniluoto 2014, 76; 2017, 3299–3300), this does not follow from the official statements of their accounts. Finally, the functional account clearly does not require progressive theories to be epistemically justified either. Indeed, the functional account distinguishes itself from all other accounts in not even requiring factivity for scientific progress – neither the problems nor their solutions must be grounded in reality in order for the 'solutions' to such 'problems' to count as progressive (see, e.g., Laudan 1977, 16).

It is important to understand that accounts that do not *require* justification for progress may still find an important role for what we might call *justificatory activities*, such as making observations, gathering data, presenting arguments, debating the plausibility of theories, and accepting or rejecting theories based on whether they meet certain epistemic standards. By the lights of any factive account of scientific progress, such as the truthlikeness account or the noetic account, justificatory activities will be integral to the progress of science – not because they are *constitutive* of progress, as per the epistemic account, but because they *promote* progress. Indeed, justificatory activities promote progress to an extent that is hard to exaggerate. In their absence, scientific progress would be a matter of pure guesswork, and should occur only in those rare instances in which we happen to chance upon a correct theory. Because of justificatory activities, however, science is one of the most successful enterprises in the history of humankind.

So, the point of disagreement between the epistemic account and other factive accounts (e.g., the noetic and truthlikeness accounts) is not about *how important justification is* to progress. Proponents of these accounts can all agree that justificatory activities are of the greatest importance to scientific progress. Rather, the difference concerns *how justification is important* to progress. On the noetic and truthlikeness accounts, its importance is instrumental: justification is an important means to making scientific progress. On the epistemic account, by contrast, justification is constitutive of progress: it is a necessary condition for knowledge, the accumulation of which is identified with progress.

What type of arguments might be provided for or against the justification requirement on progress? Thus far, the most prominent arguments in the literature have tended to appeal to *intuitions* in a very direct way. In particular, Bird's influential argument for favoring his epistemic account over the truthlikeness account is that, in certain hypothetical cases, requiring justification for progress 'accords with the verdict of intuition', whereas not doing so 'conflicts with what we are intuitively inclined to say' (Bird 2007, 66).[4] However, some philosophers have contested Bird's intuitions about these cases and/or presented cases in which, it is claimed, our intuitions point in the other direction (Dellsén 2016; Rowbottom 2008). Moreover, empirical investigations into the folk concept of progress have at best delivered ambiguous results (Mizrahi and Buckwalter 2014; Rowbottom 2015, 103).

In my view, however, the most important weakness in Bird's argument is its direct appeal to our intuitive inclinations about specific hypothetical cases as grounds for rejecting or accepting accounts of scientific progress. In light of the normative nature of the question of scientific progress, it is more appropriate to consult our reflective judgments regarding robust aspects of scientific practice and methodology. This is the approach I take in the following three sections. I will argue that there are important categories of cases in which the justification requirement (and thus the epistemic account) delivers verdicts about scientific progress that are not just counterintuitive in hypothetical cases, but also go against well-founded truisms about scientific practice, scientific methodology, and the success of science.

3 Progress and Turnover

To introduce the first problem for the justification requirement on scientific progress, consider first an argument that will be familiar from debates about scientific realism, *the pessimistic meta-induction* (see, e.g., Hesse 1976; Laudan 1981; Poincaré 1952/1905). The simplest version of this argument infers by enumerative induction from the empirical premise that most successful theories that were accepted in the past have turned out to be false, to the conclusion that currently accepted theories will probably turn out to be false as well. For our purposes, it is worth noting that if this argument were successful, it would not so much establish that current theories are in fact false (since they *might* yet be true by pure chance, for example); rather, it would establish that we are *not justified in believing* that they are true. Current theories might still *be* true, but they would not be *known* to be true.

There are various convincing responses to the pessimistic meta-induction, both to this simple form of the argument and to more sophisticated variants (e.g., Fahrbach 2017; Roush 2010). However, there is a key thought behind the pessimistic meta-induction that remains unchallenged (and rightly so): *in principle*, a sufficiently dismal track record regarding our past efforts to theorize about some phenomenon might undermine one's justification for believing current theories. If one knows that scientists in some discipline have in the past consistently produced and accepted radically false theories about some topic, and that one's epistemic situation has not changed significantly for the better since this last occurred, then surely this would undermine one's justification for believing the discipline's current theories on the topic. After all, the disciplines' poor track record would serve as a kind of higher-order evidence that the first-order evidence in favor of the theory is misleading or insufficiently probative. The upshot is that even a theory for which scientists have produced plenty of (first-order) scientific evidence might fail to be justified – either at all, or to the extent required for knowledge.

What implications does this have for the justification requirement on scientific progress? Consider a discipline whose track record regarding some specific phenomenon is sufficiently poor. That is, the discipline has produced and accepted so many theories about this phenomenon that have turned out to be radically false by our current lights that its current theories about the phenomenon would fail to be epistemically justified, even in cases where the first-order evidence for its theories would otherwise make them justified. Thus, by virtue of its track record alone, the discipline would be unable at present to accumulate knowledge by adopting new theories about this phenomenon, for the justification for any such theory would immediately be undermined by the discipline's poor track record. Given the justification requirement on scientific progress, the new theory would therefore contribute nothing to scientific progress – no matter how much more accurate the new theory is, no matter how much understanding it would facilitate, and no matter how many problems it would solve (and so on for other potential necessary conditions for scientific progress).

This implication is more than just counterintuitive. It means that scientists who are seeking to maximize their contribution to scientific progress should look backward to the history of their discipline before deciding whether to work on a given project. Specifically, they should try to ascertain whether their discipline's track record would undermine justification for future theories, including the yet-to-be-discovered theories they themselves hope to contribute to their field. These scientists should avoid studying phenomena on which past theorizing has been radically mistaken, since there would be no hope of making progress on such topics according to the justification requirement. The methodological implication of the justification requirement in these cases is thus, to put it bluntly, to avoid research on topics that would require groundbreaking research – research that goes against all previous theories about the phenomenon in question – and instead to focus on researching phenomena for which our past and current theories are already believed to be on the right track.

It is here that the normative nature of the concept of scientific progress comes to the fore. As I emphasized in the previous section, we should choose between different accounts of scientific progress based not on whether the theories or their implications accord with 'what we are intuitively inclined to say'; rather, we should consider whether such accounts agree with our reflective judgments about scientific practice and methodology. In this respect, an account of scientific progress is no different than a theory in normative ethics, which should be accepted or rejected based on whether we are prepared, on reflection, to accept its normative implications. So, what is our reflective judgment about whether our account of scientific progress ought to imply that it would be impossible for scientists to make progress regarding phenomena on which the relevant discipline's track record is sufficiently poor? Or, equivalently for our purposes, what is our reflective judgment about whether progress-seeking scientists should avoid groundbreaking research of the type described above?

In my view, progress-seeking scientists should not avoid this type of research, and, consequently, we should conceive of scientific progress in a way that does not require justification. On the contrary, it is precisely on the topics where we believe previous theories to be radically mistaken that we need scientists to do more research in order to replace previous theories. Indeed, it is worth noting that various major science funding agencies, such as the National Science Foundation (NSF) and the European Research Council (ERC), have recently adopted policies that are meant to steer scientists toward 'transformative' research, which is likely to disrupt existing scientific paradigms, and away from 'safe' research that merely builds upon previous theories and results (see Stanford 2019). The justification requirement on scientific progress implies, implausibly, that these policies are likely to hinder scientific progress, because the 'transformative' theories that scientists are being encouraged to develop would imply that previous theories are radically mistaken, which would in turn undermine, via the route sketched above, the justification for the newly developed theories.

4 Progress and Unification

I turn now to a second problem for the justification requirement on progress, which has to do with the subsumption of several previous theories under a single 'unified' theory.

The problem I have in mind is closely analogous to the so-called *preface paradox* (Makinson 1965; see also Christensen 2004). Suppose I wrote a long book containing a number of distinct factual claims $P_1,...,P_n$. I have fact-checked each claim, so I am justified in believing each P_i. But am I justified in believing the conjunction of these claims, $(P_1 \& ... \& P_n)$? Surely not, since that amounts to being justified in believing that my book is completely error-free – an amazing feat, which I have absolutely no reason to think I am fortunate enough to have accomplished. Indeed, it seems that I would be justified in believing that $(P_1 \& ... \& P_n)$ is false, and that I might accordingly say or imply as much in the preface to my book ('the errors herein are all mine'). Several lessons have been drawn from this type of case. One of the least controversial such lesson is that justification is not closed under conjunction: one can be justified in believing a number of claims and yet fail to be justified in believing their conjunction (even when one is also justified in taking the latter to follow logically from the former).

This failure of justification to be closed under conjunction can be elegantly explained on the assumption that justification requires some minimum level of probability (where 'probability' may be given any standard interpretation, e.g., in terms of rational degrees of confidence). For any two propositions, P_1 and P_2, neither of which entails the other, it is a theorem of the probability axioms that the probability of $(P_1 \& P_2)$ is lower than the probability of each of P_1 and P_2 separately (and *a fortiori* for longer conjunctions of non-entailing propositions). It follows that conjoining non-entailing propositions will, sooner or later, result in a conjunction whose probability is as low as you like.[5] So if being justified in believing P requires that the probability of P exceed some threshold $t > 0$ (e.g., 90%), then a conjunction of justified claims may itself be unjustified. In other words, justification would not be closed under conjunction. Indeed, it is not hard to see that failures of justification closure are rather commonplace in this probabilistic framework, even for maximally short conjunctions (and *a fortiori* for longer conjunctions). For example, if we set the probability threshold for justification at $t = 90\%$, and assume P_1 and P_2 to be probabilistically independent, then justification fails to be closed under their conjunction even when P_1 and P_2 each have 94% probability.

So much for the preface paradox. Now consider the scientific practice of subsuming theories about the same phenomenon under a 'unified' theory that entails each of the subsumed theories. The example with which I will operate here concerns various gas laws that were proposed and accepted in the seventeenth, eighteenth, and nineteenth centuries, relating two or more quantities of a gas:

- *Boyle's Law*, discovered in 1662, holds that for a given gas sample at a fixed temperature, pressure (P) is inversely proportional to volume (V): $P \propto 1/V$.
- *Charles's Law*, discovered around 1780, holds that in a given gas sample with fixed pressure, volume is proportional to temperature (T): $V \propto T$.
- *Avogadros's Law*, discovered in 1811, states that for fixed temperature and pressure, volume is proportional to the amount of gas molecules in the sample, measured in moles (n): $V \propto n$.
- *The Ideal Gas Law*, first formulated in 1834, subsumes all of these laws, and much else besides, under one equation: $PV = nRT$ (where R is the universal gas constant).

Now, it is not hard to see that the Ideal Gas Law entails each of Boyle's Law, Charles's Law, and Avogadros's Law, and that none of the latter (or, indeed, their conjunction) entails the former. Thus, it is plausible that at some point in history, such as just after the Ideal Gas Law was proposed in 1834, these theories constituted a case in which justification fails to be closed under conjunction. After all, each of the three sub-sumed laws (i.e., Boyle's, Charles's, and Avogadros's), were presumably justified at some point when the Ideal Gas Law failed to be so, for the simple reason that there are many more ways for the Ideal Gas Law to be false than for each of Boyle's Law, Charles's Law, and Avogadros's Law to be false. Put in terms of probability, the probability of the Ideal Gas Law is necessarily a great deal lower than the probability of each of the three subsumed laws.

Admittedly, whether this particular case counts as one in which jus-tification closure fails will depend on historical details of the case that I would not presume to know for certain. Was the Ideal Gas Law already justified in 1834? And were each of Boyle's Law, Charles's Law, and Avo-gadros's Law definitely justified at that time? If the answer to the first question is 'yes', or if the answer to the second is 'no', then this case does not exemplify the failure of justification closure. In that case, we would have to look elsewhere for a historical example to illustrate the point, or imagine a nearby possible world in which Boyle's, Charles's, and Avoga-dros's Laws were justified at some point when the Ideal Gas Law failed to be so. No matter. The general, philosophical point here is that we can surely find some case in which previous theories $T_1,...,T_n$, which are individually justified, are subsumed under a 'unified' theory T_U, which fails to be justified due to its having a (much) lower probability than any of the previous theories $T_1,...,T_n$.

The problem with all of this for the justification requirement is that, in such a case, the step from accepting each one of the subsumed theories, $T_1,...,T_n$, to (also) accepting the subsuming theory, T_U, would not count as progressive – even when all the theories involved would otherwise qualify for making progress. For example, according to the justification requirement, the formulation and acceptance of the Ideal Gas Law might very well (depending on the details mentioned above) not have been pro-gressive when it was formulated in 1834, even if the discovery of each of Boyle's Law, Charles's Law, and Avogadros's Law constituted progress. This implication is hard to swallow, since the discovery of the Ideal Gas Law seems paradigmatic of scientific progress. Indeed, it seems to be pre-cisely because the Ideal Gas Law subsumes several previously accepted gas laws that it contributes so much to progress. More generally, the justification requirement entails, implausibly, that progress-seeking sci-entists should avoid subsuming previous theories under a unified theory in cases where justification is not closed under conjunction. That cannot be right.

5 Progress and Disagreement

A third and final problem with the idea that scientific progress requires epistemic justification concerns the relationship between progress and disagreement in science.

In recent epistemology, there has been much interest in *peer disagreement*. Two or more agents count as 'peers' in the relevant sense just in case they are (roughly) equally well informed with respect to a given proposition, and (roughly) equally competent reasoners with respect to that proposition. A much-discussed question has been how (if at all) one should modify one's beliefs upon encountering disagreement from someone one recognizes as one's epistemic peer. A bewildering number of views on this issue have been defended, most of which entail that, in one way or another, the parties to a peer disagreement should 'conciliate', that is, move closer to the opinion(s) of those with whom they disagree. The 'should' here is *epistemic*: it concerns how an agent would have to modify their beliefs in order to ensure that they remain epistemically justified after becoming aware of the disagreement. So, most of the views on offer hold that a revealed peer disagreement undermines the relevant agents' justification for their initial beliefs.

Although much of the discussion about peer disagreement appeals to idealized cases, many of the lessons drawn from them apply to real disagreements in science.[6] After all, most scientists working in the same field have access to roughly the same evidence (because they share their results in a systematic way), and they are roughly equally competent in reasoning from that evidence (because they have been selected based on similar competencies, and because they have received a similar type of training). So, although not all scientists within a field will be 'peers' in the strictest sense of the term, there is reason to think that what is true of disagreement among peers will largely be applicable to disagreement among scientists within the same field. And although the parties to scientific disagreements tend to be groups of multiple scientists as opposed to individual agents, this also is no obstacle to learning from the peer disagreement debate, at least not when the groups of scientists are of comparable sizes (as in a 47/53% split, for example).

Now consider a garden-variety scientific disagreement that arises and evolves over time. Initially, a theory T_1 is overwhelmingly accepted by the scientists working in the relevant field. Then, at some later time, a rival theory T_2 is proposed, where T_2 is superior to T_1 in whatever way is required for progress (e.g., by being more truthlike). Slowly but steadily, T_2 wins over the proponents of T_1; moreover, new generations of scientists tend to favor T_2 over T_1, so that T_2 grows in popularity 'one funeral at a time'. Eventually, T_2 becomes overwhelmingly accepted, just as T_1 had been previously. But this does not happen overnight; it's a gradual process. So there is a period of time, Δt, during which proponents of T_1 and T_2 are of comparable sizes. During Δt, those on either side of the T_1/T_2 divide can look over to the other side and see that the number of fellow scientists who disagree with them is about as large as the number of those who agree.

Now consider whether the proponents of either theory would be justified in believing their preferred theory during Δt. Strictly speaking, it is impossible to give a definitive answer without further information about the case. But one thing that can be said definitively is that, according to most views of peer disagreement (i.e., any conciliatory view), one of the factors that is relevant to determining whether these beliefs are justified is that there is so much disagreement within their field. Specifically, their awareness of this widespread disagreement on T_1/T_2 would undermine their justification for believing whichever of T_1 and T_2 they actually believe.

In order to home in on how this affects scientific progress, let us consider a variation of the cases described above in which each scientist's first-order evidence for the theory they prefer is just barely sufficient during Δt to make them epistemically justified (i.e., justified to the extent required for knowledge) in believing that theory. In such cases, the justification-undermining effect of the disagreement would prevent scientists from being epistemically justified in believing either of T_1 and T_2 during Δt. Of course, the T_1/T_2-disagreement's potency to block epistemic justification in this way would disappear once the disagreement abates, that is, once T_2 starts becoming significantly more popular than T_1. But for some period of time when the disagreement is sufficiently evenly split (i.e., during Δt), the relevant scientists would not be epistemically justified in their beliefs about T_1 and T_2.

Now consider what this means for scientific progress. If justification is required for progress, the disagreement during Δt would not just prevent scientists from being justified in their beliefs regarding T_1 and T_2; it would also prevent their beliefs regarding T_1 and T_2 from contributing toward scientific progress in the way they otherwise would (e.g., through constituting accumulated knowledge). Thus, although we might have thought that the gradual replacement of T_1 by T_2 before, during, and after Δt was simply a case in which there was gradual progress as an increasing number of scientists came to accept a superior theory,[7] the justification requirement implies that the disagreement during Δt blocks any of the progress with respect to T_1 or T_2 that would otherwise occur during that period. Moreover, given that scientists' beliefs regarding T_1 were (barely) epistemically justified before Δt, the justification requirement implies that there is a sudden drop in progress at the beginning of Δt. It is as if disagreement casts a paralyzing spell that temporarily

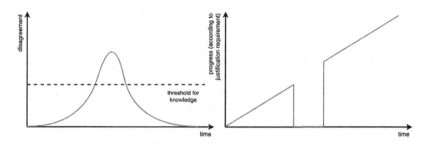

Figure 29.1 A simple illustration of how disagreement would temporarily block progress according to the justification requirement.

destroys previous progress on the topic and blocks any further progress during the period. This spell is then lifted when the disagreement has abated sufficiently so as to no longer undermine justification as it once did (see Figure 29.1 for a simple illustration).[8]

To see just how absurd these implications of the justification requirement are, it may be helpful to apply them to a historical case. In the eighteenth and early nineteenth centuries, the dominant theory of the nature of light was Isaac Newton's 'corpuscular' theory, according to which light consists of tiny particles emitted from light sources. Newton's theory went mostly unchallenged until Augustin-Jean Fresnel's formulation of his transverse wave theory in 1815–1818, according to which light consists of waves that oscillate in a direction perpendicular to its movement. Fresnel's theory enjoyed considerable empirical success shortly after it was formulated. For example, the theory accurately predicted that a bright spot would appear at the center of the shadow formed by shining light from a single source on an opaque disk. In the early 1820s, this fact (together with various other observations) had already convinced some scientists, especially Fresnel's French compatriots, that his theory was correct. At the same time, many other scientists, especially Newton's British compatriots, remained skeptical of Fresnel's theory and loyal to Newton's. During the latter half of the 1820s, however, Fresnel's theory grew steadily in popularity, even in Britain, and by the 1830s there were few supporters of Newton's theory left on the scene.

For the sake of the argument, let us suppose that Fresnel's theory is superior to Newton's in whatever way is required for a replacement of the latter by the former to constitute scientific progress. Suppose also that, at some point in the 1820s, the scientific community of optical physicists was sufficiently evenly split so that none, or at least relatively few, of them would have been epistemically justified in believing either theory according to most views of peer disagreement.[9] During this period, then, the justification requirement evidently implies that there was no, or at least little, scientific progress with respect to theories of light. Progress on the topic would have been blocked by the spell of widespread disagreement about which theory is correct. (Although once Fresnel's theory became sufficiently popular in comparison to Newton's – in the late 1820s or early 1830s, say – this would have lifted the spell so that scientific progress could again go back to normal.)

Of course, a proponent of the justification requirement could respond by rejecting any view in the epistemology of disagreement on which peer disagreement has the capacity to undermine epistemic justification. There are two related issues with this response. First, it ties the plausibility of the justification requirement (and thus any account of scientific progress that incorporates it) very tightly to particular, minority views in the epistemology of disagreement. Although many views reject the idea that one loses all justification upon discovering that one is in a peer disagreement, there are much fewer views on which awareness of peer disagreement does not to any extent undermine one's justification for one's initial beliefs in a way that could prevent the belief from being epistemically justified. Second, the few views that have been proposed on which disagreement does not undermine justification at all are arguably quite underdeveloped.[10] Indeed, it is noteworthy that one early proponent of such a view (Kelly 2005) later argued for a view on which some conciliation is often called for in cases of peer disagreement (Kelly 2010).

Better then, I say, to simply reject the justification requirement on scientific progress. Having done so, disagreement presents no obstacle to scientific progress. In a case where T_1 gradually loses out to T_2, there will come a time at which neither the proponents of T_1, nor the proponents of T_2, are justified in believing their preferred theory to be true, due to the widespread disagreement between them. Although this puts working scientists in an epistemological conundrum regarding which theory to believe and pursue, it is no hindrance to scientific progress. If the replacement of T_1 by T_2 is otherwise progressive, for example, by virtue of increasing truthlikeness or conveying more understanding, then this is simply a case of gradual scientific progress. No spell is cast, no progress blocked.

6 Conclusion

The justification requirement on scientific progress holds that in order for the adoption or replacement of a theory to qualify as progressive, scientists must be epistemically justified in believing the new theory. I have argued against this requirement by considering various commonplace scientific situations in which justification is undermined or absent. To deny that there is scientific progress in these situations is to condemn much of ordinary scientific practice as non-progressive.

Notes

1 It is also meant to cover other types of cognitive changes, such as discarding a (mistaken) theory, discovering a new phenomenon, and developing a new explanation for a familiar phenomenon using an extant theory. But for current purposes, we can set such cases aside and focus on cases of adopting/replacing theories.

2 Depending on one's theory of justification, one may also take it to be necessary to add a special condition that rules out Gettier cases. We will not be concerned with Gettier cases here, however, so I will ignore this complication in what follows.

3 I say 'roughly speaking' because a proponent of the epistemic account might argue that the new theory could still count as progressive if it merely *implies* a (previously unknown) proposition that becomes known. In that case, only the implied proposition needs to be epistemically justified. This maneuver arguably has problematic consequences for the epistemic account – for example, it implies that the introduction of any number of falsehoods still counts as progressive provided that a single unknown proposition becomes known in the process. For simplicity's sake, however, I do not take issue with it in what follows.

4 In one of Bird's cases, we are asked to consider René Blondlot's claim to have observed 'N-rays', a type of radiation supposedly similar to X-rays. N-rays do not exist, so Blondlot must have somehow hallucinated or fabricated his experimental results. It follows that Blondlot's claims about N-rays are unjustified. Now suppose N-rays did exist, so that Blondlot's claims were true by pure luck. In that case, says Bird, it is still intuitively wrong to say that Blondlot's discovery constituted progress (Bird 2007, 67).

5 Well, almost: it will never get to 0 unless one of the conjuncts is a contradiction or contradicts the rest of the conjunction.

6 A common complaint about the peer disagreement debate is that it deals only with a very idealized form of disagreement, so that little if anything can be learned from it about more realistic cases. This complaint is largely based on a confusion. No one is suggesting that we should simply extrapolate from idealized to realistic cases, as in some sort of enumerative induction from a single case. Rather, the point of focusing the debate on idealized cases is to home in on the factors that might be relevant to whether, and how, a disagreement undermines epistemic states such as justified belief. Once we have discovered what those factors might be in a controlled situation in which others factors have deliberately been eliminated, we can locate those factors in more realistic cases, where several other factors will be at play as well.

7 I am assuming for simplicity that T_2 is superior to T_1 in other respects (e.g., by being more truthlike). If, by contrast, T_1 is superior to T_2, then this would be a case of scientific regress, but the argument in the main text would apply *mutatis mutandis*.

8 Note that the issue here is not the epistemic one that we would not *know* – or be *justified* in our beliefs about – whether we are making progress on the relevant phenomenon during Δt. Rather, the issue is that, according to the justification requirement, there would not *be* any progress on the relevant phenomenon during Δt.

9 Some of these scientists might have had relevant evidence that most other scientists lacked and/or been more competent than their average colleague, in which case they might not have been sufficiently close to being peers for their justification to be undermined by the disagreement.

10 For example, Hawthorne and Srinivasan (2013) are often mentioned as advocating a view on which conciliation is not called for in peer disagreements. But Hawthorne and Srinivasan do not really defend such a view as much as they simply assume it without argument in order to explore some of its consequences.

References

Bangu, Sorin. 2015. "Progress, Understanding, and Unification." In *Romanian Studies in Philosophy of Science*, ed. Ilie Pârvu, Gabriel Sandu, and Iulian D. Toader, 239–253. Cham: Springer.

Bird, Alexander. 2007. "What Is Scientific Progress?" *Nous* 41: 64–89. https://doi.org/10.1111/j.1468-0068.2007.00638.x.

Bird, Alexander. 2008. "Scientific Progress as Accumulation of Knowledge: A Reply to Rowbottom." *Studies in History and Philosophy of Science* 39: 279–281. https://doi.org/10.1016/j.shpsa.2008.03.019

Bird, Alexander. 2016. "Scientific Progress." In *The Oxford Handbook in Philosophy of Science*, ed. Paul Humphreys, 544–565. Oxford: Oxford University Press. https://doi.org/10.1093/oxfordhb/9780199368815.013.29.

Cevolani, Gustavo, and Luca Tambolo. 2013. "Progress as Approximation to the Truth: A Defence of the Verisimilitudinarian Approach." *Erkenntnis* 78: 921–935. https://doi.org/10.1007/s10670-012-9362-y.

Christensen, David. 2004. *Putting Logic in Its Place: Formal Constraints on Rational Belief*. Oxford: Oxford University Press. https://doi.org/10.1093/0199263256.001.0001.

Dellsén, Finnur. 2016. "Scientific Progress: Knowledge versus Understanding." *Studies in History and Philosophy of Science* 56: 72–83. https://doi.org/10.1016/j.shpsa.2016.01.003.

Dellsén, Finnur. 2017. "Understanding Without Justification or Belief." *Ratio* 30: 239–254. https://doi.org/10.1111/rati.12134.

Dellsén, Finnur. 2018a. "Deductive Cogency, Understanding, and Acceptance." *Synthese* 195: 3121–3141. https://doi.org/10.1007/s11229-017-1365-4.

Dellsén, Finnur. 2018b. "Scientific Progress: Four Accounts." *Philosophy Compass* 13: e12525. https://doi.org/10.1111/phc3.12525.

Dellsén, Finnur. 2021. "Understanding Scientific Progress: The Noetic Account." *Synthese* 199: 11249-11278.

Fahrbach, Ludwig. 2017. "Scientific Revolutions and the Explosion of Scientific Evidence." *Synthese* 193: 5039–5072. https://doi.org/10.1007/s11229-016-1193-y.

Hawthorne, John and Amia Srinivasan. 2013. "Disagreement Without Transparency: Some Bleak Thoughts." In *The Epistemology of Disagreement: New Essays*, eds. David Christensen and Jennifer Lackey, 9–30. New York: Oxford University Press. https://doi.org/10.1093/acprof:oso/9780199698370.003.0002.

Hesse, Mary. 1976. "Truth and the Growth of Scientific Knowledge." *PSA: Proceedings of the Biennial Meeting of the Philosophy of Science Association*, vol. 2, 261–280. Chicago: University of Chicago Press. https://doi.org/10.1086/psaprocbienmeetp.1976.2.192385.

Hills, Alison. 2016. "Understanding Why." *Nous* 50: 661–688. https://doi.org/10.1111/nous.12092.

Kelly, Thomas. 2005. "The Epistemic Significance of Disagreement." In *Oxford Studies in Epistemology*, vol. 1, eds. Tamar Gendler and John Hawthorne, 167–196. Oxford: Oxford University Press.

Kelly, Thomas. 2010. "Peer Disagreement and Higher Order Evidence." In *Disagreement*, eds. Richard Feldman and Ted Warfield, 111–174. New York: Oxford University Press. https://doi.org/10.1093/acprof:oso/9780199226078. 003.0007.

Khalifa, Kareem. 2017. *Understanding, Explanation, and Scientific Knowledge*. Cambridge: Cambridge University Press. https://doi.org/10.1017/9781108164276.

Kuhn, Thomas S. 1970. *The Structure of Scientific Revolutions*, 2nd edition. Chicago: University of Chicago Press.

Laudan, Larry. 1977. *Progress and Its Problems: Towards a Theory of Scientific Growth*. London: Routledge and Kegan Paul.

Laudan, Larry. 1981. "A Confutation of Convergent Realism." *Philosophy of Science* 48: 19–49. https://doi.org/10.1086/288975.

Lawler, Insa. 2016. "Reductionism about Understanding Why." *Proceedings of the Aristotelian Society* 116: 229–236. https://doi.org/10.1093/arisoc/aow007.

Makinson, David C. 1965. "The Paradox of the Preface." *Analysis* 25: 205–207. https://doi.org/10.1093/analys/25.6.205.

Mizrahi, Moti and Wesley Buckwalter. 2014. "The Role of Justification in the Ordinary Concept of Scientific Progress." *Journal for General Philosophy of Science* 45: 151–166. https://doi.org/10.1007/s10838-014-9243-y.

Needham, Paul. 2020. *Getting to Know The World Scientifically: An Objective View*. Cham: Springer. https://doi.org/10.1007/978-3-030-40216-7.

Niiniluoto, Ilka. 1984. *Is Science Progressive?* Dordrecht: D. Reidel. https://doi.org/10.1007/978-94-017-1978-0.

Niiniluoto, Ilka. 2014. "Scientific Progress as Increasing Verisimilitude." *Studies in History and Philosophy of Science* 46: 72–77. https://doi.org/10.1016/j.shpsa.2014.02.002.

Niiniluoto, Ilka. 2017. "Optimistic Realism about Scientific Progress." *Synthese* 194: 3291–3309. https://doi.org/10.1007/s11229-015-0974-z.

Niiniluoto, Ilka. 2019. "Scientific Progress." In *Stanford Encyclopedia of Philosophy* (Winter 2019 edition), ed. Edward N. Zalta. https://plato.stanford.edu/archives/win2019/entries/scientific-progress/.

Park, Seungbae. 2017. "Does Scientific Progress Consist in Increasing Knowledge or Understanding?" *Journal for General Philosophy of Science* 48: 569–579. https://doi.org/10.1007/s10838-017-9363-2.

Poincaré, Henri. 1905/1952. *Science and Hypothesis*. Repr. New York: Dover.

Popper, Karl R. 1963. *Conjectures and Refutations: The Growth of Scientific Knowledge*. London: Hutchinson. https://doi.org/10.1063/1.3050617.

Roush, Sherrilyn. 2010. "Optimism about the Pessimistic Induction." In *New Waves in Philosophy of Science*, eds. P. D. Magnus and Jacob Busch, 29–58. New York: Palgrave MacMillan. https://doi.org/10.1007/978-0-230-29719-7_3.

Rowbottom, Darrell P. 2008. "N-rays and the Semantic View of Progress." *Studies in History and Philosophy of Science* 39: 277–278. https://doi.org/10.1016/j.shpsa.2008.03.010.

Rowbottom, Darrell P. 2015. "Scientific Progress Without Increasing Verisimilitude: In Response to Niiniluoto." *Studies in History and Philosophy of Science* 51: 100–104. https://doi.org/10.1016/j.shpsa.2015.01.003.

Sliwa, Paulina. 2015. "Understanding and Knowing." *Proceedings of the Aristotelian Society* 115: 57–74. https://doi.org/10.1111/j.1467-9264.2015.00384.x.

Stanford, P. Kyle. 2019. "Unconceived Alternatives and Conservatism in Science: The Impact of Professionalization, Peer-Review, and Big Science." *Synthese* 196: 3915–3932. https://doi.org/10.1007/s11229-015-0856-4.

30 The Significance of Justification for Progress

A Reply to Dellsén

C. D. McCoy

Dellsén claims that epistemic justification is not necessary for scientific progress. Each of his three arguments for this claim centers on a type of scientific scenario in which intuitive progress lacks apparent justification. However, Dellsén maintains that his conclusion does not rest on intuitive judgments on cases alone. Indeed, he argues that reliance on mere intuitive judgments of cases is insufficient to decide the issue. Thus, he supplies further argumentation in each of the three cases, which is intended to show that a justification condition on scientific progress should be rejected, for the reason that certain alleged normative consequences of this condition would countermand "truisms about scientific practice, scientific methodology, and the success of science" (this volume, 374). Dellsén maintains that this further argumentation involves the consultation of our "reflective judgments" (this volume, 375) rather than mere intuitive judgments, with the rough idea being that the former judgments are a surer basis for drawing conclusions in this context given the normative character of scientific progress.

In each of the scenarios that Dellsén identifies, however, these alleged normative consequences depend on the scenarios being cases of scientific progress that lack justification, and Dellsén's judgment that they are such is merely on an intuitive basis. Hence, intuitive judgments about cases do, in fact, remain the essential grounds of Dellsén's arguments.[1] The basic problem with relying on intuitive judgments of cases in this context is that they are inevitably indecisive, since whether one regards some case as a counter-example to the thesis (that justification is necessary for scientific progress) will necessarily depend on whether one judges that the case evinces progress, and that is something on which intuitions will vary. In every scenario proffered by Dellsén, it is interpretively open to identify the "moment of scientific progress" somewhat later in the episode than he does, in particular at the point when an adequate justification has been obtained (supposing that it is obtained). When the case is interpreted in line with this alternative judgment of progress, no counter-intuitive results need obtain, nor countermands of principles of scientific methodology.

DOI: 10.4324/9781003202905-34

Dellsén's first scenario is one where a research community investigating some phenomenon has had a "dismal track record" (this volume, 374) of theorizing about that phenomenon, so dismal that it undermines any justification the community might have for believing the current theory. By failing to meet the assumed-to-be-necessary justification requirement, the current theory would not contribute to scientific progress, even if it were a true one that would otherwise be justified by empirical evidence or the theoretical virtues it possesses. Dellsén finds this conclusion counter-intuitive, for it would seem that the otherwise justified and theoretically virtuous theory could represent a great deal of progress. What he finds more concerning, though, is the normative consequence of the justification failure. He claims that awareness of how justification can be undermined in scenarios like this would indicate to scientists that they should either avoid or abandon research programs with dismal track records, due to the risk that future justification may never be obtainable. This consequence he thinks cannot be accepted, since he holds that it is a truism that scientists should be motivated to pursue groundbreaking research in order to make progress.

If this consequence were genuine, then I would agree that it is not scientifically acceptable. However, it is not true that a research program's dismal track record precludes a theory from being justified, no matter how bad it is. Although I do agree with Dellsén that research track records are evidentially relevant (particularly as in Dawid's (2013) conception of meta-empirical theory assessment), it must be acknowledged that evidence of this kind is highly defeasible on its own, and, moreover, that we simply do not have a good grasp of how likely a dismal track record of research is given the truth and falsity of a particular theory in that research program. In any case, regardless of how bad a research program's track record is, empirical evidence obtained through observation or experiment that subsequently supports the current theory in that program will steadily overcome (often quickly, because of the power of empirical evidence) the previous dismal track record. Therefore, there is no serious risk of a successful theory failing to meet a reasonable standard of justification merely because of its poorly performing predecessors. Further ameliorating the long-term impact of the dismal track record would be the recognition of precisely why scientists were so off the mark previously and why they are now on the right track (this evidence being a defeater of the dismal track record evidence). Given the potential for overcoming a bad track record, the supporter of the justification requirement would be well within her rights to regard the achievement of progress as un-attained until such time as adequate justification is obtained. As for the contrary intuitions that the introduction of the theory itself represents progress, this, the supporter could reply, is simply a sign of potential progress yet unsecured.

Dellsén next envisions a scenario where a number of theories about some phenomenon are individually justified (but only barely). When a unifying theory that subsumes them is then developed, he supposes that this unifying theory will be unjustified, since it entails a conjunction of the (independent) predecessor theories, implying that its probability is less than or equal to the product of the component theories' probabilities, each of which is only barely justified. Dellsén points to the historical case of the unification of various gas laws into the ideal gas law as a case in point, saying that the implication that progress was not achieved in the discovery of the ideal gas law, due to its lack of justification, is "hard to swallow ... since [its discovery] seems paradigmatic of scientific progress" (this volume, 378). Building on his intuition that progress was achieved in the unification itself, Dellsén lays down the unacceptable further normative implication that the risk of a loss of justification by proposing unifying theories would lead scientists to avoid doing so (at least in cases involving the entailment of conjunctions of justified hypotheses).

Much like the previous case, there is nothing in the present scenario that bars later justification of the theory in question. As empirical evidence subsequently accrues for the unifying theory, its justification would improve until at some point it becomes sufficiently justified in the normal scientific way. One might then legitimately intuit this point as the point of progress rather than the earlier point of discovery. Intuitions of progress of the latter kind can, as before, be explained away as intuitions of the potential for progress rather than of the real thing. As for the normative consequence that he identifies, there is simply no reason for scientists to shy away from proposing unifying theories, since future justification of the theory remains possible if the evidence exists to ground it.

Finally, Dellsén envisions a scenario where there is community disagreement over whether two theories should be adopted. Citing recent literature on peer disagreement, which he says generally supposes that parties to disagreement should conciliate, he argues that at the height of the community disagreement, when the opposing scientists' empirical evidence just barely justifies them in belief in the respective theories, justification for either theory would be undermined by their awareness of the disagreement. He claims that this would suggest, counter-intuitively, that progress comes to a standstill for a period of time, until enough supporters of one theory switch allegiance to the other side, thereby diminishing the conciliating pressure of peer disagreement. Dellsén's opposing intuition is that the gradual swing in allegiance from one theory to the other represents a matter of continuous progress, with no progress-less period in the middle.

Let me grant, for the sake of argument, that community disagreement should have this normative consequence of conciliation for the individual reasoner, and also that there would be a period where the justification of either theory is lacking at the community level. This scenario constitutes a counter-example, then, only if there is clearly progress during that period. Unfortunately, I myself cannot conjure the intuition that progress is "continuous" during this period, as Dellsén has it. Is it merely the incremental changes in allegiance that are behind the intuition? If so, then it can only be a judgment with the benefit of hindsight, for we happen to know how the story ends. One who thinks that justification is essential to progress could simply reject the judgment, saying that the intuition of progress here is instead merely a matter of knowing which theory is the progressive one and measuring progress by popularity. Moreover, they could counter with the view that what actually makes the one theory progressive is that it is (eventually) justified, something that only happens at the community level after enough members have switched sides. Depending on what one takes the function from member beliefs to community beliefs to be, there may, in fact, be a period where progress comes to a standstill on this interpretation, but that may be just as it should.

Both sides can throw water into their own mills, but we will not see whose flour is better like this. How we comprehend scientific progress is ultimately a matter for interpretation, not the outcome of a clash of intuitions. For this reason, I do find Dellsén's interest in moving toward reflective judgments apt; what must be done, though, is removing the dependence of these judgments on unreliable intuitions. The best interpretation of scientific progress will then be the one that makes the best sense of scientific practice, in all its variegated modes.

Like Dellsén's other work (2016, 2021) on the topic, my contribution to this volume represents an attempt to make sense of scientific practice by characterizing scientific progress as a matter of understanding. I argue that understanding a phenomenon is the ability to use a theory to describe or explain it. The greater the ability, the greater the progress. One can ask Dellsén's question of my account: is justification necessary for progress? For both Dellsén and I, there is a clear sense in which the answer is "no," since it seems that one can, at least in principle, have an understanding without that understanding being justified. Yet, I would myself be quite unsatisfied with leaving the issue at that, for I cannot ignore the fact that I am not always in a position to know that I understand, just as I am not always in a position to know that I know some fact. In the situation where one knows not whether one understands, it seems essential to me that one should adopt a critical attitude, seek justification, and test one's understanding. In a way, the need to carry out these activities is already built in as an integral component of my view, based as it is on the critical activity of problem-solving. Solving problems is precisely the needed test of cognitive ability.

Dellsén is right, of course, to distinguish what "constitutes progress" (this volume, 371) from what is merely related to progress, and, to be sure, what remains ultimately valuable, according to me, is the progress-constituting understanding, not the knowledge thereof. However, in the further elaboration of this important distinction, he mentions that "justificatory activities" (this volume, 373) are integral to the progress of science only because they promote it. That they certainly do, especially when they show us that our understanding is lacking and motivate us to improve, but equally well they inform us whether an understanding has been achieved. This latter function brings out more clearly the essential critical function of justification vis-à-vis understanding. In my view, this role of justification is so intimately involved in the genesis of understanding that I would be quite reluctant to say that there is scientific progress without it. An analysis that does not acknowledge at least a place-marker for justification's genetic role will have as a likely consequence, I expect, an unfortunate surfeit of theory, dislocated from scientific practice.

Acknowledgments

I would like to thank Insa Lawler for comments that helped to improve this chapter and Vera Matarese for checking over the original manuscript.

Note

1 Dellsén himself appears to acknowledge that his argument involves both intuitive judgments and reflective judgments (2016, 7).

References

Dawid, Richard. 2013. *String Theory and the Scientific Method.* Cambridge: Cambridge University Press. doi: 10.1017/CBO9781139342513.

Dellsén, Finnur. 2016. "Scientific Progress: Knowledge Versus Understanding." *Studies in History and Philosophy of Science Part A* 56:72–83. doi: 10.1016/j.shpsa.2016.01.003.

Dellsén, Finnur. 2021. "Understanding Scientific Progress: The Noetic Account." *Synthese* 199:11249–11278. doi: 10.1007/s11229-021-03289-z.

Dellsén, Finnur. this volume. "Scientific Progress Without Justification".

31 Scientific Progress without Problems

A Reply to McCoy

Finnur Dellsén

1 Introduction

In the course of developing an account of scientific progress, C. D. McCoy appeals centrally to *understanding* as well as to *problem-solving*. On the face of it, McCoy's account could thus be described as a kind of hybrid of the understanding-based account that I favor (Dellsén 2016, 2021) and the functional (a.k.a. problem-solving) account developed most prominently by Laudan (1977; see also Kuhn 1970; Shan 2019). In this commentary, I offer two possible interpretations of McCoy's account and explain why I do not find it compelling on either interpretation.

2 Problem-Solving as Promoting Progress?

Like other understanding-based accounts, such as the noetic account (Dellsén 2016, 2021), McCoy identifies scientific progress with an 'increase in understanding' (this volume, 363). What is distinctive about McCoy's account as compared with other understanding-based accounts, however, is what he says about the type of understanding that increases as science progresses:

> [...] progress in understanding simply involves improvement in theoretical understanding, which is a matter of improved ability to explain and describe phenomena, the degree of which is something evaluable by a problem-solving standard.
>
> (this volume, 362)

In this passage, McCoy appears to be appealing to elements from the functional (a.k.a. problem-solving) account of scientific progress in characterizing increases in scientific understanding. In particular, McCoy says that degree of understanding is 'evaluable by a problem-solving standard'. But what exactly does this mean?

DOI: 10.4324/9781003202905-35

Earlier in the paper, McCoy suggests that the primary function of problem-solving is twofold, depending on the stage of development the scientific discipline in question finds itself in. In 'normal science', that is, as scientists are working within a Kuhnian paradigm, the main point of problem-solving is 'to evaluate the degree of understanding so far achieved in the current, established paradigm' (this volume, 358). By contrast, in 'revolutionary science', that is, as scientists are replacing one paradigm with another, problem-solving serves primarily 'to evaluate the potential for understanding that may be achieved through solving problems with the new resources furnished by the future paradigm under development' (this volume, 359). In both cases, then, problem-solving serves primarily to *evaluate* how much understanding is, or could be, achieved within a paradigm.

Why would scientists need to evaluate how much understanding has or will be achieved within different paradigms? The only sensible answer that I can think of is that, in evaluating how much understanding has been, and will be, achieved within paradigms, scientists will be better placed to choose to adopt those paradigms in which they will achieve the most understanding – and, consequently, the most scientific progress. After all, scientists would not know whether to abandon their current paradigm P_c in favor of a newfangled paradigm P_n unless they knew whether P_c or P_n is more likely to better serve the scientific aim of increasing understanding. So, a perfectly straightforward rationale for spending time and resources evaluating paradigms, through problem-solving, is to maximize the amount of scientific progress made at any given time.

But if that is the role of problem-solving in scientific progress on McCoy's view, then I am not sure whether problem-solving really is an essential part of McCoy's account of scientific progress. To explain why, let me introduce the distinction between *constituting* and *promoting* progress. A scientific development *constitutes progress* if and only if it is an improvement in and of itself, that is, regardless of what other developments are thereby brought about (or are made more likely to be brought about). By contrast, a scientific development *promotes progress* if and only if it is an improvement because and in so far as it brings about other developments that constitute scientific progress (or makes it more likely that they are brought about). So, a progress-promoting episode would contribute to progress only to the extent that it leads to – or is likely to lead to – scientific progress at a later time; whereas a progress-constituting episode would do so regardless of its actual or probable causal effects.

Given this distinction, it seems quite clear what the relationship between problem-solving, understanding, and scientific progress is within McCoy's account: problem-solving enables scientists to evaluate the extent to which paradigms foster understanding, thereby causing or probabilifying increases in understanding, which, in turn, constitutes progress. Thus, the problem-solving aspect of McCoy's account should, I believe, be seen as a view of what promotes, rather than what constitutes, scientific progress. This is important for how McCoy's account is situated among other accounts of scientific progress, because other accounts of scientific progress are not intended to be accounts of what promotes progress; rather, they are accounts only of what constitutes progress (see Bird 2008, 280; Dellsén 2018, 73).

Indeed, there is reason to be pessimistic about the prospects of providing any general philosophical account of what promotes progress. The point applies to any attempt at giving an account of what promotes scientific progress, but for specificity let us consider the claim (which I am tentatively attributing to McCoy) that problem-solving promotes scientific progress. If this is meant to be an exceptionless generalization, then it seems empirically unsubstantiated. After all, it seems at least possible for solutions to problems to lead researchers into blind alleys, prolonged fruitless debates, and so forth, which do not promote progress in the long (or short) run. So, if we are to accept as an exceptionless generalization that problem-solving promotes progress, it would at the very least need to be backed up by substantial empirical data. If, by contrast, the claim that problem-solving promotes scientific progress is meant to be a more qualified type of generalization (e.g., that solving problems *often* promotes progress), then it is not clear that anyone would ever have disagreed. Moreover, there are surely lots of *other* activities that promote progress as well, such as collecting data, developing scientific concepts, and refining theoretical arguments. So why single out problem-solving specifically as the progress-promoting activity to highlight in discussions of scientific progress?

3 Problem-Solving as Constituting Progress?

In this section, I will consider a different possible role for problem-solving in an otherwise understanding-based account of scientific progress. To be clear (and fair), I do not think the following is particularly plausible as an interpretation of McCoy's chapter. But the discussion below may be of some general interest in so far as it develops and criticizes a genuinely hybrid account in which both understanding and problem-solving play central roles.

Consider an account of what constitutes scientific progress wherein progress consists in increasing understanding, and where understanding is itself constituted by problem-solving. On this (hypothetical) account, what it is to understand something is to have (or have the ability to provide) solutions to scientific problems, and one's understanding increases precisely to the extent that one has (or has the ability to provide) a greater number of such solutions. If this account is meant to combine elements of understanding-based and standard functional accounts of scientific progress, the notion of 'problem-solving' used here must bear some close relation to that used by proponents of the latter, such as Kuhn and Laudan.[1]

So, what notion of 'problem-solving' do we find in the standard functional accounts of scientific progress? Let us focus on Laudan (1977, 1981), who provides the most detailed characterization of what a 'problem' and a corresponding 'solution' would be. According to Laudan (1977, 11–69), there are two distinct kinds of problems: *empirical problems* are questions concerning the objects or entities that a particular scientific theory is meant to explain or account for, and *conceptual problems* are questions about the theories themselves or how they relate to other theories. Since there are clearly infinitely many questions of either kind in logical space, Laudan holds that only some of these constitute genuine scientific problems such that solving them would constitute progress on his functional account. But which ones? And which ones of the infinitely many answers to a given problem-question count as genuine solutions?

In each case, Laudan's view is that this is entirely determined by what he calls a *research tradition* (corresponding roughly to Kuhn's notion of a paradigm). For Laudan (1977, 78–95), a research tradition is a set of assumptions about the entities and processes in some domain and the appropriate methods for studying them. So which questions count as problems, and which answers count as solutions, is fully determined by these assumptions (i.e., by the research tradition). There is no requirement here that these assumptions are in any way true, truthlike, or otherwise anchored in some type of objective reality (Laudan 1977, 16–17, 24–25). This departure from 'scientific realism' (in one sense of the term) is not a coincidence or oversight from Laudan; on the contrary, the whole point of developing his functional account of scientific progress was to get away from the supposedly 'utopian' idea that scientific progress must consist in revealing an objective reality (see, e.g., Laudan 1981, 145). Laudan's research traditions are, in effect, what replaces objective reality as the determiner of whether, and the extent to which, some scientific activity constitutes progress.

Now, do these conceptions of scientific 'problems' and their 'solutions' help us get a grip on what constitutes increased understanding of the type that might in turn be taken to constitute scientific progress? Can we define understanding in terms of having (or having the ability to provide) 'solutions' to scientific 'problems', in the above sense of the latter terms? I think not. The main problem, as I see it, is that this would fail to make sense of one of the most basic facts about understanding, namely, that it is possible to *mis*-understand, that is, to have a mistaken understanding of something. Relatedly, it also fails to make sense of the possibility of scientists being *mistaken* about whether a given scientific development is progressive.

Consider a concrete example of something that Laudan himself classified as a 'problem' relative to a once-dominant research tradition in medicine, namely, 'that bloodletting cured certain diseases' (Laudan 1977, 16). According to Laudan's account, medical theories that answer questions about why bloodletting (allegedly!) cures certain diseases – for example, by appealing to some version of the humoral theory – provide 'solutions' to this 'problem'. Thus, if we identify understanding with problem-solving in Laudan's sense, humoral explanations of the (alleged) benefits of bloodletting would count as increasing our understanding, and thus, on an understanding-based account of scientific progress, as constituting progress. Apart from being wildly counterintuitive, this seems to me to make a mockery of the distinction between understanding and mis-understanding. Humoral explanations of bloodletting were *attempts* to understand what cures certain diseases, but these attempts were mistaken. They were mis-understandings. Relatedly, this muddles the distinction between (genuine) progress and what scientists at the time *believed* was progress. The doctors who developed humoral medicine certainly thought they were making progress with their explanations of how bloodletting (allegedly) cures diseases, but we now know that they were not.

Now, you might agree with the argument of the last two paragraphs and yet think that *a* notion of problem-solving might still be used to define (increasing) understanding – and, consequently, progress. The idea would be to adopt a *factive*, or at least *quasi-factive*, notion of problem-solving, where genuine problems are required to have some basis in reality (e.g., in being questions with approximately true presuppositions) and similarly for genuine solutions (e.g., in that they must appeal to approximately true theories). However, I fail to see how this maneuver constitutes a step forward in our thinking about scientific progress. After all, as I noted above, the whole point of introducing the notion of problem-solving was, at least for Laudan, to *replace* objective reality as a determiner of whether something counts as progress or not. If we add requirements to the effect that genuine problem-solving must be anchored in objective reality, then why not cut out the middleman and define understanding and progress directly in terms of having a more accurate representation of some aspect of that objective reality?

References

Bird, Alexander. 2008. "Scientific Progress as Accumulation of Knowledge: Reply to Rowbottom." *Studies in History and Philosophy of Science* 39: 279–281. https://doi.org/10.1016/j.shpsa.2008.03.019.

Dellsén, Finnur. 2016. "Scientific Progress: Knowledge versus Understanding." *Studies in History and Philosophy of Science* 56: 72–83. https://doi.org/10.1016/j.shpsa.2016.01.003.

Dellsén, Finnur. 2018. "Scientific Progress, Understanding, and Knowledge: Reply to Park." *Journal for General Philosophy of Science* 49: 451–459. https://doi.org/10.1007/s10838-018-9419-y.

Dellsén, Finnur. 2021. "Understanding Scientific Progress: The Noetic Account." *Synthese.* https://doi.org/10.1007/s11229-021-03289-z.

Kuhn, Thomas S. 1970. *The Structure of Scientific Revolutions*, 2nd edition. Chicago: University of Chicago Press.

Laudan, Larry. 1977. *Progress and Its Problems: Towards a Theory of Scientific Growth*. London: Routledge and Kegan Paul.

Laudan, Larry. 1981. "A Confutation of Convergent Realism." *Philosophy of Science* 48: 19–49. https://doi.org/10.1086/288975.

McCoy, Casey D. this volume. "Understanding the Progress of Science."

Shan, Y. 2019. "A New Functional Approach to Scientific Progress." *Philosophy of Science* 86: 739–758. https://doi.org/10.1086/704980.

Note

1 I do not here consider the possibility of appealing to the notion of problem-solving used by more recent proponents of functional accounts, such as Shan (2019), since McCoy refers only to Kuhn and Laudan in his contribution.

Index

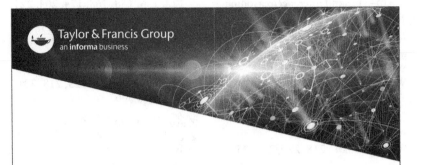

Printed in the United States
by Baker & Taylor Publisher Services